Fu-Sheng Chu

朱福盛

Digital Signal Processing

DIGITAL SIGNAL PROCESSING

THEORY, DESIGN, AND IMPLEMENTATION

ABRAHAM PELED
IBM Corporation

BEDE LIU
Princeton University

ISBN 0-471-01941-0

發行人：楊鏡秋
發行所：儒林圖書有限公司
地址：台北市重慶南路一段一一一號
電話：三一四〇一一一號
三八一二三〇二號
郵政劃撥：一〇六七九二號
印刷所：吉豐印製有限公司
板橋市三民路二段正隆巷46弄7號
電話：九六一四三四八號
中華民國七十二年月
行政院新聞局局版台業字第一四九二號

To our families

Acknowledgment

We thank the IBM Corporation for its generous support of this undertaking. The manuscript was prepared by means of interactive text editing and formatting facilities in conjunction with a computer controlled experimental printer at the IBM Thomas J. Watson Research Center. These services have reduced significantly the time lag between the completion of the manuscript and its publication. Consequently, we were able to include more up-to-date material than otherwise would be possible. In our opinion, this is especially important in this relatively new field. In this connection we thank C. Thompson, N. Badre, and P. Capek for their special assistance in helping us to effectively use these facilities.

We also thank the Electrical and Optical Communications Program, Division of Engineering National Science Foundation and the Directorate of Mathematical and Information Sciences of the Air Force Office of Scientific Research for their continuing support of research at Princeton University. This support enabled one of us (B. L.) to engage in research in the exciting field of digital signal processing. Some of the results obtained during these research programs are included here.

There are a number of individuals to whom we are especially indebted. One of us (B. L.) thanks Professor M. E. Van Valkenburg for his encouragement and advice during this project. The other of us (A. P.) thanks Dr. J. S. Birnbaum for his continuing encouragement and guidance, which made working under him technically exciting and a personal pleasure. Both of us have benefited greatly from technical interaction with our distinguished colleagues, especially J. Cocke, T. Kaneko, H. Silverman, K. Steiglitz and S. Winograd. We especially thank T. Thong, who contributed the material in Appendix 5.1, and A. Hochberg, who contributed the material in Appendix 5.2, and to C. Tappert for the proofreading of the manuscript.

Finally, we thank Mrs. B. A. Smalley for diligent typing, design of the text layout, and endless corrections of style and punctuation.

Abraham Peled

Bede Liu

Preface

Res sunt futurae digitales

The emergence of ***digital signal processing*** as a major discipline began in the mid 1960s when high speed digital computers became widely available for serious research and development work. Many concepts that form the theoretical basis of digital signal processing, such as the Z-transform and the Fourier analysis, had been familiar, however, to engineers for a long time. In the ensuing years, this field has matured considerably. Its development is intimately tied with advances in the computer field.

The past decade has been marked with phenomenal progress in computer technology. With each stride forward, computers became more accessible and more affordable to an everincreasing user community, and the users discovered more new applications, generating new demands for even more sophisticated technology. These developments have had a profound impact on almost all scientific disciplines, and the field of digital signal processing benefitted greatly from these developments.

In digital signal processing, we deal with signals and systems that are the discrete-time counterpart of the more familiar continuous-time systems. The field may be subdivided into two interrelated areas: digital filtering and spectral analysis. Digital filters can perform the same function that analog filters do, while the analog approach, in some cases, may be difficult or unfeasible to implement practically. The use of digital filters offers important engineering advantages, such as perfect reproducibility and a guaranteed level of performance, the increased ease in changing the filter characteristics, and the possibility of time sharing the same hardware system among a multiplicity of filtering functions. These advantages alone would, in many cases, make digital filtering an attractive alternative to analog processing. There is a further important advantage: the possible modularized hardware for customized large scale integration. Digital spectral analysis has been given a tremendous boost by the introduction of the fast Fourier transform (FFT). These computationally efficient algorithms have gained widespread use in many diverse scientific disciplines, making possible accuracies and resolution that could not even be contemplated before with an analog approach.

The proliferation of the use of digital signal processing can be witnessed by its appearance in a variety of areas of scientific endeavor such as biomedical engineering, seismic and geophysical research, image processing and pattern recognition, radar and sonar detection and countermeasures, acoustics and speech research, and telecommunications. In many of these applications, there is a real need for the signal processor to operate at sufficiently high speed as to permit real time processing. At the same time, these processors must be economically competitive to be within the reach of a large user community. The field is expanding rapidly, creating a need for graduating engineers with some exposure and skill in the theory, design, and implementation of digital signal processors. There are a growing number of schools offering a course on digital signal processing; however, a suitable textbook, with emphasis on practical design and implementations, is still lacking. In addition, since the area of digital signal processing is relatively new, many of the practicing engineers today may find themselves thrust into a job that requires considerable knowledge on digital signal processing, but have only a limited time to acquire it. The book by Gold and Rader [1], published in 1969, served these needs for some time. However, there have been many new developments since its publication. The more recent books are directed primarily at the graduate level [2—5].

In this textbook, we have attempted to present a balanced blend of theory and hardware implementation techniques which, in our opinion, constitute the essential body of knowledge in digital signal processing. The material included will enable the reader to enter this important field, and to follow the published literature and the new developments in this area. We have directed this book at both the undergraduate engineering students and the practicing engineers. It can serve as a textbook for a one semester senior course on digital signal processing with a practical bent. The inclusion of the two actual projects in Appendixes 5.1 and 5.2 should help students, who are engaged in similar projects, gain valuable *hands on* experience in this area. The practicing engineer will find in this text the basic theory that he needs for a better understanding of this topic, as well as a large number of specific references to a more detailed treatment of the various subjects. Furthermore, the computer programs included and the detailed hardware implementation discussions should prove to be useful and directly applicable to some of his current problems at work.

The six chapters of the book are organized with minimum interdependence, so that each can be used separately as a reference by persons working in the field for the various topics that these chapters discuss. Chapter 1 contains the essential theoretical background needed for the understanding of the main aspects of digital signal processing, and their relation to the more familiar analog signal processing. Chapter 2 presents the main design methods of digital filters, including some computer programs for their design. In addition to the design of *standard* filters, we also consider the design of digital filters for interpolation and decimation, a process with no counterpart in analog signal processing. Through a number of examples, we demonstrate the effect that finite word length of the filter coefficients has on the filter characteristics, and alert the reader to this important aspect of digital filtering. Chapter 3 is devoted to the fast Fourier transform and its application to power spectra measurement and to performing linear convolution. Chapter 4 deals with the problems of the hardware implementation of general purpose digital signal processors, which are essentially specialized computers. Chapter 5, on the other hand, deals with the hardware implementation of dedicated hardware special purpose digital signal processors which, although less flexible, offer a high degree of cost effectiveness. Finally, in Chapter 6, we discuss some additional implementation considerations, arising from the use of finite word length, such as scaling and limit cycles, which have both theoretical and practical importance. These should be understood and appreciated in order to design a successful processor.

In summary, this book presents, in a concise yet reasonably complete and directly usable form, the main body of knowledge in the area of digital signal processing. It is mainly aimed at the undergraduate engineering student to ease his entrance into this expanding field, and at the practicing engineer to facilitate his ever more difficult job of staying abreast his field of endeavor.

Abraham Peled
Bede Liu

1. B. Gold and C. Rader, *Digital Processing of Signals*, McGraw-Hill Inc., N.Y., 1969.
2. A. V. Oppenheim and R. W. Schafer, *Digital Signal Processing*, Prentice Hall Inc., Engelwood Cliffs, N.J., 1975.
3. L. R. Rabiner and B. Gold, *Theory and Application of Digital Signal Processing*, Prentice-Hall Inc., Engelwood Cliffs, N.J., 1975.
4. W. D. Stanley, *Digital Signal Processing*, Reston Publishing Company Inc., Reston, VA., 1975.
5. B. Liu (editor), *Digital Filters and the Fast Fourier Transform*, Dowden, Hutchinson Ross, Inc., Stroudsburg, PA., 1975.

Contents

CHAPTER 1

An Introduction to Digital Signal Processing

1.1 INTRODUCTION

In this chapter we introduce the reader to the basic theory of digital signal processing, and we also review briefly those topics in signal and system theory that are pertinent to the study of this field. The material included in this chapter is intended to provide the necessary background essential for an understanding of the digital processing of signals that are derived from analog waveforms. The signals and systems treated in this chapter are entirely deterministic in nature. There is a parallel body of knowledge concerning stochastic signals. Readers who are familiar with them will have no difficulty to expand in that direction.

In Section 1.2 we review briefly some aspects of continuous time linear system theory and proceed to define a discrete time system counterpart. In Section 1.3 we introduce the Z-transform, which plays a role similar to that of the Laplace transform in continuous systems. Section 1.4 contains an introduction to digital filtering. This is followed by Section 1.5, where the digital filtering of analog signals is discussed. In Section 1.6 we consider the problems of performing the digital filtering with limited accuracy and alert the reader to some of the problems arising there. Section 1.7 is devoted to the discrete Fourier transform, its properties, and its relationship to the familiar Fourier transform.

The theoretical treatment in this chapter is, by necessity, brief. Although it will provide the reader with a sufficient background for the understanding of the rest of the book and the central area of digital signal processing, additional reading on some of the topics treated in this chapter is advised, should a deeper understanding be desired.

1.2 CONTINUOUS-TIME AND DISCRETE-TIME SIGNALS AND SYSTEMS

This section begins with a review of the elementary properties of analog and digital signals, and the response of linear systems to these signals. *Analog signals* are those operating in the continuous-time domain, while *digital signals* operate in discrete-time. The term *digital* also implies that these signals have values limited to discrete levels. The conversion of analog signals to digital signals is taken up in a later section. Only deterministic or nonrandom signals will be discussed in this section.

A continuous-time or analog signal may be described by a function of time, say f(t). Under quite general conditions that are almost always met in engineering practice, we may take the Fourier transform of f(t). That is, the integral

$$F(\omega) = \int_{-\infty}^{\infty} f(t)e^{-j\omega t}\, dt \tag{1-1}$$

can be evaluated for all, or almost all, real values of ω, thus defining a function of ω. Often $F(\omega)$ is referred to as the *spectrum* of the signal f(t); or more precisely, $|F(\omega)|^2$ is called the *power spectrum* of the signal. There is an inverse relationship to Eq. 1-1,

$$f(t) = (1/2\pi) \int_{-\infty}^{\infty} F(\omega)e^{j\omega t}\, d\omega \tag{1-2}$$

The two functions f(t) and $F(\omega)$ form a Fourier transform pair. The analytic evaluation of the integrals is facilitated sometimes by regarding ω or t as a complex variable and using techniques of contour integration. The subject of numerical computation of $F(\omega)$ from a given f(t) is discussed in Section 1.7.

Consider next the response of a linear time-invariant system to a signal as depicted in the block diagram of Figure 1.1. The output g(t) and the input f(t) are related by a superposition integral of the form

$$g(t) = \int_{-\infty}^{\infty} h(\tau)f(t-\tau)\, d\tau \tag{1-3}$$

where h(t) is the unit impulse response, which is the response of the system at time t due to a unit impulse input at time 0. With a simple change of variable $\tau' = t - \tau$, Eq. 1-3 can also be written as

$$g(t) = \int_{-\infty}^{\infty} f(\tau')h(t-\tau')\,d\tau' \qquad (1\text{-}4)$$

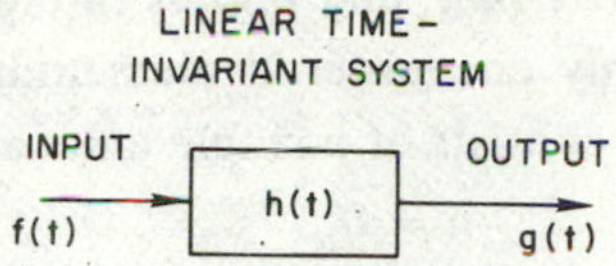

Figure 1.1 Continuous-time signal and system.

The right hand side of Eq. 1-3 or Eq. 1-4 is commonly called the convolution of the two functions f(t) and h(t), often denoted by f(t)*h(t). By a well-known theorem in Fourier integrals, we have

$$G(\omega) = H(\omega)\,F(\omega) \qquad (1\text{-}5)$$

where $G(\omega)$, $H(\omega)$, and $F(\omega)$ are respectively the Fourier transform of g(t), h(t), and f(t). $H(\omega)$ is called the transfer function of the system.

It can be shown that for an input signal of the form $f(t) = A\,e^{j\omega_1 t}$ the output is simply

$$g(t) = A\,H(\omega_1)\,e^{j\omega_1 t} \qquad (1\text{-}6)$$

Since $e^{j\omega_1 t}$ corresponds to a sinusoidal signal of frequency ω_1, we see that the response of a linear time-invariant system to a sinusoidal input is a sinusoidal signal of the same frequency. The amplitude of the output sinusoid is $A\,|\,H(\omega_1)\,|$, which is equal to the input amplitude multiplied by the magnitude of the complex number $H(\omega_1)$, and the phase by which the output lags the input is simply the argument *Arg* $H(\omega_1)$. $H(\omega)$ is seen to characterize completely the response of the system to pure sinusoidal signals, and is, therefore, called the *frequency response* of the system.

We now turn our attention to discrete-time signals. A *discrete-time signal* is a sequence of numbers, $\{x_n\}$, where the index n may vary over a finite or an infinite range. When it is desired to display explicitly the range of n, say $N \le n \le M$, we shall use the notation $\{x_n\}_{N,M}$ or $\{x_n\}_{n=N,M}$. Although the most commonly encountered discrete-time signals are the samples of analog signals at uniform intervals, these are by no means the only types of discrete-time signals. The arrival times of automobiles at a toll booth, a record stored on a memory device in a computer system, and the attendance at successive home games of the New York Mets during a season are all examples of discrete-time signals. Other common terminologies used for these signals are sampled data signals and digital signals. However, strictly speaking, the term *digital* carries with it the implication that each sample value x_n is also digitized or quantized to a discrete set. In this book, we shall use *discrete-time* and *digital* interchangeably.

A number of commonly encountered elementary signals are illustrated in Figure 1.2. The first signal consists of a single unit sample at n=0. That is,

$$x_n = \begin{cases} 1 & n = 0 \\ 0 & n \neq 0 \end{cases} \tag{1-7}$$

This particular signal is called a unit sample signal or, more popularly, a unit impulse signal, even though there is no impulse in the signal. The second signal is a unit step,

$$x_n = \begin{cases} 1 & n \geq 0 \\ 0 & n < 0 \end{cases} \tag{1-8}$$

The last one is a sinusoidal signal

$$x_n = \sin(an+b) \qquad -\infty < n < \infty \tag{1-9}$$

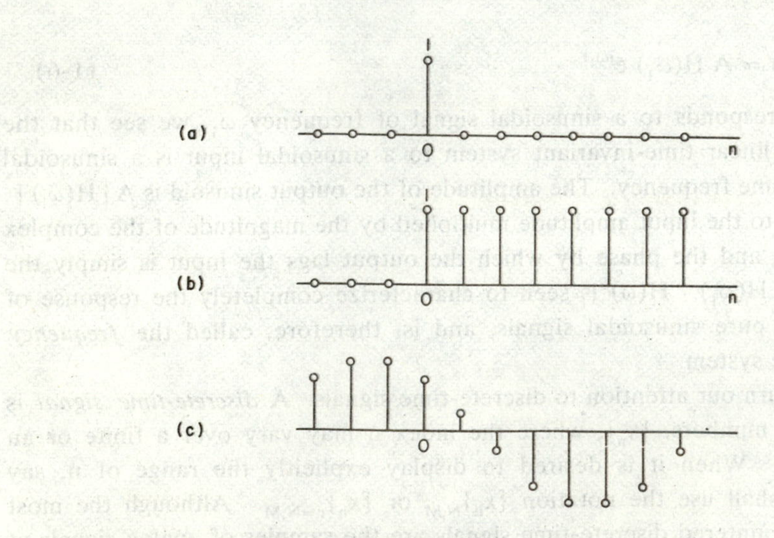

Figure 1.2 Three commonly encountered digital signals (a) unit impulse, (b) unit step, (c) sinusoidal signal.

A discrete-time system operates on an input digital signal to produce an output digital signal. A linear time-invariant discrete-time system can be described by the input-output relationship

$$y_n = \sum_{m=-\infty}^{\infty} x_m h_{n-m} \tag{1-10}$$

where $\{x_n\}$ and $\{y_n\}$ are, respectively, the input and output signals, and $\{h_n\}$ is the impulse response of the system. That is, h_n is the response of the system at n, due to a unit sample input at 0. By letting $n-m=k$, so $m=n-k$, Eq. 1-10 becomes

$$y_n = \sum_{k=-\infty}^{\infty} x_{n-k} h_k \tag{1-11}$$

This is depicted in Figure 1.3. The right hand side of Eqs. 1-10 and 1-11 is called the convolution sum of the two sequences $\{x_n\}$ and $\{h_n\}$. When the sequence $\{h_n\}$ has only a finite number of nonzero terms, we say the system has a *finite impulse response* (FIR). Otherwise, the system is said to possess an *infinite impulse response* (IIR). If $h_n = 0$ for $n \leq 0$, we say the system is *causal* or *physically realizable*.

A large class of linear time-invariant discrete-time systems can also be described by the linear constant coefficient difference equation

$$y_n = \sum_{k=0}^{M} a_k x_{n-k} - \sum_{k=1}^{L} b_k y_{n-k} \tag{1-12}$$

where $\{x_n\}$ is the input, $\{y_n\}$ is the output, and $a_0, a_1, \ldots, a_M, b_1, \ldots, b_L$ are constants that determine the characteristics of the system. It is possible to convert Eq. 1-12 to an equation of the form of Eq. 1-10. However, we shall defer the discussion of this topic for the time being.

The reader is undoubtedly familiar with the important role that Fourier and Laplace transforms play in continuous-time signals and systems. In an analogous

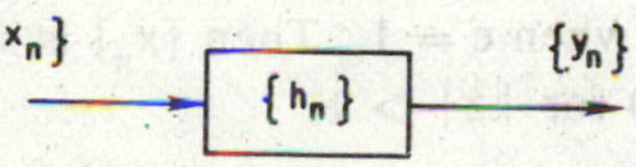

$\{x_n\}$ → $\{h_n\}$ → $\{y_n\}$

Figure 1.3 A linear time invariant discrete-time system.

manner, the study of discrete-time signals and systems is facilitated by the Z-transform. This is the subject of the next section.

1.3 THE Z-TRANSFORM

(a) *Definition and Some Examples*

Given a sequence $\{x_n\}$, its Z-transform is defined by

$$X(z) = \sum_{n=-\infty}^{\infty} x_n z^{-n} \tag{1-13}$$

where z is a complex variable and plays a role similar to that of the variable s in the Laplace transform. The series on the right side of Eq. 1-13 converges, when z takes on values in a certain region in the complex plane. For many commonly encountered signals the series can be summed in close form. $\{x_n\}$ is called the inverse Z-transform of X(z).

Let us illustrate this with a few examples.

Example 1

$$x_n = \begin{cases} c^n & n \geq 0 \\ 0 & n < 0 \end{cases} \tag{1-14}$$

According to Eq. 1-13

$$X(z) = \sum_{n=0}^{\infty} c^n z^{-n} = \sum_{n=0}^{\infty} (cz^{-1})^n$$

which is a geometric series. It converges if $|cz^{-1}| < 1$, or

$$|z| > |c| \tag{1-15}$$

If Eq. 1-15 is satisfied, then the convergent series has a closed form expression

$$X(z) = \frac{1}{1 - cz^{-1}}, \quad |z| > |c| \tag{1-16}$$

A special case of interest is when c = 1. Then $\{x_n\}$ is simply the unit step, Figure 1.2*b*, and $X(z) = 1/(1-z^{-1})$ for $|z| > 1$.

Example 2

$$x_n = c^{|n|} \qquad -\infty < n < \infty \tag{1-17}$$

with $|c| < 1$. We break the infinite summation of Eq. 1-13 as follows

$$X(z) = \sum_{n=-\infty}^{-1} c^{|n|} z^{-n} + \sum_{n=0}^{\infty} c^{|n|} z^{-n}$$

$$= \sum_{n=-\infty}^{-1} c^{-n} z^{-n} + \sum_{n=0}^{\infty} c^{n} z^{-n}$$

$$= \sum_{n=1}^{\infty} c^{n} z^{n} + \sum_{n=0}^{\infty} c^{n} z^{-n} \tag{1-18}$$

The first series converges if $|cz| < 1$ or $|z| < 1/|c|$, and the second series converges if $|z| > |c|$, according to the previous example. Since $|c| < 1$ as given, $|c| < 1/|c|$. Thus, if z lies in the ring $|c| < |z| < 1/|c|$, both infinite series in Eq. 1-18 will converge, and we have

$$X(z) = \frac{cz}{1 - cz} + \frac{1}{1 - cz^{-1}}$$

$$= \frac{1 - c^2}{(1 - cz)(1 - cz^{-1})}, \quad |c| < |z| < 1/|c| \tag{1-19}$$

Example 3

$$x_n = \begin{cases} 1 & n = 0 \\ 0 & n \neq 0 \end{cases} \tag{1-20}$$

This is the unit impulse sequence, Figure 1.2*a*. We have in this case,

$$X(z) = 1 \tag{1-21}$$

Example 4

$$x_n = \begin{cases} 1 & n = N \\ 0 & n \neq N \end{cases} \tag{1-22}$$

This is a shifted unit impulse sequence. Again, straightforwardly,

$$X(z) = z^{-N} \tag{1-23}$$

Example 5

$$x_n = \begin{cases} a^n \sin(nb) & n \geq 0 \\ 0 & n < 0 \end{cases} \tag{1-24}$$

We write sin(nb) as $(e^{jnb} - e^{-jnb})/2j$. By Eq. 1-13

$$X(z) = \sum_{n=0}^{\infty} a^n \sin(nb)\, z^{-n}$$

$$= \sum_{n=0}^{\infty} (a^n e^{jnb} z^{-n} - a^n e^{-jnb} z^{-n})/2j$$

$$= [\sum_{n=0}^{\infty} (ae^{jb}z^{-1})^n - \sum_{n=0}^{\infty} (ae^{-jb}z^{-1})^n]/2j \tag{1-25}$$

Both series converge if $|ae^{jb}z^{-1}| = |ae^{-jb}z^{-1}| < 1$ or $|z| > |a|$. Thus,

$$X(z) = \frac{1}{2j} [\frac{1}{1 - ae^{jb}z^{-1}} - \frac{1}{1 - ae^{-jb}z^{-1}}]$$

$$= \frac{a \sin(b)\, z^{-1}}{(1 - ae^{jb}z^{-1})(1 - ae^{-jb}z^{-1})}$$

$$= \frac{a \sin(b)\, z^{-1}}{1 - 2a \cos(b)\, z^{-1} + a^2 z^{-2}} \qquad |z| > |a| \tag{1-26}$$

Example 6

$$x_n = \begin{cases} a^n \cos(nb) & n \geq 0 \\ 0 & n < 0 \end{cases} \tag{1-27}$$

Following a similar approach as in the previous example, we obtain

$$X(z) = \frac{1 - a \cos(b)\, z^{-1}}{1 - 2a \cos(b)\, z^{-1} + a^2 z^{-2}} \tag{1-28}$$

Example 7

$$x_n = \begin{cases} nc^n & n \geq 0 \\ 0 & n < 0 \end{cases} \tag{1-29}$$

The Z-transform is given by

$$X(z) = \sum_{n=0}^{\infty} n\, c^n z^{-n} \tag{1-30}$$

To sum up the series, we note that it can be rewritten slightly differently,

$$X(z) = z^{-1} \frac{d}{dz^{-1}} \left(\sum_{n=0}^{\infty} c^n z^{-n} \right) \tag{1-31}$$

and interchanging the order of summation and the term by term differentiation can be mathematically justified. We recall from Example 1,

$$\sum_{n=0}^{\infty} c^n z^{-n} = \frac{1}{1 - cz^{-1}}$$

The differentiation with respect to z^{-1} of the righthand side can be carried out straightforwardly. The reader can easily verify that Eq. 1-31 becomes

$$X(z) = \frac{cz^{-1}}{(1 - cz^{-1})^2} \tag{1-32}$$

(b) *Elementary Properties*

A number of elementary properties of the Z-transform can easily be derived from the definition, Eq. 1-13. The reader will recognize the analogous properties in Laplace and Fourier transforms. In what follows we shall denote the Z-transforms of $\{x_n\}$, $\{y_n\}$, and $\{w_n\}$ by Z(z), Y(z), and W(z), respectively; also, a and b are constants, and k an integer.

1. *Linearity*: The Z-transform of $\{ax_n + by_n\}$ is $aX(z) + bY(z)$.
2. *Shifts*: The Z-transform of $\{x_{n+k}\}$ is $z^k X(z)$.
   These two properties follow directly from Eq. 1-13.
3. *Convolution*: Let $\{w_n\}$ be the convolution of $\{x_n\}$ and $\{y_n\}$. That is,

$$w_n = \sum_{m=-\infty}^{\infty} x_m y_{n-m} = \sum_{k=-\infty}^{\infty} x_{n-k} y_k \tag{1-33}$$

   Then $W(z) = X(z)\,Y(z)$. This can be shown as follows:

$$W(z) = \sum_{n=-\infty}^{\infty} w_n z^{-n}$$

$$= \sum_{n=-\infty}^{\infty} \left[\sum_{m=-\infty}^{\infty} x_m y_{n-m} \right] z^{-n}$$

$$= \sum_{m=-\infty}^{\infty} x_m \left[\sum_{n=-\infty}^{\infty} y_{n-m} z^{-(n-m)} \right] z^{-m}$$

$$= \sum_{m=-\infty}^{\infty} x_m Y(z) z^{-m}$$

$$= Y(z) \sum_{m=-\infty}^{\infty} x_m z^{-m} = X(z)\, Y(z) \qquad (1\text{-}34)$$

(c) *Calculation of the Inverse Z-Transform.*

From complex variable theory, it can be shown that each individual x_n in the series of Eq. 1-13 is related to the function X(z) through

$$x_n = \frac{1}{2\pi j} \oint_C X(z) z^n \frac{dz}{z} \qquad (1\text{-}35)$$

where C is an appropriately chosen contour in the region of the complex plane where the series converges. However, in practice, one usually finds that other techniques are easier to apply. We shall discuss, by means of examples, the use of partial fraction expansion and long division, which is applicable when X(z) is a rational function in z. Unless otherwise stated, we shall assume that X(z) corresponds to a sequence $\{x_n\}$ that is bounded as n approaches $\pm\infty$, This means that the unit circle $|z| = 1$ is either in or forms the boundary of the region of convergence for X(z). As a first step in applying these methods, X(z) is written as the sum of two components

$$X(z) = X_1(z) + X_2(z)$$

where $X_1(z)$ is analytic inside the unit circle $|z| = 1$ and $X_2(z)$ analytic outside. That is, all the poles of $X_1(z)$ are outside $|z| = 1$ and all the poles of $X_2(z)$ are inside. Thus, $X_1(z)$ is the Z-transform of a one-sided sequence that vanishes for positive index n, and $X_2(z)$ is the Z-transform of another one-sided sequence that vanishes for negative n.

Example 1

$$X(z) = \frac{1}{1-0.5z^{-1}}$$

This X(z) is analytic outside the unit circle $|z|=1$, since its pole is at 0.5. A long division of the denominator into the numerator would result in an infinite series in powers of z^{-1}.

$$\begin{array}{r|l} & 1 + 0.5z^{-1} + 0.25z^{-2} + 0.125z^{-3} + \ldots \\ 1-0.5z^{-1} & 1 \\ & 1 - 0.5z^{-1} \\ \hline & \quad 0.5z^{-1} \\ & \quad 0.5z^{-1} - 0.25z^{-2} \\ \hline & \qquad 0.25z^{-2} \\ & \qquad 0.25z^{-2} - 0.125z^{-3} \\ \hline & \qquad\qquad 0.125z^{-3} \\ & \qquad\qquad \vdots \end{array}$$

Thus we see

$$X(z) = 1 + 0.5z^{-1} + 0.25z^{-2} + 0.125z^{-3} + \ldots$$

By comparing this series with the definition, Eq. 1-13, we see that

$$x_n = \begin{cases} 0 & n < 0 \\ 1, 0.5, 0.25, 0.125, \ldots, & n \geq 0 \end{cases}$$

The reader will recognize that $x_n=(0.5)^n$, for $n \geq 0$.

Example 2

$$X(z) = \frac{1}{(1-0.5z^{-1})(1+0.8z^{-1})} = \frac{1}{1+0.3z^{-1}-0.4z^{-2}}$$

As in Example 1, X(z) is analytic outside $|z|=1$ its two poles being at 0.5 and and at −0.8, so $\{x_n\}$ is a one sided sequence that vanishes for $n < 0$. A long division can be carried out as follows:

$$\begin{array}{r|l} & 1 - 0.3z^{-1} + 0.31z^{-2} - 0.029z^{-3} + \ldots \\ 1+0.3z^{-1}-0.4z^{-2} & 1 \\ & 1 + 0.3z^{-1} - 0.4z^{-2} \\ \hline & \quad -0.3z^{-1} + 0.4z^{-2} \\ & \quad -0.3z^{-1} - 0.09z^{-2} + 0.12z^{-3} \\ \hline & \qquad 0.31z^{-2} - 0.12z^{-3} \\ & \qquad 0.31z^{-2} + 0.91z^{-3} - 0.124z^{-4} \\ \hline & \qquad\qquad -0.029z^{-3} + 0.124z^{-4} \\ & \qquad\qquad \vdots \end{array}$$

Thus, $x_n=0$ for $n<0$, and $\{x_n\} = \{1,-0.3,0.31,-0.029,...\}$ for $n \geq 0$.

Suppose we break X(z) into partial fraction,

$$X(z) = \frac{0.5/1.3}{1-0.5z^{-1}} + \frac{0.8/1.3}{1+0.8z^{-1}}$$

The first term is recognized as the Z-transform of the sequence $x_n^{1} = (0.5/1.3)(0.5)^n$, $n \geq 0$, and the second term is the Z-transform of $x^2_n = (0.8/1.3)(-0.8)^n$, $n \geq 0$. So the inverse Z-transform is

$$x_n = \begin{cases} 0 & n<0 \\ [0.5^{n+1} - (-0.8)^{n+1}]/1.3 & \end{cases}$$

The reader can easily verify that this result agrees with the one obtained from the long division.

Example 3

$$X(z) = \frac{z^2}{1-0.5z^{-1}}$$

A long division of the denominator into the numerator gives

$$X(z) = z^2 + 0.5z^1 + 0.25 + 0.125z^{-1} + 0.0625z^{-2} + \ldots$$

From which we can make the identification $x_n=0$, $n < -2$, $x_{-2} = 1$, $x_{-1} = 0.5$, $x_0 = 0.25$, $x_1 = 0.125$, . . . or $x_n = (0.5)^{n+2}$, $n \geq -2$. This result can also be obtained from Example 1 by using the shifting property discussed in (b).

Example 4

$$X(z) = \frac{1}{(1+0.5z^{-1})(1-1.5z^{-1})}$$

This X(z) has one pole at -0.5 and another at 1.5, so its inverse is no longer one sided. To find $\{x_n\}$, we first expand X(z) into a partial fraction

$$X(z) = \frac{0.25}{1+0.5z^{-1}} + \frac{-0.5z}{1-z/1.5}$$

The first term is analytic outside $|z|=1$, corresponding to a sequence vanishing for negative index. The second term is analytic inside $|z|=1$, so its inverse vanishes for positive n. Using long division, the first term is expanded into a power series in z^{-1} and the second term is expanded into a power series in z.

$$\frac{0.25}{1 + 0.5z^{-1}} = 0.25(1 - 0.5z^{-1} + 0.25z^{-2} - 0.125z^{-3} + \ldots)$$

$$\frac{-0.5z}{1 - z/1.5} = -0.5(z + z^2/1.5 + z^3/1.5^2 + \ldots)$$

Thus the inverse is

$$x_n = \begin{cases} 0.25(-0.5)^n & n \geq 0 \\ -0.5/1.5^n & n < 0 \end{cases}$$

Example 5

$$X(z) = \frac{z^{-3}}{(1 - 0.5z^{-1})(1 + 0.8z^{-1})(1 - 2z^{-1})}$$

Leaving aside the factor z^{-3} which corresponds to a simple shift, the rest can be expanded into a partial fraction

$$\frac{-0.1282}{1 - 0.5z^{-1}} + \frac{-1.7778}{1 + 0.8z^{-1}} + \frac{0.9524}{1 - 2z^{-1}}$$

The inverse Z-transform of each of these terms can be easily obtained. On combining them, we have

$$x_n = \begin{cases} -1.9060, 1.3581, -1.1698, 0.8942, -0.7362, \ldots & n \geq -3 \\ -0.4762, -0.2381, -0.1190, -0.05952, \ldots & n < -3 \end{cases}$$

From these examples, we see that the determination of the inverse Z-transforms from a rational function in z^{-1} involves the following steps:

1. Set aside any factor of the form z^k, k integer.
2. Expand the remaining part into a partial fraction.
3. Obtain the inverse Z-transform of each term in the partial fraction expansion.
4. Combine the results and make the necessary shift due to the factor z^k taken out in step (1).

1.4 INTRODUCTION TO DIGITAL FILTERS

A linear time invariant discrete-time system, described by Eq. 1-12, is commonly called a digital filter. These filters form one of the most important classes of digital signal processing systems and have found important applications in a number of diverse fields of science and engineering.

To facilitate the discussion, we repeat Eq. 1-12 below as Eq. 1-36

$$y_n = \sum_{k=0}^{M} a_k x_{n-k} - \sum_{k=1}^{L} b_k y_{n-k} \tag{1-36}$$

where $\{x_n\}$ is the input signal, $\{y_n\}$, the output signal, and $a_0, a_1, ..., a_M, b_1, ..., b_L$ are constants. Effective methods have been developed to enable a designer to choose the filter coefficients, namely the a_k's and b_k's, so that the filter possesses certain desired characteristics for the processing of digital signals. These design methods are discussed in Chapter 2. It is also possible to use a digital filter to process analog signals by connecting an analog-to-digital converter (A/D) preceding the filter, and following it with a digital-to-analog converter (D/A). This subject is discussed in the next section.

The filter of Eq. 1-36 can obviously be programmed or simulated on a general purpose digital computer. It can also be implemented by using digital electronics components (registers, adders, multipliers, etc.).

There are two outstanding advantages of using digital filters for signal processing. First, they are highly reliable and have a highly stable and predictable behavior because they are constructed with electronic circuits that operate with only two states (zero and one). Secondly, they permit a high degree of flexibility. For example, their characteristics can be changed easily by simply reading in from memory a new set of filter coefficients. In conjunction with a time sharing operation, this means that a single filter structure can serve a multiplicity of input and output signals.

There is, however, an inherent limitation on the accuracy of digital filters due to the quantized nature of digital signals and the finite register length used in the implementation. This and other practical considerations will be discussed later in the book.

Let us take the Z-transform of Eq. 1-36.

$$\sum_{n=\infty}^{\infty} y_n z^{-n} = \sum_{n=-\infty}^{\infty} [\sum_{k=0}^{M} a_k x_{n-k} - \sum_{k=1}^{L} b_k y_{n-k}] z^{-n} \tag{1-37}$$

The left side is simply Y(z), the Z-transform of $\{y_n\}$. The right side can be evaluated with the help of the shift property: the Z-transform of $\{y_{n-k}\}$ is $z^{-k}Y(z)$ and that of $\{x_{n-k}\}$ is $z^{-k}X(z)$ where X(z) is the Z-transform of $\{x_n\}$. Thus, Eq. 1-37 becomes

$$Y(z) = [\sum_{k=0}^{M} a_k z^{-k}] X(z) - [\sum_{k=1}^{L} b_k z^{-k}] Y(z) \tag{1-38}$$

which can be put in a more convenient form,

$$Y(z) = X(z)\,H(z) \tag{1-39}$$

where

$$H(z) = \frac{\sum_{k=0}^{M} a_k z^{-k}}{1 + \sum_{k=1}^{L} b_k z^{-k}} \tag{1-40}$$

H(z) is called the transfer function of the filter; it is the ratio of the Z-transform of the output Y(z) to that of the input, X(z).

As seen from Eq. 1-40, H(z) is a rational function in z^{-1}. Its numerator is

$$N(z) = \sum_{k=0}^{M} a_k z^{-k} \tag{1-41}$$

and its denominator is

$$D(z) = 1 + \sum_{k=1}^{L} b_k z^{-k} \tag{1-42}$$

The roots of N(z) and D(z) are, respectively, the zeros and poles of H(z).

As discussed in Section 1.2, the impulse response $\{h_n\}$ is the output, when the input $\{x_n\}$ is a unit impulse. In this case X(z) is simply 1 and Y(z) = H(z). So we see that the impulse response $\{h_n\}$ is the inverse Z-transform of the transfer function H(z).

The digital filter of Eq. 1-36 calculates the n-th output sample y_n by using the previous L output samples, $y_{n-1}, y_{n-2}, \ldots, y_{n-L}$ as well as the current and the previous M input samples, $x_n, x_{n-1}, \ldots, x_{n-M}$. Since no future samples from either the input or the output are needed to calculate the current output sample, the filter is physically realizable or causal. In this case the impulse response $\{h_n\}$ has the property $h_n = 0$, $n < 0$. We shall assume the filter to be stable. That is, $h_n \to 0$ as $n \to \infty$. So all the poles lie inside the unit circle.

The value of h_n is given by (See Eq. 1-35)

$$h_n = \frac{1}{2\pi j} \oint_{|z|=1} H(z)\, z^n \frac{dz}{z} \tag{1-43}$$

Since H(z) is a rational function in z^{-1}, the impulse response $\{h_n\}$ can be calculated by the method discussed in the previous section. It is a sum of *damped sinusoids.* From Eq. 1-39, we see that the output $\{y_n\}$ is the convolution of the input $\{x_n\}$ and the impulse response $\{h_n\}$

$$y_n = \sum_{m=0}^{\infty} h_m x_{n-m} \tag{1-44}$$

The summation starts with m=0 because $h_m=0$ for m<0. Thus, Eqs. 1-40 and 1-43 define the relations between $\{h_n\}$, the impulse response of the filter and the coefficients $\{a_k\}$ and $\{b_k\}$ of the linear difference equation that characterizes the filter.

Digital filters can be classified as either the recursive or the nonrecursive type. When the b_k's are not all zero, the calculation of y_n requires the value of some output samples that have already been calculated. In this case the filter is said to be of the recursive type. On the other hand, if all the b_k's are zero, that is, $b_1=b_2=...b_L=0$, then the calculation of y_n does not require the use of previously calculated samples of the output, and the filter is said to be of the nonrecursive type. For nonrecursive filters, the transfer function H(z) is seen to reduce to a polynomial in z^{-1}. Furthermore, with $b_k=0$, the right side of Eq. 1-36 is recognized as a convolution sum of the form of Eq. 1-44. Therefore, the impulse response is identical to the coefficients $\{a_k\}$, that is,

$$h_n = \begin{cases} a_n & 0 \leq n \leq M \\ 0 & \text{otherwise} \end{cases} \tag{1-45}$$

For this reason, a nonrecursive filter is also a finite impulse response (FIR) filter, and a recursive filter is an infinite impulse response (IIR) filter.

Despite these differences, filters of either the recursive type or the nonrecursive type may be designed to achieve almost any desired performance characteristic. The implementation, of course, is quite different. The selection of a particular type must be based on the specific application considered, as well as on some other considerations which we will discuss in later chapters.

It is a simple matter to verify that the input-output relationship of Eq. 1-36 can be represented by the block diagram of Figure 1.4. This block diagram, commonly known as the *direct form*, is but one of the many ways in which a digital filter can be implemented. Another direct form of realization is shown for the case that L=M in Figure 1.5. From the practical point of view, the two direct forms of realizing a transfer function of order higher than 2 are usually undesirable. Two alternative forms, known as the *parallel* and *cascade* forms, are preferred. To realize a filter in the parallel form the transfer function H(z) is written as a sum by using partial fraction expansion

$$H(z) = \sum_{i=1}^{K} H_i(z) \tag{1-46}$$

where each $H_i(z)$ is a first or second order transfer function. The filter can then be realized via a parallel connection of lower order filters as depicted in Figure 1.6. Similarly, the transfer function H(z) can be written as a product of lower order transfer function,

$$H(z) = \prod_{i=1}^{K} H_i^*(z) \tag{1-47}$$

and the entire filter can be realized as a cascade connection of simpler filters as illustrated in Figure 1.7. Usually, these simpler filters $H_i(z)$ and $H_i^*(z)$ are of order two or one.

Consider now the response of a digital filter with a transfer function H(z) to a pure sinusoidal signal

$$x_n = \begin{cases} A\sin(cn) & n \geq 0 \\ 0 & n < 0 \end{cases} \tag{1-48}$$

Its Z-transform can be obtained as in Example 5 of Section 1.3

$$X(z) = \frac{Az^{-1}\sin(c)}{(1 - e^{jc}z^{-1})(1 - e^{-jc}z^{-1})} \tag{1-49}$$

The Z-transform of the output, given by Eq. 1-39, may be written as a partial fraction

$$Y(z) = \frac{AH(e^{jc})}{2j(1 - e^{jc}z^{-1})} - \frac{AH(e^{-jc})}{2j(1 - e^{-jc}z^{-1})} + [\text{terms due to poles of } H(z)] \tag{1-50}$$

Let us examine the limiting behavior of y_n as $n \to \infty$. When the inverse Z-transform of Y(z) is taken, the terms due to the poles of H(z) give rise to damped sinusoids that tend to zero as $n \to \infty$. The first two terms of Eq. 1-50 can be rewritten as

$$\frac{Az^{-1}\sin(c)[H(e^{jc}) + H(e^{-jc})] - j[1 - z^{-1}\cos(c)][H(e^{jc}) - H(e^{-jc})]}{2[1 - 2\cos(c)z^{-1} + z^{-2}]}$$

which has the inverse Z-transform of

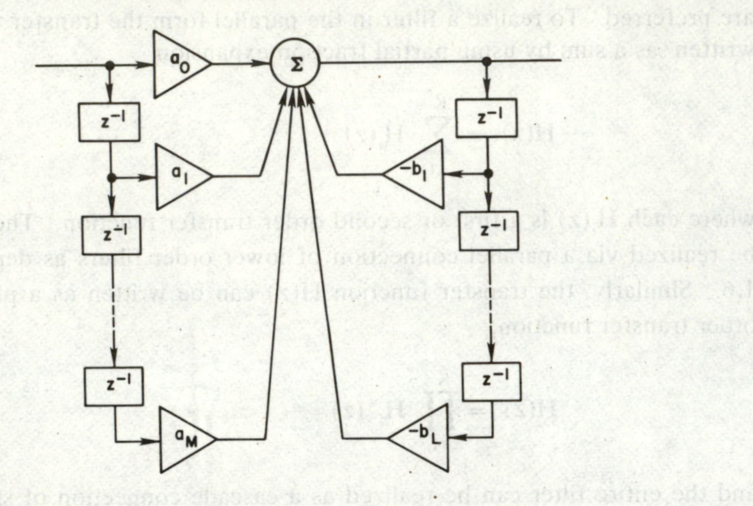

Figure 1.4 Direct form realization of the digital filter of Eq. 1-36.

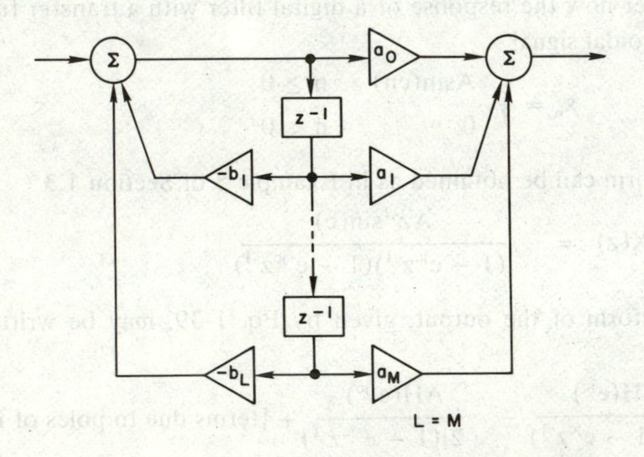

Figure 1.5 Canonical form realization of Eq. 1-36.

$$\frac{H(e^{jc}) + H(e^{-jc})}{2} \sin(nc) + \frac{H(e^{jc}) - H(e^{-jc})}{2j} \cos(nc)$$

$$= A\,|H(e^{jc})|\sin[nc + Arg H(e^{jc})] \qquad (1\text{-}51)$$

Thus, the steady state response of a stable digital filter to a sinusoidal input of frequency c is a sinusoid of the same frequency. Its amplitude is that of the input

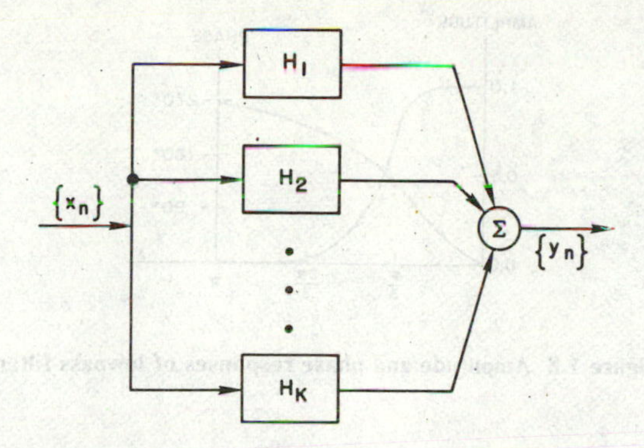

Figure 1.6 Parallel form realization.

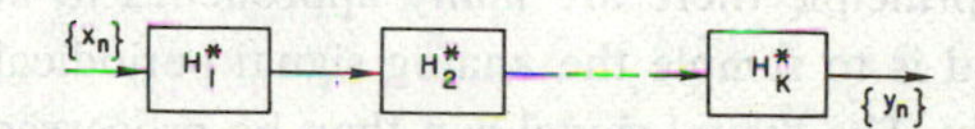

Figure 1.7 Cascade form realization of digital filters.

scaled by $|H(e^{jc})|$, the *gain* of the filter, and its phase is changed by $Arg H(e^{jc})$. The behavior of H(z) on the unit circle $|z|=1$ is, therefore, of special interest. We call $H(e^{j\lambda})$ the *frequency response*. Its amplitude $|H(e^{j\lambda})|$ with $-\pi \leq \lambda < \pi$ is the *amplitude response* of the digital filter, and its argument $Arg H(e^{j\lambda})$, the *phase characteristic*. λ is the *digital frequency*.

Shown in Figure 1.8 is the amplitude and phase characteristic of a digital filter. It is a lowpass filter with a passband of $0 \leq \lambda \leq \pi/3$. The transfer function of this particular filter is

$$H(z) = \frac{0.10073(1 + z^{-1})^3}{(1 - 0.42265z^{-1})(1 - 0.69783z^{-1} + 0.39566z^{-2})}$$

The design of this and other types of digital filters is discussed in Chapter 2.

The zeros play no role in the stability of the filter. They are, however, important in determining the frequency response. For example, if there is a zero on the unit circle, say at $z = e^{jb}$, then the transmission at the digital frequency b is zero, or the loss at b is infinite.

A filter is of the allpass type, when the amplitude characteristic $|H(e^{j\lambda})|$ is a constant for all λ. The poles and zeros of these filters are reciprocals of each other. That is, for each pole at p_i, there is a zero at b_i such that $p_i\, b_i = 1$.

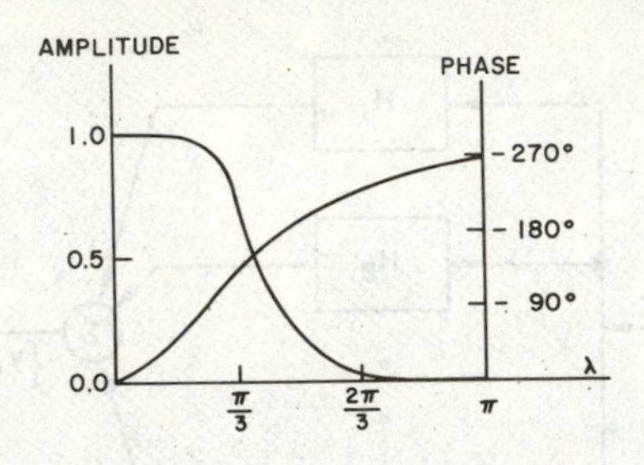

Figure 1.8 Amplitude and phase responses of lowpass filter.

1.5 DIGITAL FILTERING OF ANALOG SIGNALS

To process an analog signal by digital techniques, the signal must first be digitized. Although in principle there are many approaches to accomplish this, the only practical method is to sample the analog signal periodically and to quantize and code each sample. The digital signal can then be processed by a digital filter. After processing, the signal may be converted back into the analog form, if so desired. Thus, the most common digital processor for analog signals is that shown in Figure 1.9. In this section we will derive the basic relationships governing the operation of the various components of such a system.

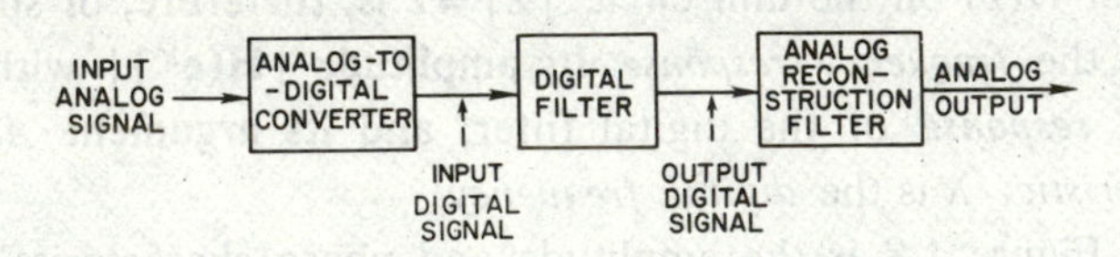

Figure 1.9 Processing of analog signal by digital filter.

(a) *Sampling of Analog Signals (Analog to Digital Conversion).*

The first component of the system shown in Figure 1.9 is the Analog to Digital Converter simply A/D, which is basically a sample and hold circuit followed by a quantizer and coder. We are concerned here only with the sampler; quantization is discussed in the next section.

Suppose that an analog signal x(t) is sampled every T seconds as shown in Figure 1.10. At the output of the sampler, we obtain a digital signal $\{x_n\}$ where

$$x_n = x(nT) \tag{1-52}$$

Let the Z-transform of the digital sequence $\{x_n\}$ be denoted by X(z) and the

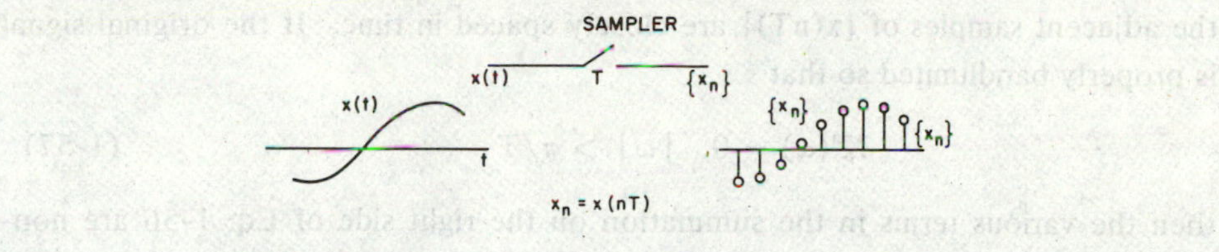

Figure 1.10 Sampling of an analog signal to obtain a digital signal.

Fourier transform of the analog signal x(t) be denoted by $X^a(\omega)$, where the superscript 'a' is used to designate analog. Then

$$X(z) = \sum_{n=-\infty}^{\infty} x_n z^{-n} \tag{1-53}$$

Let us set $z = e^{j\omega}$, and use the Fourier transform relation between x(t) and $X^a(\omega)$, Eq. 1-2, then we obtain

$$X(e^{j\omega}) = \sum_{n=-\infty}^{\infty} x(nT)e^{-jn\omega}$$

$$= \sum_{n=-\infty} (1/2\pi) \int_{-\infty}^{\infty} X^a(u)e^{jnTu}\, du\, e^{-jn\omega}$$

$$= (1/2\pi) \int_{-\infty}^{\infty} X^a(u) \sum_{n=-\infty}^{\infty} e^{jn(uT-\omega)}\, du \tag{1-54}$$

By using the equality

$$\sum_{n=-\infty}^{\infty} e^{jn(uT-\omega)} = (2\pi/T) \sum_{k=-\infty}^{\infty} \delta(u - \omega/T + 2\pi k/T) \tag{1-55}$$

the integration in Eq. 1-54 can be carried out to yield

$$X(e^{j\omega}) = (1/T) \sum_{k=-\infty}^{\infty} X^a[\,(\omega + 2\pi k)/T\,] \tag{1-56}$$

It is seen that $X(e^{j\omega})$ is periodic in ω with period 2π.

Figure 1.11 illustrates the relationship between the digital spectrum X(•)

and the analog spectrum $X^a(\bullet)$ for two different values of T. When T is small, the adjacent samples of $\{x(nT)\}$ are closely spaced in time. If the original signal is properly bandlimited so that

$$X^a(\omega) = 0, \quad |\omega| > \pi/T \tag{1-57}$$

then the various terms in the summation on the right side of Eq. 1-56 are non-overlapping. This is illustrated in Figure 1.11*b*. As T increases, the adjacent samples become farther apart, and the terms in the summation of Eq. 1-56 move closer together and eventually overlap causing an *aliasing error*. This is depicted in Figure 1.11*c*. Frequencies that are separated by multiples of $2\pi/T$ are called aliases of one another. If Eq. 1-57 is satisfied, then the various components of the summation in Eq. 1-56 do not overlap, and the digital spectrum $X(e^{j\omega})$, $|\omega| < \pi$, determines the spectrum of the analog signal $X^a(\omega)$. This results in the familiar *sampling theorem*. The minimum sampling frequency $1/T$, for which Eq. 1-57 holds, is called the Nyquist frequency. It should also be noted that even if $X^a(\omega)$ is not strictly bandlimited, that is, it has some negligible energy outside π/T, a small enough T can be chosen so that the overlap of the components of the summation in Eq. 1-56 is below a prescribed level. This is very important when a sampling frequency is to be selected for a particular signal.

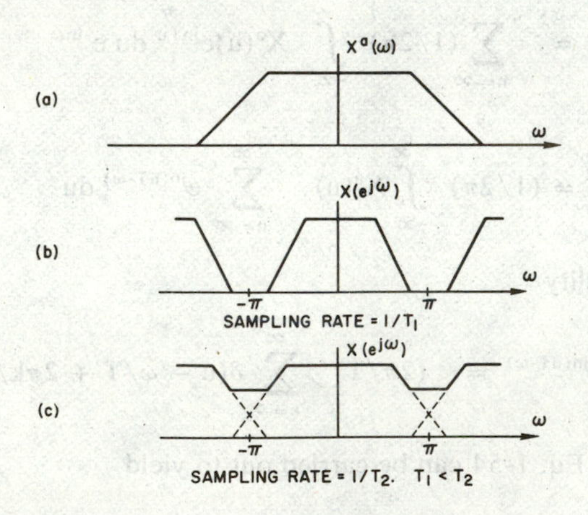

Figure 1.11 Spectra of analog signal and digital signal derived from it; (a) analog signal, (b) digital signal for sampling rate $1/T_1$, (c) digital signal for sampling rate $1/T_2$.

(b) *Digital Signal Applied to an Analog System (Data Reconstruction, Digital to Analog Conversion).*

The last component of the system shown in Figure 1.9 is an analog filter that takes the digital signal at the output of the digital filter and reconstructs it into an analog signal. This step is called digital to analog conversion, or simply D/A. As illustrated in Figure 1.12, a digital signal $\{y_n\}$ is applied to an analog reconstruction filter with an impulse response g(t). The output, denoted by y(t), is given by

$$y(t) = \sum_{n=-\infty}^{\infty} y_n g(t-nT) \tag{1-58}$$

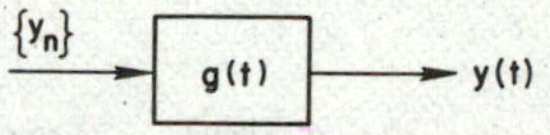

Figure 1.12 Reconstruction of analog signal from the digital samples (D/A).

One can visualize that the sequence of numbers $\{y_n\}$ is read out every T seconds to form a train of very narrow pulses which is then applied to the input of the analog system.

One of the most common filters used in practice is the zero order hold circuit that has a g(t) as shown in Figure 1.13. Other commonly used reconstruction filters are RC filters and lowpass filters.

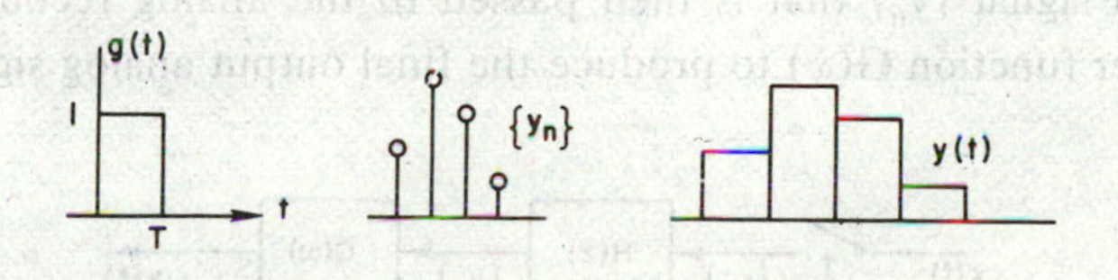

Figure 1.13 Reconstruction of y(t) using zero order hold circuit.

Let Y(z) denote the Z-transform of the digital signal and $Y^a(\omega)$ denote the Fourier transform of the output analog signal, that is,

$$Y(z) = \sum_{n=-\infty}^{\infty} y_n z^{-n} \tag{1-59}$$

$$Y^a(\omega) = \int_{-\infty}^{\infty} y(t)e^{j\omega t}dt \tag{1-60}$$

Substituting Eq. 1-58 into Eq. 1-60 yields

$$Y^a(\omega) = \int_{-\infty}^{\infty} [\sum_{n=-\infty}^{\infty} y_n g(t-nT)] \, e^{-j\omega t} \, dt$$

$$= \sum_{n=-\infty}^{\infty} y_n G(\omega) e^{-j\omega nT} = G(\omega) Y(e^{j\omega T}) \tag{1-61}$$

where $G(\omega)$ is the transfer function of $g(t)$ and we have used Eq. 1-59 with $z=e^{j\omega T}$

(c) *Digital Processing of Analog Signals.*

We are now in a position to take a closer look at the problem of processing analog signals by a digital filter, as depicted in Figure 1.9. The block diagram is redrawn in Figure 1.14 in which an analog signal $x(t)$ is converted into a digital signal $\{x_n\}$ by sampling it at the rate $1/T$. The digital signal $\{x_n\}$ is processed by a linear time invariant digital filter with a transfer function $H(z)$ to produce an output digital signal $\{y_n\}$ that is then passed to the analog reconstructing filter with a transfer function $G(\omega)$ to produce the final output analog signal $y(t)$.

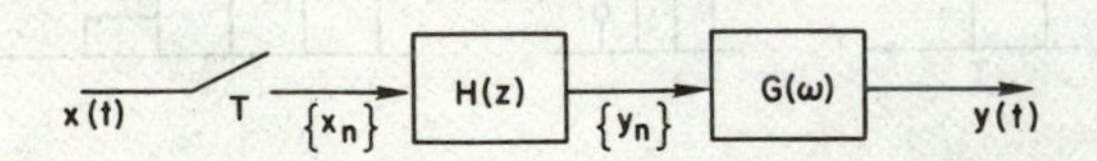

Figure 1.14 Processing of an analog signal by a digital filter.

We now examine in detail the relationships between the input and output. First, we apply Eq. 1-56 to the sampler,

$$X(e^{j\omega}) = (1/T) \sum_{k=-\infty}^{\infty} X^a[\,(\omega+2\pi k)/T\,] \tag{1-62}$$

The input output relationship of the digital filter is

$$Y(z) = H(z)\,X(z) \tag{1-63}$$

Finally, the Fourier transform of the output of the reconstruction filter is given by Eq. 1-61.

$$Y^a(\omega)=G(\omega)\,Y(e^{j\omega T})=G(\omega)\,H(e^{j\omega T})\,X(e^{j\omega})$$

$$= (1/T)G(\omega)\,H(e^{j\omega T}) \sum_{k=-\infty}^{\infty} X^a[\,(\omega + 2\pi k)/T\,] \tag{1-64}$$

The sampling rate $1/T$ is usually taken to be high enough so that the components in the summation $\Sigma X^a[(\omega+2\pi k)/T]$ do not have a significant overlap. The reconstruction filter is usually a lowpass filter whose function is to eliminate the undesirable high frequency components while introducing negligible distortion at low frequencies. Ideally,

$$G(\omega) \approx \begin{cases} 1 & |\omega| < \pi/T \\ 0 & |\omega| > \pi/T \end{cases} \tag{1-65}$$

All terms in the summation $\Sigma\, X^a[(\omega+2\pi k)/T]$ other than $k=0$ would be cut off by $G(\omega)$, and the $k=0$ term is essentially unaffected by $G(\omega)$. Eq. 1-64 thus reduces to

$$Y^a(\omega) \approx \begin{cases} (1/T)H(e^{j\omega T})X^a(\omega) & \text{if } |\omega| < \pi/T \\ 0 & \text{otherwise} \end{cases} \tag{1-66}$$

This is the fundamental relationship in the digital filtering of analog signals. It has the interpretation that the overall system of Figure 1.14 appears to have an effective transfer function of $(1/T)H(e^{j\omega T})$, and the digital filter $H(z)$ is actually performing the spectral shaping of the analog input signal $x(t)$. Eqs. 1-64 and 1-66 are illustrated in Figure 1.15.

We note that ω is the analog frequency in radian/seconds. From the discussion in Section 1.4, we recall that for the transfer function $H(z)$ with $z=e^{j\lambda}$, λ is the digital frequency. By comparison with $H(e^{j\omega T})$ in Eq. 1-66, we see that the digital frequency λ corresponds to an analog frequency ω with $\lambda=\omega T$. Since $1/T$ is the sampling frequency F_S, and $\omega=2\pi f$ where f is the analog frequency in Hz., we have

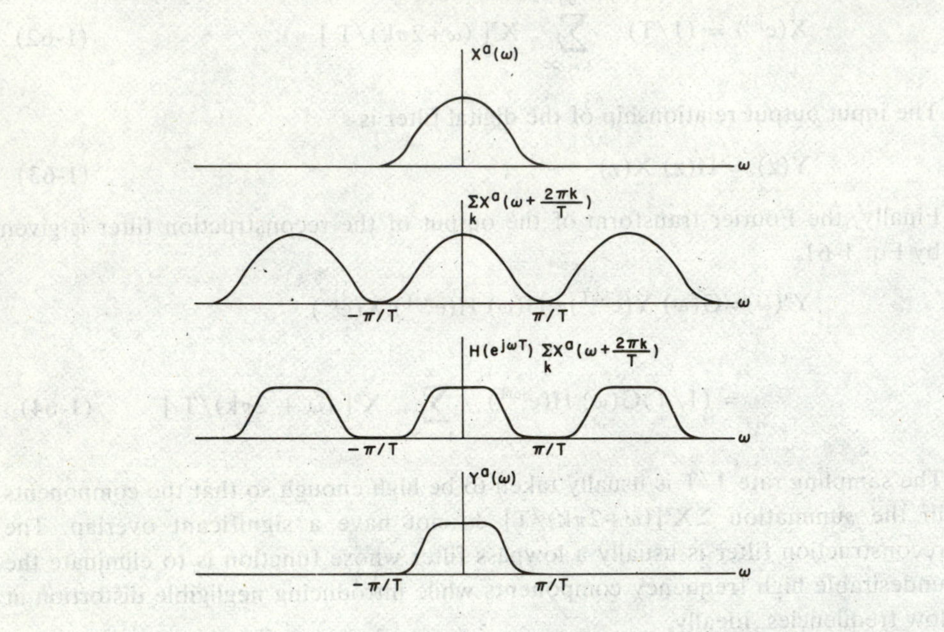

Figure 1.15 Spectra of signals in the system of Figure 1.14.

$$\lambda = 2\pi f/F_s \tag{1-67}$$

As the digital frequency ranges from 0 to π, the analog frequency f ranges from 0 to $F_s/2$, half the sampling frequency.

The function of the reconstruction filter $G(\omega)$ is primarily to eliminate frequency components that are aliases while avoiding introduction of distortion in the frequency range of interest $-(\pi/T) \leq \omega \leq (\pi/T)$. Although $G(\omega)$ should be ideally a perfect lowpass filter, practical reconstruction filters can only approximate this ideal frequency characteristic. The most commonly used reconstruction filter is perhaps the zero-order hold circuit whose impulse response is

$$g(t) = \begin{cases} 1 & 0 \leq t \leq T \\ 0 & \text{otherwise} \end{cases} \tag{1-68}$$

The use of this filter gives a piecewise constant reconstructed analog signal as shown in Figure 1.13. The corresponding $G(\omega)$ is

$$G(\omega) = [\ 2\sin(\omega T/2)/\omega\ ]\ e^{-j\omega T/2} \tag{1-69}$$

It has a linear phase which corresponds to a constant delay of T/2. Its amplitude response is plotted in Figure 1.16 together with the ideal characteristic, for comparison. The deviation from the ideal characteristic introduces a distortion in the passband and at frequencies beyond π/T. The effect of such a distortion can be investigated by using Eq. 1-64 or 1-66. If desirable, this distortion can be compensated for either by incorporating it in the design of the digital filter H(z) or by adding an analog equalizer after the reconstruction filter.

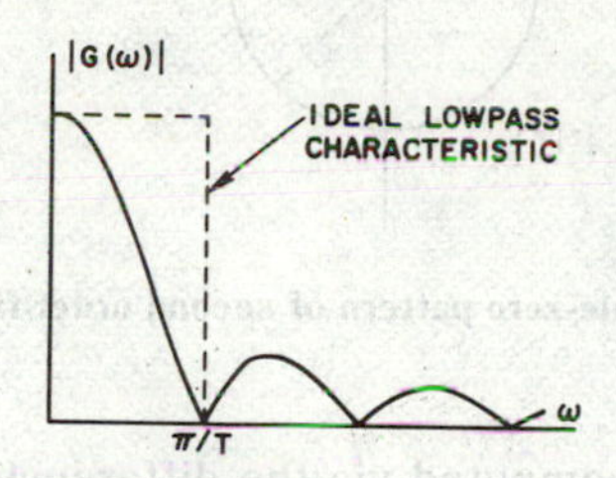

Figure 1.16 Amplitude response of zero order hold reconstruction filter.

The other commonly used reconstruction filter is the RC filter which has an impulse response

$$g(t) = \begin{cases} e^{-t/a} & t \geq 0 \\ 0 & t < 0 \end{cases} \tag{1-70}$$

and a frequency response

$$G(\omega) = \frac{1}{j\omega + a} \tag{1-71}$$

The parameter a is selected by the designer to match the signal.

1.6 IMPLEMENTATION OF DIGITAL FILTERS WITH FINITE REGISTER LENGTH

As mentioned in Section 1.4, there is an inherent limitation on the accuracy of digital signal processors due to the fact that all digital computational elements operate with a finite number of bits. In this section we shall examine the various effects due to this finite word length.

We choose to illustrate these various effects through a simple example. Consider a second order bandpass filter with zeros at ± 1 and poles at $0.9e^{\pm j\pi/4}$ or $0.6363961 \pm j0.6363961$ (Figure 1.17). The transfer function of this filter is:

$$H(z) = \frac{1 - z^{-2}}{1 - 1.2727922z^{-1} + 0.81z^{-2}} \tag{1-72}$$

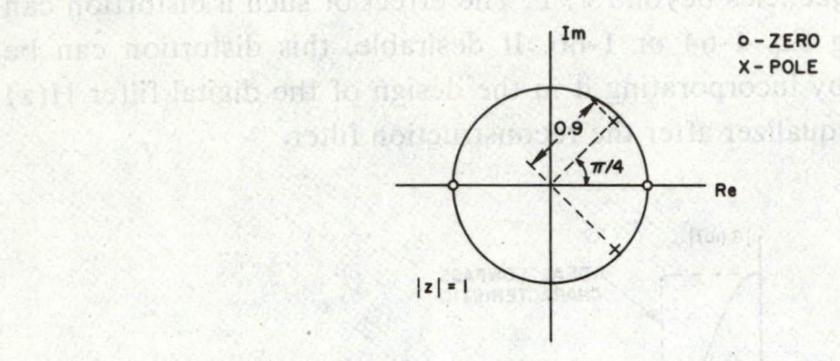

Figure 1.17 Pole-zero pattern of second order filter in example.

The n-th output sample is computed via the difference equation

$$y_n = x_n - x_{n-2} + 1.2727922y_{n-1} - 0.81y_{n-2} \tag{1-73}$$

To build this filter we use the basic arrangement shown in Figure 1.18. It consists of four parts: a memory for storing the filter coefficients, a set of data registers for storing the input and output samples, an arithmetic unit to perform the computation according to Eq. 1-73, and a control unit (not shown) for providing the timing signals.

The question of hardware implementation of digital filters will be discussed in detail in Chapters 4 and 5. Here we examine briefly only some aspects of the problem in order to demonstrate the effects of finite word length. A more detailed treatment of this topic can be found in Chapter 6.

To begin with, the numbers we are dealing with must be represented in binary notation in order to be stored, manipulated, and operated upon by digital computational elements. Consider the coefficient 0.81. It can be written in base 2 as 0.110011110101... . That is,

$$0.81 = (1/2)^1 + (1/2)^2 + (1/2)^5 + (1/2)^6 + ...$$

An infinite amount of bits are needed to represent this coefficient exactly. Since all practical memory circuits have a finite number of bits for each word, the infinite binary string must be modified. Suppose that the memory used for this filter implementation has 6 bits for each word. A simple way to store our number is to keep only the 6 most significant bits, that is, 0.11001, as the approximate value for 0.81. However, 0.11001 in base 2 represents the number 0.78125. Thus, we see that our filter when implemented will use the coefficient 0.78125,

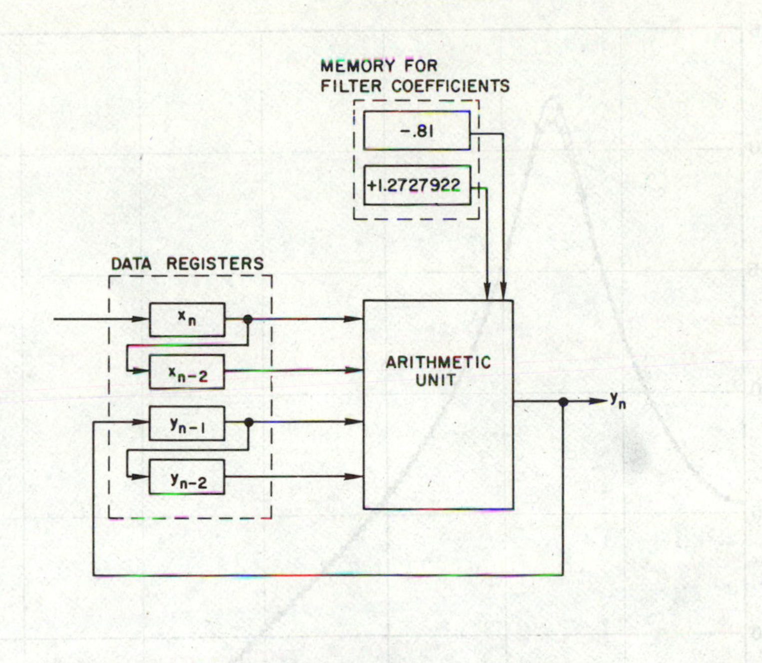

Figure 1.18 Basic parts of digital filter realization.

instead of the true coefficient 0.81. A difference or error of 0.02875 is introduced. Similarly, 1.2727922 is 1.010001011101... and a 6 bit approximation is 1.01000 or 1.25, resulting in an error of 0.0227922. On the other hand, a coefficient of ± 1 or 0 can be represented exactly. Thus instead of Eq. 1-73 the input output relationship of the filter actually implemented is

$$y_n = x_n - x_{n-2} + 1.25y_{n-1} - 0.78125y_{n-2} \quad (1\text{-}74)$$

and the transfer function is modified accordingly. While the zeros of the filter remain unchanged in this example, the poles of this filter have moved to $0.625 \pm j0.625$. Consequently, the frequency response differs slightly from the response of the original filter. For comparison, the amplitude response of the filter before and after the quantization of the coefficients is plotted in Figure 1.19.

Another source of error is the quantization of the input data. Suppose that the filter input is a sinusoidal signal, and consider the following input segment, ..., 0.2955, 0.5564, 0.8912, 0.9320,.... These numbers must be stored in registers for the arithmetic unit to perform the necessary calculations. Again, only a finite

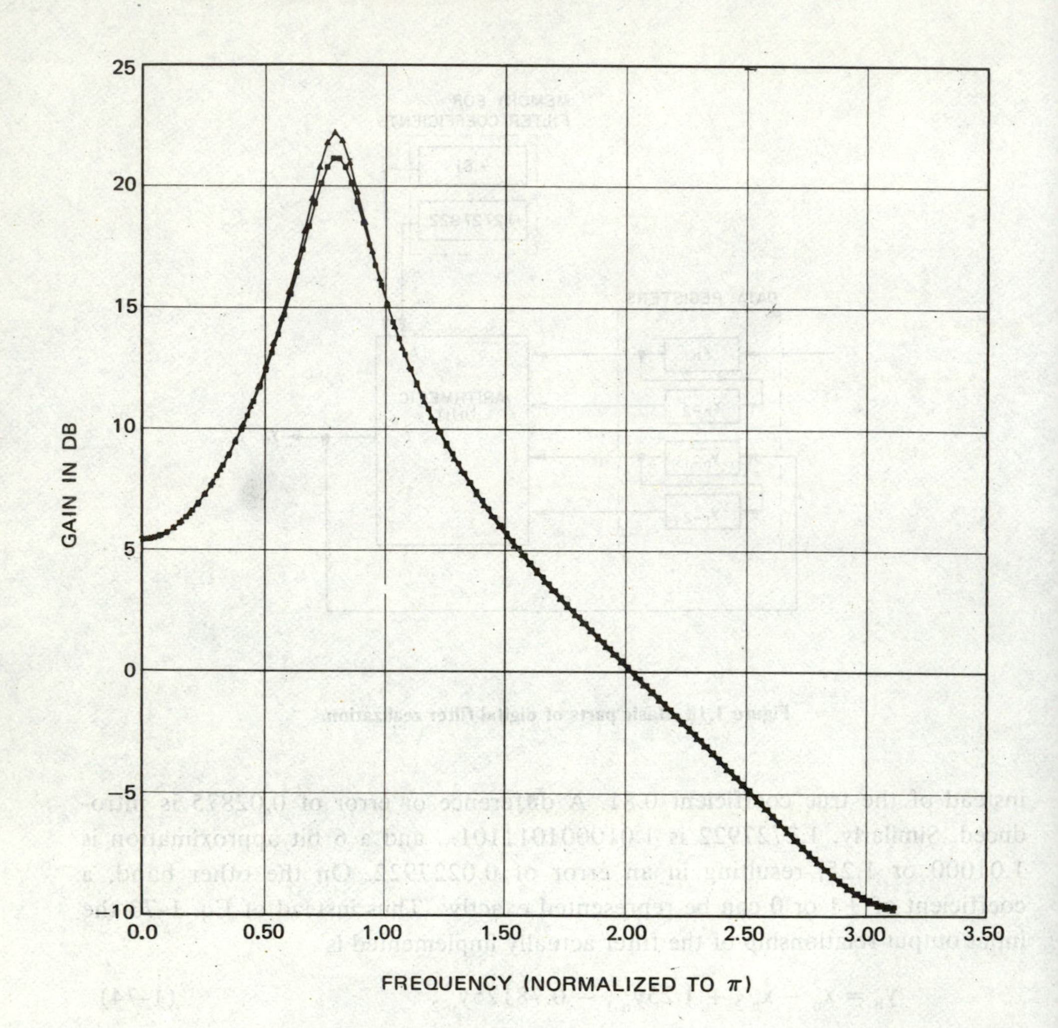

Figure 1.19 Amplitude response of filter before (triangles) and after coefficient quantization (squares).

number of bits can be accommodated. Suppose the data registers are 8 bit, then the input data in binary form, after it was truncated to 8 bits, is

. . . ,0.0100101, 0.1000111, 0.111010, 0.1110111, . . .

This corresponds to the values

. . . ,0.2890625, 0.5546875, 0.890625, 0.9265, . . .

which obviously differ from the actual samples of the sinusoidal signal.

The third type of error is due to the limited accuracy with which arithmetic operations are performed. Consider the term $-0.78125\ y_{n-2}$ appearing in Eq. 1-74. This is the product of a 6 bit number and an 8 bit number. In general it will have 14 significant bits that must be shortened to 8 bits, so that the result will fit in the 8 bit data register. The error thus committed is known as round-off error. Furthermore, the previously computed output samples are used via Eq. 1-74 to compute later output samples, and this has a cumulative effect.

To summarize, for a digital filter as specified by Eq. 1-36, there are three common error sources due to the finite word length:

1. The quantization of the input sequence $\{x_n\}$ into a set of discrete levels.
2. The modification of the filter coefficients $\{a_k\}$ and $\{b_k\}$ due to the use of a finite number of bits to represent them.
3. The accumulation of round-off errors committed in arithmetic operations.

We address these problems in more detail throughout the later chapters in this book.

1.7 THE DISCRETE FOURIER TRANSFORM

Consider the problem of calculating numerically the Fourier integral of a function f(t) as given in Eq. 1-1, which is repeated below.

$$F(\omega) = \int_{-\infty}^{\infty} f(t)e^{j\omega t}\,dt \tag{1-75}$$

Suppose the set of values of ω, for which the integral is to be calculated, is equally spaced in an interval of the ω-axis. The usual approach is to pick an interval on the t-axis, outside which f(t) is negligibly small. A uniform grid of points in this interval is chosen, and the integral of Eq. 1-75 is approximated by a summation,

$$F(\omega) \approx \Delta \sum_{k} f(k\Delta)e^{j\omega k\Delta} \tag{1-76}$$

where Δ is the step size on the t-axis. For uniformly spaced values of ω, the right side of Eq. 1-76 defines what is known as the *Discrete Fourier Transform* (DFT) or the finite Fourier transform. The discrete Fourier transform is very similar to the familiar Fourier integral and Fourier series. Yet, it has some distinct features of its own. This is the subject studied in this section. We first give the definition of the DFT and discuss some of its elementary properties. Then the calculation of the Fourier transforms, Eq. 1-75, using the DFT is treated. The computation of the DFT, using a fast algorithm known as the fast Fourier transform or FFT, is the subject of Chapter 3. Chapters 4 and 5 take up the question of hardware implementation of FFT processors.

(a) *Definition of the Discrete Fourier Transform*

Let $\{x_n\}$, $n=0,1,...,N-1$, be N complex numbers. We form for each $p=0,1,...,N-1$, the sum

$$A_p = \sum_{n=0}^{N-1} x_n (W_N)^{pn} \tag{1-77}$$

where $W_N = e^{-2j\pi/N}$. The set of N complex numbers so obtained, $\{A_p\}$, is called the discrete Fourier transform (DFT) of the sequence $\{x_n\}$. The term *N-point DFT* is also used when it is desirable to indicate explicitly the length of the data (and transform) sequence.

Each x_k can be obtained from the transform sequence $\{A_p\}$ by

$$x_k = (1/N) \sum_{p=0}^{N-1} A_p (W_N)^{-pk}$$

$$k = 0,1,...,N-1 \tag{1-78}$$

This relationship can be verified readily by inserting Eq. 1-77 into Eq. 1-78, changing the order of summation, and using the relationship

$$\sum_{p=0}^{N-1} (W_N)^{pn} (W_N)^{-pk} = \begin{cases} N \text{ if } k = n \\ 0 \text{ if } k \neq n \end{cases} \tag{1-79}$$

$\{x_n\}$ is called the inverse discrete Fourier transform (IDFT) of $\{A_p\}$.

If in the direct transform relationship, Eq. 1-77, p is allowed to take on values outside the range $\{0,1,...,N-1\}$, A_p would repeat itself periodically, that is,

$$A_{p+mN} = A_p, \quad m = 0, \pm 1, \pm 2,.... \tag{1-80}$$

Similarly, from Eq. 1-78, we have

$$x_{k+mN} = x_k, \quad m = 0, \pm 1, \pm 2,.... \tag{1-81}$$

We shall often take $\{x_n\}$ to mean the periodic extensions according to Eq. 1-81, and $\{A_p\}$ its periodic extension according to Eq. 1-80.

(b) *Elementary Properties*

A number of elementary properties of the DFT can be easily derived from its definition, Eq. 1-77. These are listed below. The reader will recognize the analogous properties of Fourier series and Fourier integral. We denote by $\{A_p\}$ the DFT of $\{x_n\}$; both are taken to mean the periodic extensions when appropriate. The asterisk denotes complex conjugate.

1. The DFT of $\{A_p\}$ is $\{Nx_{-n}\}$.
2. DFT is *linear*, that is, the DFT of $\{ax_n+by_n\}$ is $\{aA_p+bB_p\}$, where a and b are constants, and $\{A_p\}$ and $\{B_p\}$ are respectively the DFT of $\{x_n\}$ and $\{y_n\}$.
3. If $\{x_n\}$ has even symmetry, that is, $x_n=x_{-n}$ or equivalently, $x_n=x_{N-n}$, then its DFT is also even, $A_p=A_{-p}=A_{N-p}$.
4. If $\{x_n\}$ has odd symmetry, that is, $x_n=-x_{-n}$ or equivalently $x_n=-x_{N-n}$, then its DFT is also odd, $A_p=-A_{-p}=-A_{N-p}$.
5. If $\{x_n\}$ is real, then its DFT has even real part and odd imaginary part, that is, $A_p{}^*=A_{-p}=A_{N-p}{}^*$.
6. If $\{x_n\}$ is imaginary, then its DFT has odd real part and imaginary even part, that is, $A_p=-A_{-p}{}^*=-A_{N-p}{}^*$.
7. The DFT of $\{x_{-n}\}$ is $\{A_{-p}\}$.
8. The DFT of $\{x_n{}^*\}$ is $\{A_{-p}{}^*\}$.
9. The DFT of $\{x_{n-k}\}$ is $\{(W_N)^{pk}A_p\}$.
10. The DFT of $\{(W_N)^{-nk}x_n\}$ is $\{A_{p-k}\}$.

Let us work out some examples that are analytically tractable.

Example 1

$$x_n = e^{-an}, \quad n=0,1,....,N-1$$

From the definition of Eq. 1-77, the DFT is given by:

$$A_p = \sum_{n=0}^{N-1} e^{-an} \, (e^{-2j\pi/N})^{pn}$$

which is a geometric progression. It can be summed in close form

$$A_p = \frac{1 - e^{-aN}}{1 - e^{-a} \, e^{-2jp\pi/N}} \qquad p = 0,1,2,...,N-1$$

The sequences $\{x_n\}$ and $\{A_p\}$ are sketched in Figure 1.20 for a=0.5 and N=15.

Example 2

$$x_n = \sin(an), \quad n = 0,1,...,N-1$$

To compute its DFT we write sin(an) as $(e^{jan}-e^{-jan})/2j$ and use the linearity property of the DFT,

$$A_p = (1/2j) \sum_{n=0}^{N-1} e^{jan}\, e^{-2j\pi pn/N} - (1/2j)\sum_{n=0}^{N-1} e^{-jan}\, e^{-2j\pi pn/N}$$

$$= \frac{1 - e^{jaN}}{2j(1 - e^{j(a-2p\pi/N)})} - \frac{1 - e^{-jaN}}{2j(1 - e^{-j(a+2p\pi/N)})},$$

assuming $a \neq 2p\pi/N$ for integer p. Consider the first term. For those values of p such that $2p\pi/N$ is closer to a, the denominator would be small, so that the first term would be comparatively large, indicating a discrete spectral component there. The second term is essentially a mirror image of the first term. In Figure 1.21, we have plotted $\{ | A_p | \}$ for a=1 and N=30.

(c) *Periodic Convolution*

Let $\{x_n\}$ and $\{y_n\}$ be two periodic sequences of period N. Define

$$u_n = \sum_{k=0}^{N-1} x_{n-k} y_k = \sum_{k=0}^{N-1} x_k y_{n-k} \qquad (1\text{-}82)$$

$$n = 0,1,2,...,N-1$$

Then $\{u_n\}$ is also periodic with period N. It is called the convolution of $\{x_n\}$ and $\{y_n\}$. The terms periodic or circular convolution are also used. Suppose that $\{x_n\}$ and $\{y_n\}$ have DFTs $\{A_p\}$ and $\{B_p\}$, respectively. Then the DFT of $\{u_n\}$ is $\{A_p B_p\}$. The proof is very similar to that for Fourier integral and Fourier series and is accomplished by straightforward substitution. Eq. 1-82 is the familiar time domain convolution. There is also a frequency domain convolution defined as follows. Let the sequence $\{u_n\}$ be given by

$$u_n = x_n\, y_n \qquad (1\text{-}83)$$

and let the DFT of $\{u_n\}$ be denoted by $\{C_p\}$. Then

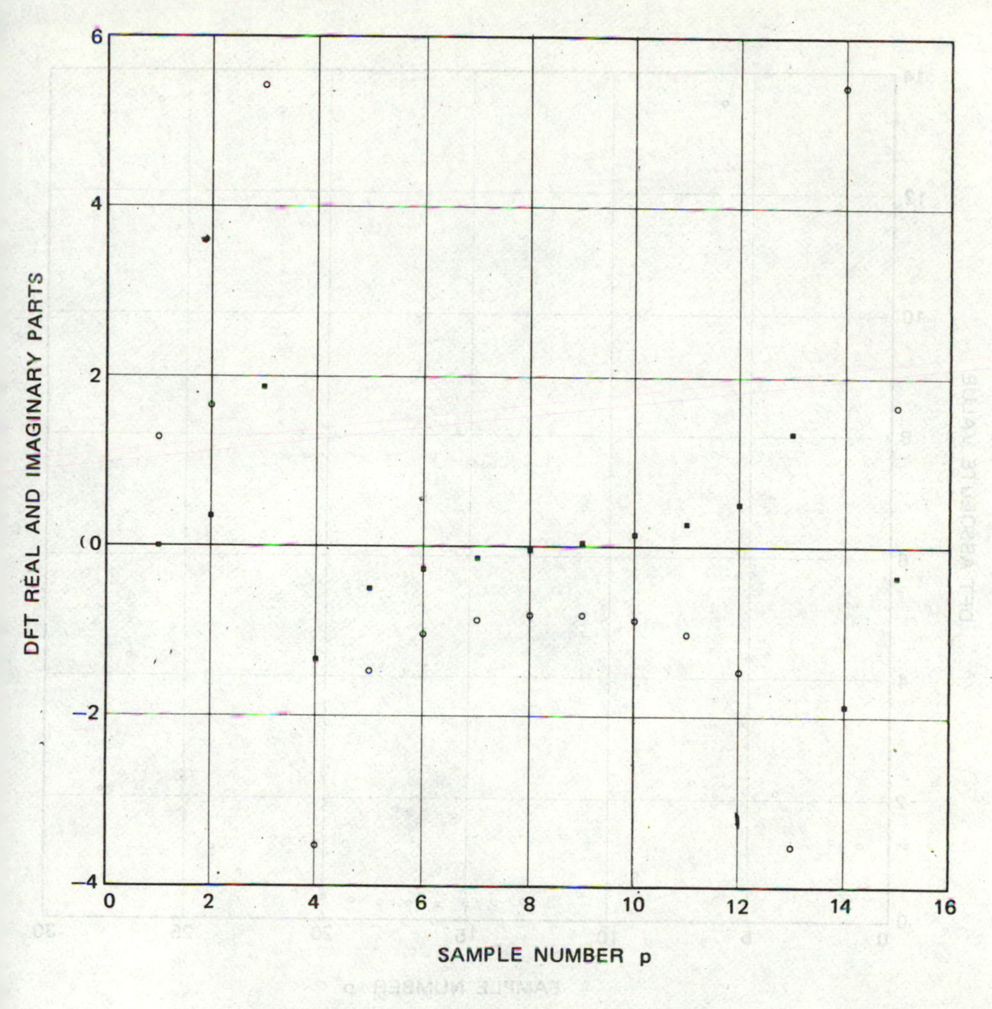

Figure 1.20 The DFT of the signal in Example 1. Real part shown in circles, imaginary part in squares.

$$C_p = \sum_{k=0}^{N-1} A_k B_{p-k} = \sum_{k=0}^{N-1} A_{p-k} B_k \tag{1-84}$$

$$p = 0,1,2,\ldots,N-1$$

The important Parseval theorem for DFT relates the *energy* of $\{x_n\}$ and $\{A_p\}$, and states

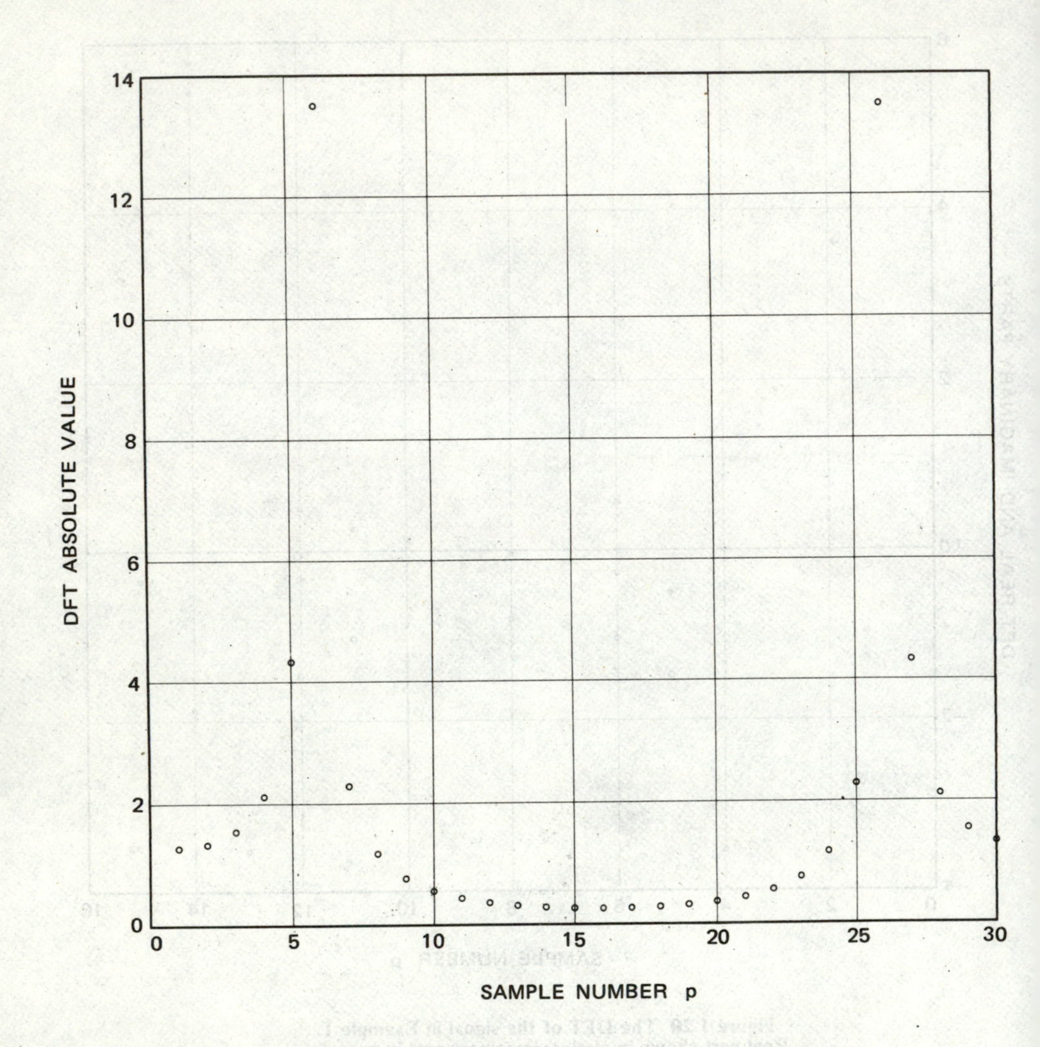

Figure 1.21 The magnitude of the DFT sequence in Example 2.

$$N \sum_{n=0}^{N-1} |x_n|^2 = \sum_{n=0}^{N-1} |A_p|^2 \qquad (1\text{-}85)$$

It can be proven either by direct substitution or by using the results on convolution.

(d) *The Calculation of the DFT*

To calculate the DFT of a sequence of N data points, it is seen from Eq. 1-77 that N complex multiplications and (N-1) complex additions are needed for one Fourier coefficient. Thus, a total of approximately N^2 complex multiplications and N(N-1) complex additions are needed to compute all the numbers in the transform. In many applications this amount of computation is excessive to be practical. We shall discuss in Chapter 3 a way of calculating the DFT that requires a total of $(N/2)\log_2 N$ complex multiplications and $N\log_2 N$ complex additions. Although this approach was known much earlier, its recent rediscovery is credited to Cooley and Tukey, and the method often bears their names as the *Cooley-Tukey algorithm*. The commonly accepted name, and most widely used, is the *Fast Fourier Transform* or simply FFT.

The savings in computer time of FFT over the brute force method is almost a factor of $N/(\log_2 N)$, which for large N is quite substantial, and the saving increases as N gets larger. When N=1024, the saving is approximately 100 to 1. The need to apply Fourier analyses to long sequences arises frequently. For example, in picture processing 1024 points corresponds to a grid of only 32×32 samples. In many problems, savings of the kind introduced by the FFT can mean the difference whether or not the solution is practical. Thus the impact of FFT can be easily appreciated.

(e) *Calculation of the Fourier Transform Using the DFT*

We illustrate here the use of the DFT in the evaluation of Fourier integrals, Eq. 1-1. Since the DFT consists of a finite number of points, we must represent f(t) and $F(\omega)$ each by a finite sequence. Suppose f(t) is represented by its samples spaced T apart, as shown in Figure 1.22*a*. Similarly, $F(\omega)$ is represented by its sample values Ω apart, as illustrated in Figure 1.22*b*. T and Ω represent respectively the *resolution* in the time and frequency domains.

We define, for $0 \leq n \leq N-1$,

$$f_n = \sum_{k=-\infty}^{\infty} f(nT + kNT) \tag{1-86}$$

and

$$F_n = \sum_{k=-\infty}^{\infty} F(n\Omega + kN\Omega) \tag{1-87}$$

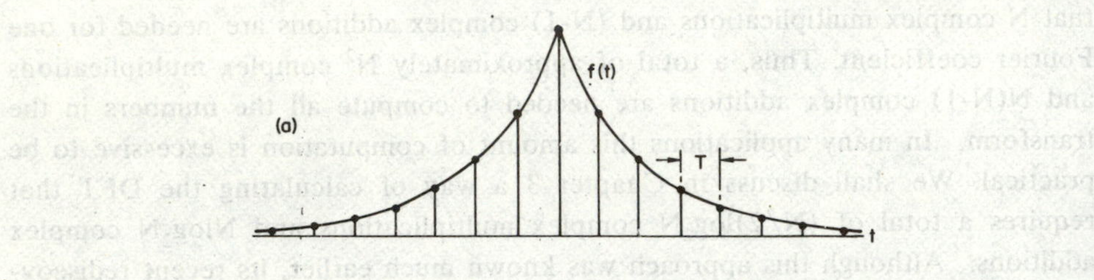

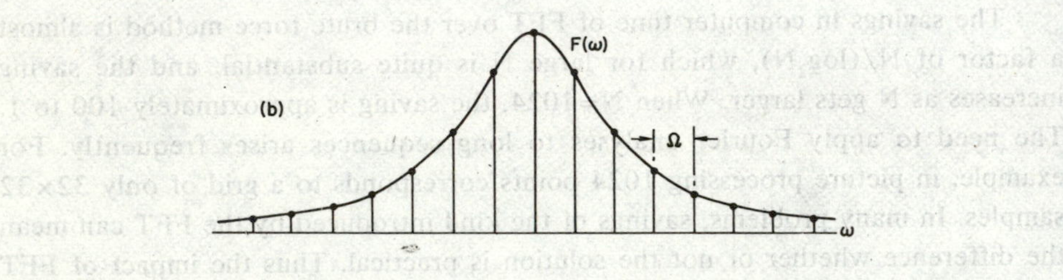

Figure 1.22 Calculation of the Fourier integral through the DFT.

Then it can be shown that if $T\Omega = 2\pi/N$, the DFT of $\{Tf_n\}$ is $\{F_n\}$. That is,

$$F_p = T \sum_{n=0}^{N-1} f_n e^{-2j\pi pn/N} \qquad (1\text{-}88)$$

The inverse DFT relationship is

$$T f_n = (1/N) \sum_{n=0}^{N-1} F_p e^{2j\pi pn/N} \qquad (1\text{-}89)$$

If NT and $N\Omega$ are sufficiently large, then only one term in each of the summations of Eqs. 1-86, 1-87 needs to be included. In other words, f_n and F_n are nearly equal to the actual samples of f(t) and $F(\omega)$, so that the DFT of Eq. 1-88 would enable us to calculate the samples of the Fourier transform $F(\omega)$ from the samples of f(t).

To prove Eq. 1-88, we first express f(nT) as follows:

$$f(nT) = (1/2\pi) \int_{-\infty}^{\infty} F(\omega)e^{jnT\omega}\, d\omega$$

$$= (1/2\pi) \sum_{k=-\infty}^{\infty} \int_{k2\pi/T}^{(k+1)2\pi/T} F(\omega)e^{jnT\omega}\, d\omega$$

$$= (1/2\pi) \sum_{k=-\infty}^{\infty} \int_{0}^{2\pi/T} F(\omega+2\pi k/T)e^{jnT(\omega+2\pi k/T)}\, d\omega$$

$$= (1/2\pi) \int_{0}^{2\pi/T} [\ \sum_{k=-\infty}^{\infty} F(\omega+2\pi k/T)\ ]\ e^{jnT\omega}\, d\omega \qquad (1\text{-}90)$$

Since $\Sigma\ F(\omega+2\pi k/T)$ is periodic in ω with period $2\pi/T$, it has a Fourier series expansion;

$$\sum_{k=-\infty}^{\infty} F(\omega+2\pi k/T) = \sum_{n=-\infty}^{\infty} a_n e^{-jnT\omega} \qquad (1\text{-}91)$$

The Fourier coefficients a_n are given by

$$a_n = (T/2\pi)\int_{0}^{2\pi/T} [\ \sum_{k=-\infty}^{\infty} F(\omega+2\pi k/T)\ ]\ e^{jnT\omega}\, d\omega$$

$$= T\, f(nT) \qquad (1\text{-}92)$$

in view of Eq. 1-90. On setting $T\Omega=2\pi/N$ and $\omega=p\Omega$, p integer, Eq. 1-91 becomes

$$\sum_{k=-\infty}^{\infty} F(p\Omega+kN\Omega) = T \sum_{n=-\infty}^{\infty} f(nT)e^{-jpn2\pi/N} \qquad (1\text{-}93)$$

The left side is simply F_p. We now replace n on the right side by m+kN with $m=0,1,...,N-1$, so that the sum over n becomes a double sum, one over m from 0 to $N-1$ and another over k from $-\infty$ to ∞. The right side of Eq. 1-93 then becomes

$$T \sum_{m=0}^{N-1} \sum_{k=-\infty}^{\infty} f(mT+kNT)e^{-j(m+kN)p2\pi/N}$$

$$= \sum_{m=0}^{N-1} [T\sum_{k=-\infty}^{\infty} f(mT+kNT)\,]e^{2j\pi pm/N} = T\sum_{m=0}^{N-1} f_m e^{-2j\pi pm/N} \qquad (1\text{-}94)$$

which is Eq. 1-88.

It should be noticed that if f(t) is of finite duration and its samples taken every T seconds are f(0), f(T),....,f((N-1)T), then

$$f_n = f(nT),\ \ n=0,1,....,N-1$$

We now give a simple example to illustrate this calculation. Let f(t) be the one sided exponential function

$$f(t) = \begin{cases} e^{-t} & t > 0 \\ 0.5 & t = 0 \\ 0 & t < 0 \end{cases}$$

It is well known that its Fourier transform is

$$F(\omega) = \frac{1}{1 + j\omega}$$

Suppose we pick T = 0.5 and N = 20. So $\Omega = 2\pi/NT = \pi/5$. We note that for t > NT = 10, f(t) can be regarded as negligible, so that f(nT) can be taken as f_n for n=0,1,....,19. We also note that f(0) = 0.5, because of the discontinuity of f(t) at t = 0. The sequence $\{f_n\}$ is 0.50000, 0.60653, 0.36788, 0.22313, 0.13534, 0.08208, 0.04979, 0.03020, 0.01832, 0.01111, 0.00674, 0.00409, 0.00248, 0.00150, 0.00091, 0.00055, 0.00034, 0.00020, 0.00012, 0.00008. Its DFT $\{F_n\}$ is tabulated below for n=0,1,....,10. For n>10, $F_n = F^*_{19-n}$. This symmetry is due to the real $\{f_n\}$. Because of the periodicity of the DFT, $F_{19} = F_{-1}$, $F_{18} = F_{-2}$, etc. That is, F_{19}, F_{18}, ... , F_{11} are the approximation to $F(\omega)$ for negative ω. Also tabulated are the values of $F(n\Omega)$, n=0,1,...,10. The agreement between F_n and $F(n\Omega)$ is seen to be reasonably good. For better accuracy, we must increase the value of N and decrease the value of T.

F_n	$F(n\Omega)$
1.0207 + j 0.0000	1.0000 + j 0.0000
0.7378 + j 0.4375	0.7170 + j 0.4505
0.4089 + j 0.4612	0.3877 + j 0.4872
0.2413 + j 0.3746	0.2196 + j 0.4140
0.1591 + j 0.2904	0.1367 + j 0.3435
0.1155 + j 0.2217	0.0920 + j 0.2890
0.0907 + j 0.1655	0.0657 + j 0.2478
0.0759 + j 0.1179	0.0492 + j 0.2162
0.0673 + j 0.0759	0.0381 + j 0.1914
0.0627 + j 0.0372	0.0303 + j 0.1715
0.0612 + j 0.0000	0.0247 + j 0.1552

1.8 SUMMARY

This chapter began with a brief review of some aspects of signal and system theory, both in continuous-time and in discrete-time. No attempt at completeness was made in this review; the primary purpose was to provide, in a limited number of pages, an adequate background necessary for the understanding of the topics to be discussed later in the book. For readers who have not had sufficient previous exposure to discrete-time systems, additional reading, particularly on Z-transform and related topics, is highly advisable.

An overview of the basic theory of digital filtering was then presented, and a fundamental relationship governing the processing of analog signals by digital means was derived, Eq. 1-66. The design of digital filters to meet a given set of specifications will be taken up in the next chapter. The hardware implementation of these filters will be discussed in Chapters 4 and 5. The reader is then introduced to the effects of finite word length on the accuracy of digital filters. These effects must be taken into account in any practical application of digital signal processing. Additional discussions on this important topic can be found in Chapters 2 and 6.

We concluded the chapter with an introduction to the discrete Fourier transform. In Chapter 3 a class of efficient algorithms to calculate the transform will be discussed in detail, and the implementation of special purpose hardware to perform these calculations will be presented in Chapter 5.

Digital signal processing is a broad field where active research is being pursued in many directions. It is not possible to cover in a book such as ours all the topics in this field. Among the more important topics that we shall not touch upon, we may mention the homomorphic signal processing, the Chirp Z-transform, and the processing of two dimensional signals. Furthermore, digital signal processing has many important applications in diverse fields as speech, radar, seismology, pattern recognition, and so on. We do not discuss in this book specific applications, but rather concentrate on the aspects that are common to all such systems. Readers interested in these topics may consult the bibliography at the end of this chapter for specific information on these subjects.

EXERCISES

1.1 Show that the systems described by Eq. 1-10 and the system described by Eq. 1-12 are both linear and time invariant.

1.2 Find the Z-transform of the sequences:
(a) $x_n=n,\ n\geq 0$.
(b) $x_n=1/n,\ n\geq 1$.
(c) $x_n=n\sin(an),\ n\geq 0$.
(d) $x_n=c$ for $0\leq n\leq N$ and $x_n=0$ for $n>N$.
(e) $x_n=n$ for $0\leq n\leq N$ and $x_n=0$ for $n>N$.

1.3 Let the Z-transform of $\{x_n\}$ and $\{y_n\}$ be X(z) and Y(z). Relate Y(z) to X(z) if
(a) $y_n=c^n x_n$.
(b) $y_n=nx_n$.
(c) $y_{2n}=x_n$ and $y_{2n+1}=0$.
(d) $y_{2n}=y_{2n+1}=x_n$.
(e) $y_n=x_{-n}$.

1.4 Find the inverse Z-transform of
(a) $1/(1-cz^{-1})^k$, $|z|>|c|$, k integer greater than 1
(b) $e^{c/z}$, $|z|>0$

1.5 *Final Value Theorem*. Show that

$$\lim_{n\to\infty} x_n = \lim_{z\to 1}(1-z)\,X(z)$$

1.6 *Parseval's Theorem*. Show that

$$\sum_{n=-\infty}^{\infty} x_n^{\ 2} = (1/2\pi j)\oint_C X(z)X(z^{-1})z^{-1}dz$$

where C is an appropriately chosen contour.

1.7 One definition for a system to be stable is that a bounded input produces a bounded output. Show that for the system described by Eq. 1-10, a necessary and sufficient condition is that

$$\lim_{n\to\infty} h_n = 0$$

Show that in the case of a system with rational transfer function, this condition is equivalent to the absence of poles outside the unit circle.

1.8 For the third order lowpass filter of Section 1.4:

(a) Write its time domain input-output relationship in the form of Eq. 1-36.

(b) Determine the two direct form realizations of the filter.

(c) Determine the parallel and cascade realizations of the filter.

(d) Write a computer program to calculate the first 100 samples of the filter with two input sequences: (*i*) $x_n=\sin(10n)$ and (*ii*) $x_n=\sin(n/10)$, assuming zero initial condition. Sketch both the input and output signal in these two cases.

1.9 A symmetric triangular pulse x(t) of width 10 msec is sampled at t=nT.

(a) Sketch the analog and digital spectra $X^a(\omega)$ and $X(e^{j\omega})$ for T=2 msec and for T=0.5 msec.

(b) Determine T so that x(t) has 1% of its energy beyond the frequency π/T rad/sec.

(c) Repeat (a) and (b) for a square pulse x(t) of same width.

1.10 Use Eq. 1-64 and Figure 1.9 to derive the well known Sampling theorem by showing the output y(t) is identical to the input x(t) under the following conditions: (*i*) x(t) is bandlimited, that is, $X^a(\omega)=0$ for $|\omega|>\pi/T$, (*ii*) H(z)=1, and (*iii*) $G(\omega)$ is an ideal lowpass filter. $G(\omega)=T$ for $|\omega|<\pi/T$ and $G(\omega)=0$ for $|\omega|>\pi/T$.

1.11 Suppose, in the system described by Figure 1.9, $x(t)=e^{-a|t|}$, H(z)=1, and $G(\omega)$ is the ideal lowpass filter.

(a) Calculate the output analog spectrum $Y^a(\omega)$.

(b) Calculate the mean square error

$$\int_{-\infty}^{\infty}[x(t)-y(t)]^2dt$$

and sketch this error as a function of the parameter aT.

(c) Repeat (a) and (b) for the case where $X^a(\omega)=1$ in $|\omega|<\pi/T$ and is zero otherwise, H(z)=1, and $G(\omega)$ is an RC filter with time constant 1/a.

1.12 For a *notch filter* which has a pair of zeros at $z=e^{\pm ja}$ and a pair of poles at $be^{\pm ja}$, $|b|<1$:

(a) Sketch the frequency response.

(b) If such a notch filter is applied to a signal sampled at 100 KHz in order to remove a 60 Hz pickup, what should a be.

1.13 Rework exercise 1.8 by using five bits for the coefficients and eight bits for the data. Also, calculate the frequency response of the filter using quantized coefficients, and compare with Figure 1.8.

1.14 Prove the elementary properties of DFT of Section 1.7(b).

1.15 Find the DFT of

(a) {1,1,−1,−1}.

(b) {1,j,−1,−j}.

(c) $x_n=cn$, $0\le n\le N-1$.

(d) $x_n=\sin(2\pi n/N)$, $0\le n\le N-1$.

1.16 For the sequences {1,2,3,4,5,0,0}, {1,1,1,1,0,0,0}:

(a) Find their periodic convolution.

(b) Find the convolution of the above two sequences by treating them as finite, not periodic, sequences. Compare the result with that obtained in (a).

1.17 Let $\{u_n\}$ be the periodic convolution of $\{x_n\}$ and $\{y_n\}$ as defined by Eq. 1-82. Prove the statement: "Then the DFT of $\{u_n\}$ is $\{A_pB_p\}$."

1.18 Prove the Parseval theorem for DFT, Eq. 1-85.

1.19 Let $\{A_p\}$ be the N point DFT of $\{x_n\}$. Define $\{y_n\}$, $0 \leq n \leq 2N-1$, by $y_{2n} = x_n$ and $y_{2n+1} = 0$.

(a) Find the DFT of $\{y_n\}$.

(b) Repeat (a) if $y_n = x_n$ for $0 \leq n \leq N-1$ and $y_n = 0$ for $N \leq n \leq 2N-1$.

1.20 Use the approach of Section 1.7(e) to calculate an approximation to the Fourier transform of a square pulse of width 1. Choose N=20 and T=1. Note half of the samples f(nT) would be zero.

BIBLIOGRAPHY

1. E. O. Brigham, *The Fast Fourier Transform*, Prentice-Hall, Engelwood Cliffs, New Jersey, 1974.
2. J. A. Cadzow, *Discrete-Time Systems*, Prentice-Hall, Engelwood Cliffs, New Jersey, 1973.
3. E. I. Jury, *Theory and Application of the Z-Transform Method*, John Wiley & Sons, New York, 1964.
4. A. V. Oppenheim, R. W. Schafer and T. G. Stockham, *Nonlinear Filtering of Multiplied and Convolved Signals*, Proc. IEEE, vol. 56, August 1968, pp. 1264-1291.
5. A. V. Oppenheim and R. W. Schafer, *Digital Signal Processing*, Prentice-Hall, Engelwood Cliffs, New Jersey, 1975.
6. A. Papoulis, *The Fourier Integral and its Applications*, McGraw Hill, New York, 1962.
7. L. R. Rabiner, R. W. Schafer and C. M. Rader, *The Chirp Z-Transform Algorithm*, IEEE Trans. Audio & Electroacoust., vol. AU-17, June 1969, pp. 86-92.
8. L. R. Rabiner and B. Gold, *Theory and Application of Digital Signal Processing*, Prentice-Hall, Engelwood Cliffs, New Jersey, 1975.
9. M. Schwartz and L. Shaw, *Signal Processing, Discrete Spectral Analysis, Detection and Estimation*, McGraw Hill, New York, 1975.
10. K. Steiglitz, *An Introduction to Discrete Systems*, John Wiley & Sons, New York, 1974.
11. *Digital Signal Processing*, Proc. IEEE (special issue), vol. 63, April 1975.

CHAPTER 2

Design of Digital Filters

2.1 INTRODUCTION

The design of a digital filter is the task of determining a transfer function which is a rational function in z^{-1} in the case of recursive (IIR) filters or a polynomial in z^{-1} in the case of nonrecursive (FIR) filters. The transfer function must, of course, meet certain performance specifications. In most design problems, it is also desirable to arrive at a transfer function of minimal complexity or nearly so, compatible with the given specification. We discuss in this chapter a number of techniques for designing digital filters. Two computer programs for filter design are listed in Appendices 2.1 and 2.2.

The implementation of the filter transfer function by digital hardware is discussed in Chapters 4 and 5. In the implementation, one must also take into account a number of practical considerations arising primarily from the use of finite word length. Some of these will be discussed in Section 2.8, and others are discussed in Chapter 6.

A number of excellent design methods for digital filters have been developed in recent years. The use of a particular method would depend on whether an IIR (infinite impulse response) or FIR (finite impulse response) filter is called for, and on whether the design specification is given in the frequency domain or in the time domain. In most applications, a frequency selective filter is needed. This filter lets certain frequency components pass through, but attenuates others. For this type of filter, the design specification is given in the frequency domain. We shall limit our discussion in this chapter to this class of design problems.

In the case of IIR filters, we will draw heavily upon the well developed body of knowledge of analog filter design [1,2,3]. The basic approach here is a transformation between the analog and digital frequency variables. When the specification is not suitable for the application of the transformation approach, one must resort to computer aided methods. This is discussed in Section 2.5. Appendix 2.4 contains a brief review of analog filter design methods.

For the design of FIR filters, we also discuss two methods. The first uses the Fourier series approach with *windows*, Section 2.6. The merit of this approach lies in its simplicity, requiring comparatively little computation; however, the resulting design is usually not the best possible. The second method, discussed in Section 2.7, uses a computer aided approach to arrive at an optimum design based on the minimax error criterion.

The use of FIR filters for decimation and interpolation in applications requiring multirate processing is discussed in Section 2.8. We also show that this approach can lead to an efficient use of FIR filters to implement narrow band digital filters.

Finally, in Section 2.9 we discuss the degradation in the frequency response characteristic of digital filters, when the filter coefficients are represented by finite length word (binary number) in the implementation.

2.2 DESIGN OF LOWPASS IIR FILTERS

A typical amplitude characteristic of a lowpass type filter is shown in Figure 2.1. The passband extends from $\lambda = 0$ to the cutoff frequency $\lambda = \Lambda_c$, and the stopband extends from $\lambda = \Lambda_s$ to $\lambda = \pi$. Only positive frequencies are shown, since the amplitude characteristic exhibits even symmetry. The frequency range $\Lambda_c \leq \lambda \leq \Lambda_s$ is the transition or *don't care* region. The design specifications typically include the tolerance limits δ_1 allowed in the passband and δ_2 in the stopband. The goal of the design is to determine a transfer function H(z), so that its amplitude on the unit circle $|H(e^{j\lambda})|$ satisfies

$$1 - \delta_1 \leq |H(e^{j\lambda})| \leq 1 + \delta_1 \quad \text{for } 0 \leq \lambda \leq \Lambda_c$$

and

$$|H(e^{j\lambda})| \leq \delta_2 \quad \text{for } \Lambda_s \leq \lambda \leq \pi$$

Let us relate these digital frequencies to analog frequencies. Suppose that the digital filter is intended to process an analog signal sampled at a frequency of $F_s = 1/T$, T being the time between samples. From Eq. 1-67 in Chapter 1, we see that the digital cutoff frequency Λ_c corresponds to the analog frequency of

$$f_c = \Lambda_c F_s / 2\pi \quad \text{Hz}$$

Similarly, the stopband in analog frequency begins at

$$f_s = \Lambda_s F_s / 2\pi \quad \text{Hz}$$

These are indicated in Figure 2.1.

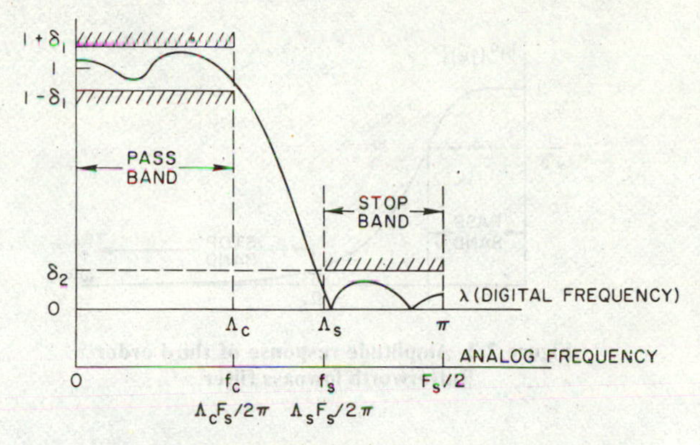

Figure 2.1 Amplitude response specification of a typical lowpass filter.

One approach to the design problem at hand is to select an appropriate analog lowpass filter and to apply a frequency transformation to its transfer function so as to obtain the desired H(z). There are three commonly encountered classes of lowpass filters: the Butterworth, Chebyshev, and Cauer or elliptic function filters. Butterworth filters have a maximally flat amplitude response, their poles being uniformly spaced on the unit circle in the s-plane. Chebyshev filters have an equiripple amplitude response in the passband (or in the stopband), their poles being located on an ellipse in the s-plane. Cauer filters have an equiripple amplitude response in both the passband and stopband. We shall assume in the following discussion a minimal familiarity on the part of the reader with these filters. Appendix 2.4 contains a brief summary of these concepts. The reader may wish to review this material before proceeding.

For the purpose of our discussion let us consider a third order Butterworth lowpass filter with the transfer function

$$H^a(s) = \frac{1}{(s+1)(s^2+s+1)} \tag{2-1}$$

Its amplitude response, shown in Figure 2.2, is

$$|H^a(j\omega)| = (1+\omega^6)^{-1/2} \tag{2-2}$$

where the 3 dB cutoff frequency has been normalized to $\omega = 1$. Suppose we take the stopband of this filter to be $\Omega_s \leq \omega < \infty$, then the amplitude response $|H^a(j\omega)|$ is less than $(1+\Omega_s{}^6)^{-1/2}$ in the stopband.

Now consider the transformation from ω to λ given by

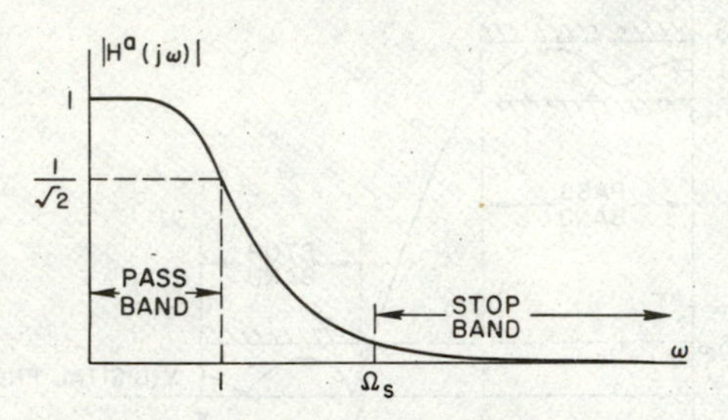

Figure 2.2 Amplitude response of third order Butterworth lowpass filter.

$$\omega \tan(\Lambda_c/2) = \tan(\lambda/2) \qquad (2\text{-}3)$$

The λ versus ω curve of this transformation is plotted in Figure 2.3, and it is seen to be monotonically increasing. With this transformation, the semi-infinite interval $0 \leq \omega < \infty$ is mapped into the interval $0 \leq \lambda < \pi$, with $\omega = 1$ corresponding to $\lambda = \Lambda_c$. The transformation of Eq. 2-3, when applied to $|H^a(j\omega)|$, yields a function in λ which exhibits a lowpass characteristic, as shown in Figure 2.3.

With $s = j\omega$ and $z = e^{j\lambda}$, Eq. 2-3 can be rewritten as

$$\begin{aligned} s &= j\cot(\Lambda_c/2)\,\frac{\sin(\lambda/2)}{\cos(\lambda/2)} \\ &= j\cot(\Lambda_c/2)\,\frac{(e^{j\lambda/2} - e^{-j\lambda/2})/2j}{(e^{j\lambda/2} + e^{-j\lambda/2})/2} \\ &= c\,\frac{z-1}{z+1} \qquad (2\text{-}4) \end{aligned}$$

where $c = \cot(\Lambda_c/2)$. The expression for $H^a(s)$ of Eq. 2-1 can now be rewritten in terms of z, using Eq. 2-4, and we obtain

$$H^a[c(z-1)/(z+1)] = \frac{(z+1)^3}{c_1(z+p_1)(z^2+b_1z+d_1)} = H(z) \qquad (2\text{-}5)$$

where the constants are given by:

$$c_1 = (1+c)(1+c+c^2) \qquad p_1 = (1-c)/(1+c)$$

$$b_1 = (2-2c^2)/(1+c+c^2) \qquad d_1 = (1-c+c^2)/(1+c+c^2)$$

This H(z) is a rational function in z, hence it may be regarded as the transfer function of an IIR digital filter. From the previous discussion, it is seen that its amplitude $|H(e^{j\lambda})|$ for $0 \leq \lambda \leq \pi$ is given by the $|H^a(j\omega)|$ of Eq. 2-2 under the transformation of Eq. 2-3.

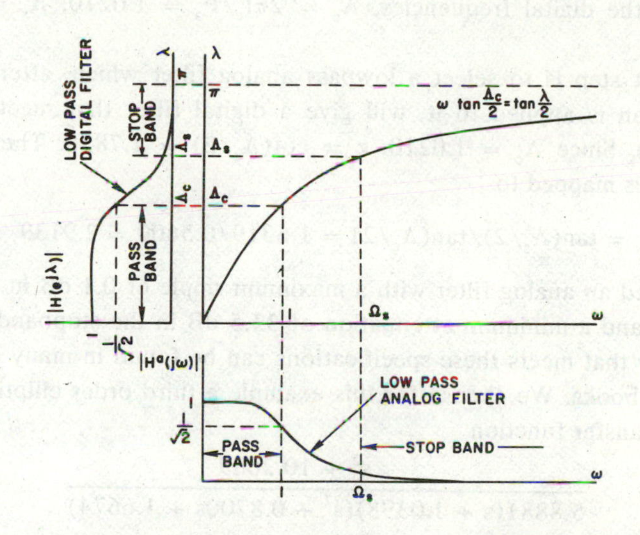

Figure 2.3 A lowpass to lowpass frequency transformation, Eq. 2-3.

The stopband of the digital filter H(z) is from $\lambda = \Lambda_s$ to $\lambda = \pi$, where Λ_s is related to the stopband of the analog filter Ω_s through Eq. 2-3, that is,

$$\Omega_s \tan(\Lambda_c/2) = \tan(\Lambda_s/2) \tag{2-6}$$

Thus in the stopband, the amplitude response satisfies

$$|H(e^{j\lambda})| \leq (1 + \Omega_s^{6})^{-1/2} \qquad \Lambda_s \leq \lambda \leq \pi \tag{2-7}$$

The transformation of Eq. 2-4 is a *bilinear transformation* between the complex variables s and z. It maps the left half of the s-plane to the interior of the unit circle $|z| = 1$ and, as noted earlier, $0 \leq \omega < \infty$ is mapped to $0 \leq \lambda < \pi$. Thus, a stable $H^a(s)$ will produce a stable H(z).

Let us now apply this approach to the design of a digital lowpass filter to process an analog signal with the following specifications:

sampling frequency (F_s) : 8 kHz
passband (f_c) : from 0 to 1.3 kHz
maximum ripple in passband (δ_1) : 0.1 dB
stopband (f_s) : beyond 2.6 kHz
minimum attenuation in stopband (δ_2) : 33.5 dB

In terms of the digital frequencies, $\Lambda_c = 2\pi f_c/F_s = 1.0210$, $\Lambda_s = 2\pi f_s/F_s = 2.0420$.

The first step is to select a lowpass analog filter which, after the bilinear transformation is applied to it, will give a digital filter that meets the design specifications. Since $\Lambda_c = 1.0210$, $c = \cot(\Lambda_c/2) = 1.7856$. The edge of the stopband Λ_s is mapped to

$$\Omega_s = \tan(\Lambda_s/2)/\tan(\Lambda_c/2) = 1.6319/0.5600 = 2.9139$$

Thus, we need an analog filter with a maximum ripple of 0.1 dB in the passband $0 \le \omega \le 1$, and a minimum attenuation of 33.5 dB in the stopband $2.9139 \le \omega < \infty$. A filter that meets these specifications can be found in many of the analog filter design books. We choose for this example a third order elliptic filter [2, p. 49] with a transfer function

$$H^a(s) = \frac{s^2 + 10.2089}{5.8881(s + 1.0398)(s^2 + 0.8700s + 1.6674)}$$

meets these specifications. The bilinear transformation is $s=1.7856(z-1)/(z+1)$. So the desired digital filter has the transfer function

$$H(z) = \frac{0.1256(1 + 1.0478z^{-1} + z^{-2})(1 + z^{-1})}{(1 - 0.2640z^{-1})(1 - 0.4748z^{-1} + 0.5153z^{-2})}$$

The amplitude response of this filter is shown in Figure 2.4*a*.

The above example illustrates the basic idea behind the frequency transformation approach to the design of digital filters. There are basically four steps in the design procedure:

1. From the sampling frequency, the specified analog cutoff frequency f_c and the specified analog stopband edge frequency f_s determine the digital cutoff frequency Λ_c and the digital stopband edge frequency Λ_s by $\Lambda_c=2\pi f_c/F_s$ and $\Lambda_s=2\pi f_s/F_s$. This step is not necessary if the specification is already given in terms of the digital frequency λ.
2. Determine the stopband edge frequency Ω_s of the analog filter from Eq. 2-6.
3. From the specifications on the amplitude response, determine an appropriate analog transfer function $H^a(s)$.

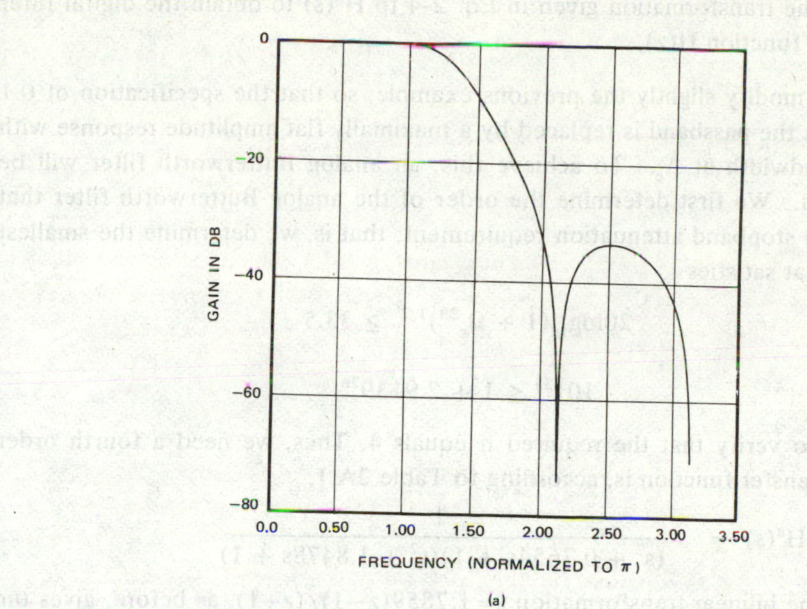

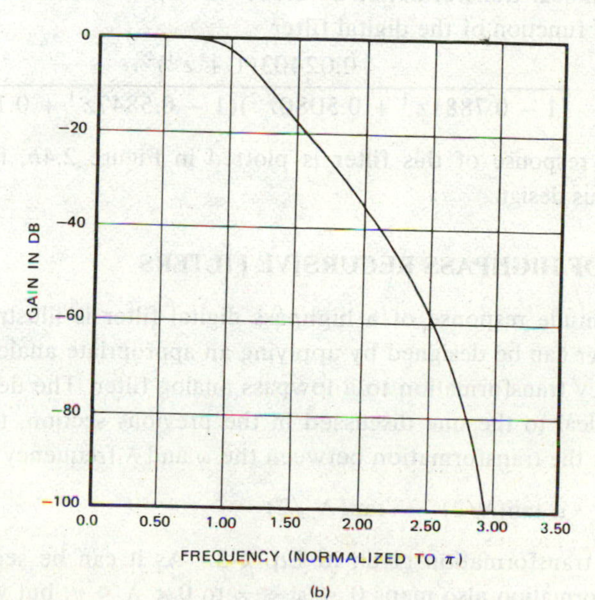

Figure 2.4 Amplitude response of the lowpass filters discussed in the example: (a) a filter designed for a specified ripple, (b) same filter designed for 3 dB bandwidth.

4. Apply the transformation given in Eq. 2-4 to $H^a(s)$ to obtain the digital filter transfer function H(z).

Let us modify slightly the previous example, so that the specification of 0.1 dB ripple in the passband is replaced by a maximally flat amplitude response with a 3 dB bandwidth at Λ_c. To achieve this, an analog Butterworth filter will be transformed. We first determine the order of the analog Butterworth filter that satisfies the stopband attenuation requirement; that is, we determine the smallest integer n that satisfies

$$20\log_{10}(1 + \Omega_s^{2n})^{1/2} \geq 33.5$$

or

$$10^{3.35} \leq 1 + 2.9139^{2n}$$

It is easy to verify that the required n equals 4. Thus, we need a fourth order filter. Its transfer function is, according to Table 2A.1,

$$H^a(s) = \frac{1}{(s^2 + 0.7654s + 1)(s^2 + 1.8478s + 1)}$$

Applying the bilinear transformation $s=1.7859(z-1)/(z+1)$, as before, gives the desired transfer function of the digital filter

$$H(z) = \frac{0.02403(1 + z^{-1})^4}{(1 - 0.7881z^{-1} + 0.5080z^{-2})(1 - 0.5847z^{-1} + 0.1188z^{-2})}$$

The amplitude response of this filter is plotted in Figure 2.4*b*, for comparison with the previous design.

2.3 DESIGN OF HIGHPASS RECURSIVE FILTERS

A typical amplitude response of a highpass digital filter is illustrated in Figure 2.5. Such a filter can be designed by applying an appropriate analog frequency to digital frequency transformation to a lowpass analog filter. The design procedure is almost identical to the one discussed in the previous section, the only difference being that the transformation between the ω and λ frequency variables is

$$\omega \tan(\lambda/2) = \tan(\Lambda_c/2) \qquad (2\text{-}8)$$

instead of the transformation given in Eq. 2-3. As it can be seen from Figure 2.6, this transformation also maps $0 \leq \omega < \infty$ to $0 < \lambda \leq \pi$, but with $0 < \omega < 1$ corresponding to $\Lambda_c < \lambda < \pi$. The edges of the stopbands Λ_s and Ω_s are related through

$$\Omega_s \tan(\Lambda_s/2) = \tan(\Lambda_c/2) \qquad (2\text{-}9)$$

with $s = j\omega$ and $z = e^{j\lambda}$, Eq. 2-8 can be rewritten as

$$s = c' \frac{z+1}{z-1} \tag{2-10}$$

where $c' = \tan(\Lambda_c/2)$. This bilinear transformation, when applied to the transfer function of a lowpass analog filter, will produce the transfer function of a highpass digital filter, that is,

$$H(z) = H^a[c'(1+z)/(1-z)] \tag{2-11}$$

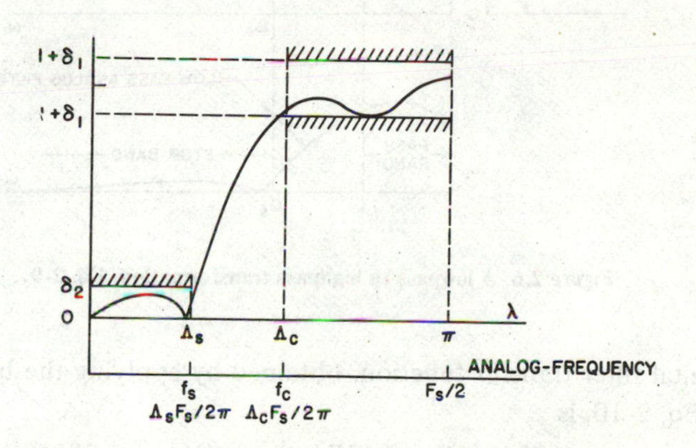

Figure 2.5 Amplitude response specification for a highpass filter.

We illustrate this design procedure with the following example. Suppose we wish to design a highpass digital filter with the following specifications:

sampling frequency (F_s) : 15 kHZ
passband (f_c) : from 4.5 to 7.5 kHz
maximum ripple in passband : 0.1 dB
stopband (f_s) : from 0 to 3.4 kHz
minimum attenuation in the stopband : 30 dB

In terms of digital frequencies, $\Lambda_c = 2\pi f_c/F_s = 1.8850$, and $\Lambda_s = 2\pi f_s/F_s = 1.4242$. With $c' = \tan(\Lambda_c/2) = 1.3764$, the edge of the stopband Λ_s is mapped into the analog frequency $\Omega_s = \tan(\Lambda_c/2)/\tan(\Lambda_s/2) = 1.5946$. A fourth order elliptic filter [2, p. 65] meets these requirements. The transfer function of the analog filter is

$$H^a(s) = \frac{(s^2 + 2.8598)(s^2 + 14.0262)}{38.2880(s^2 + 0.3702s + 1.2697)(s^2 + 1.3869s + 0.8342)}$$

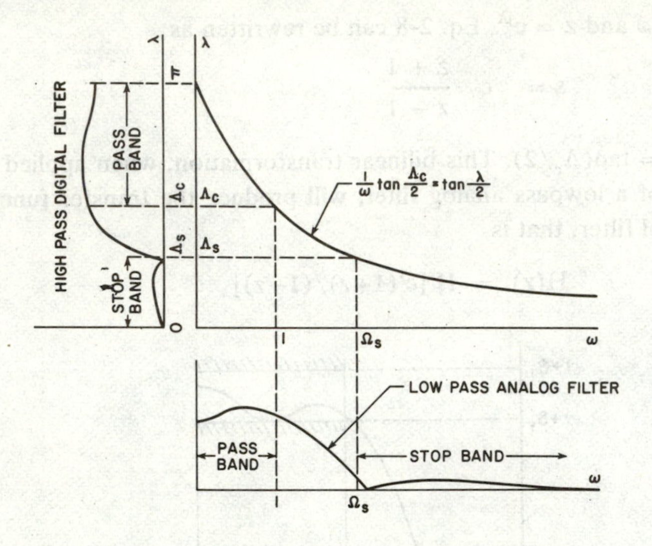

Figure 2.6 A lowpass to highpass transformation, Eq. 2-9.

and the digital filter transfer function, obtained by applying the bilinear transformation of Eq. 2-10, is

$$H(z) = \frac{0.09421(1 - 0.1174z^{-1} + z^{-2})(1 - 1.3886z^{-1} + z^{-2})}{(1 + 0.5873z^{-1} + 0.7318z^{-2})(1 + 0.6114z^{-1} + 0.2085z^{-2})}$$

Its amplitude response is plotted in Figure 2.7.

2.4 DESIGN OF BANDPASS AND BANDSTOP FILTERS

A typical amplitude response of a digital bandpass filter is shown in Figure 2.8. The passband extends from Λ_c' to Λ_c'', and the stopbands are $0 \leq \lambda \leq \Lambda_s'$ and $\Lambda_s'' \leq \lambda \leq \pi$. It is possible to construct a transformation between ω and λ which, when applied to an analog lowpass filter, will give a bandpass digital filter. One such transformation is

$$\omega = \alpha(\beta - \cos\lambda)/\sin\lambda \tag{2-12}$$

where

$$\alpha = \cot[(\Lambda_c'' - \Lambda_c')/2]$$

$$\beta = \frac{\sin(\Lambda_c' + \Lambda_c'')}{\sin\Lambda_c' + \sin\Lambda_c''} \tag{2-13}$$

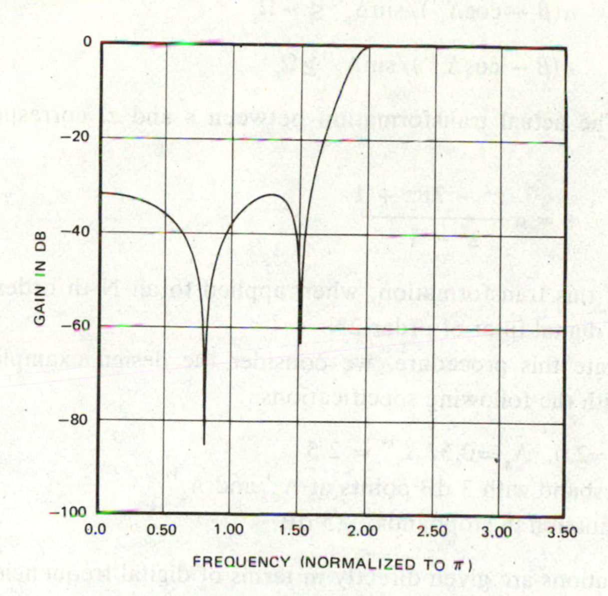

Figure 2.7 Amplitude response of the highpass filter in the example.

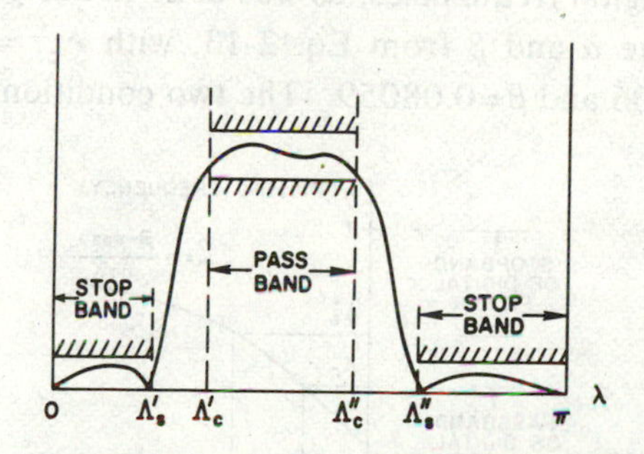

Figure 2.8 Amplitude response specification for a bandpass filter.

This transformation is plotted in Figure 2.9. Notice that the entire ω axis, $-\infty < \omega < \infty$, is shown, whereas only the positive λ axis is shown. As can be seen, it maps the passband $-1 \leq \omega \leq 1$ of the analog lowpass filter to the passband of the desired digital filter, $\Lambda_c' \leq \lambda \leq \Lambda_c''$. The edges of the stopbands, Λ_s' and Λ_s'', map to $\alpha(\beta - \cos\Lambda_s')/\sin\Lambda_s'$ and $\alpha(\beta - \cos\Lambda_s'')/\sin\Lambda_s''$, respectively. Therefore, we require both conditions:

$$\alpha(\beta - \cos\Lambda_s')/\sin\Lambda_s' \leq -\Omega_s$$

$$\alpha(\beta - \cos\Lambda_s'')/\sin\Lambda_s'' \geq \Omega_s \tag{2-14}$$

be satisfied. The actual transformation between s and z, corresponding to Eq. 2-12, is

$$s = \alpha \frac{z^2 - 2\beta z + 1}{z^2 - 1} \tag{2-15}$$

It is clear that this transformation, when applied to an N-th order analog filter, will result in a digital filter of order 2N.

To illustrate this procedure, we consider the design example of bandpass digital filter with the following specifications:

$\Lambda_c'=1.0$, $\Lambda_c''=2.0$, $\Lambda_s'=0.5$, $\Lambda_s'' = 2.5$
monotonic passband with 3 dB points at Λ_c' and Λ_c''
minimum attenuation in stopbands: 25 dB

These specifications are given directly in terms of digital frequencies. In the case of an analog application, where the specifications are initially given in analog frequencies together with a sampling frequency F_s, the first step would be to translate them into digital frequencies, as was done in our previous examples.

We first calculate α and β from Eq. 2-13, with $\Lambda_c' = 1.0$ and $\Lambda_c'' = 2.0$. The result is α=1.8306 and β=0.08059. The two conditions of Eq. 2-14 become

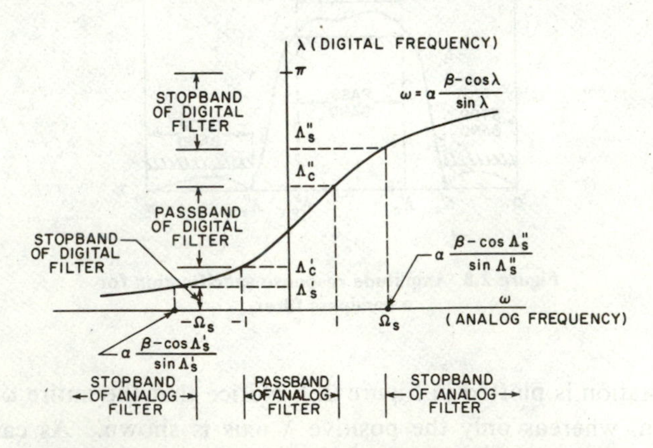

Figure 2.9 A lowpass to bandpass transformation, Eq. 2-12.

$$1.8306(0.08059 - 0.8776)/0.4794 \leq -\Omega_s \quad \text{or } \Omega_s \leq 3.0431$$

and

$$1.8306(0.08059 + 0.8012)/0.5985 \geq \Omega_s \quad \text{or } \Omega_s \leq 2.6971$$

Therefore, we pick Ω_s=2.6971. We next choose an analog lowpass filter of the Butterworth type, which will yield a 25 dB attenuation in the stopband and, therefore, is characterized by an n such that

$$20 \log_{10}(1 + \Omega_s^{2n})^{1/2} \geq 25$$

The minimum n satisfying this inequality for Ω_s=2.6971 is 3. Therefore, the analog lowpass filter has the transfer function

$$H^a(s) = \frac{1}{(s + 1)(s^2 + s + 1)}$$

and the digital filter transfer function is

$$H(z) = H^a[1.8306(z^2-0.1612z+1)/(z^2-1)]$$

$$= \frac{0.05715(1 - z^{-1})^3(1 + z^{-1})^3}{(1-0.1042z^{-1}+0.2934z^{-2})(1-0.2225z^{-1}+0.7747z^{-2}-0.1270z^{-3}+0.4077z^{-4})}$$

The amplitude response of this filter is shown in Figure 2.10. It is noted that the attenuation drops below −25 dB, before the desired cutoff $\Lambda_s' = 0.5$ demanded by the specification.

Bandstop filters, with an amplitude response as shown in Figure 2.11, can be obtained by applying the transformation

$$\omega = \frac{\alpha \sin\lambda}{\beta - \cos\lambda} \tag{2-16}$$

to a lowpass analog filter, with

$$\alpha = \frac{\cos\Lambda_c' - \cos\Lambda_c''}{\sin\Lambda_c' + \sin\Lambda_c''}$$

$$\beta = \frac{\sin(\Lambda_c' + \Lambda_c'')}{\sin\Lambda_c' + \sin\Lambda_c''} \tag{2-17}$$

Furthermore, in order for the stopband to be mapped to the stopband of the analog filter, $|\omega| > \Omega_s$, the following conditions must be satisfied:

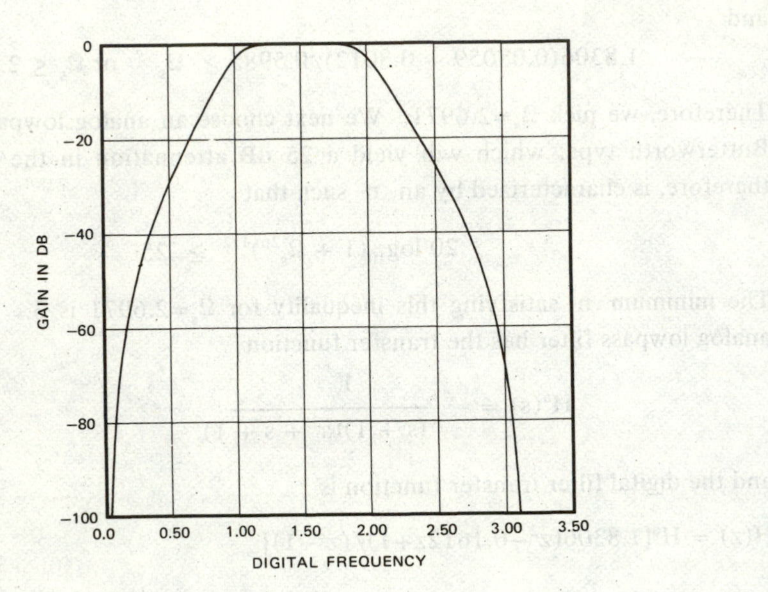

Figure 2.10 Amplitude response of bandpass filter in the example.

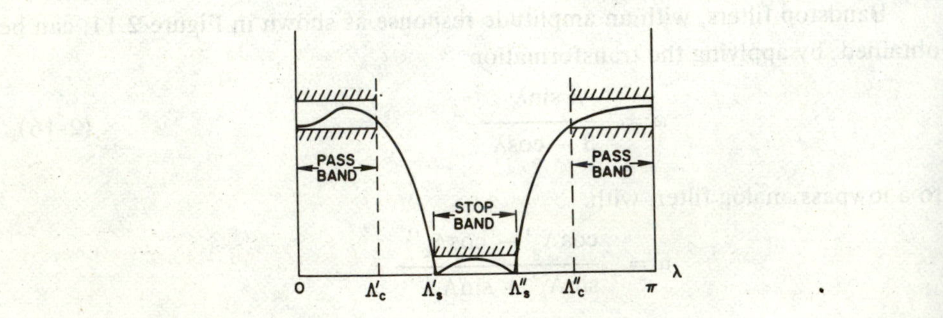

Figure 2.11 Amplitude response of bandstop filter.

$$\frac{\alpha \sin\Lambda_s'}{\beta - \cos\Lambda_s'} \leq -\Omega_s$$

$$\frac{\alpha \sin\Lambda_s''}{\beta - \cos\Lambda_s''} \geq \Omega_s \qquad (2\text{-}18)$$

The actual transformation from s to z is given by

$$s = \frac{\alpha(z^2 - 1)}{z^2 - 2\beta z + 1} \qquad (2\text{-}19)$$

We do not include a numerical design example to illustrate the design of bandstop filters, since by now the reader is probably familiar with the transformation approach.

2.5 COMPUTER AIDED DESIGN OF IIR FILTERS

Although the frequency transformation approach of designing digital filters, discussed in the last three sections, is applicable to the majority of situations encountered in practice, it can not be used if the desired frequency response has more than one passband or one stopband. In such instances, one must resort to other methods. In this section we describe a computer aided design method for IIR digital filters to achieve an arbitrarily specified amplitude response. We discuss the method proposed by Steiglitz [4], and in Appendix 2.1, the listing of the computer program for the design is given.

Let λ_n, $1 \leq n \leq N$, be a set of digital frequencies in the range $[0, \pi]$, and let A_n be the desired amplitude response at frequency λ_n. For numerical accuracy reasons, it is advantageous to write the transfer function in the factored form (cascade form) as:

$$H(z) = C \prod_{k=1}^{K} \frac{1 + f_k z^{-1} + g_k z^{-2}}{1 + c_k z^{-1} + d_k z^{-2}} \qquad (2\text{-}20)$$

where the real constants C, c_k, d_k, f_k, g_k, $1 \leq k \leq K$ are to be determined, so that the amplitude of the resulting $H(z)$ approximates A_n at $z_n = e^{j\lambda_n}$, as closely as possible.

We choose to minimize the mean square error Q

$$Q = \sum_{n=1}^{N} \left[\, |H(e^{j\lambda_n})| - A_n \,\right]^2 \qquad (2\text{-}21)$$

Let θ denote the 4K unknown constants $(c_1,d_1,f_1,g_1,\ldots,c_K,d_K,f_K,g_K)$. Then the error Q is a function of θ and the constant multiplier C. We may write Eq. 2-21 as follows:

$$Q(\theta,C) = \sum_{n=1}^{N} [C\,|H_n| - A_n]^2 \tag{2-22}$$

where for simplicity we have defined

$$H_n = H(e^{j\lambda_n})/C \tag{2-23}$$

It is obviously a function of C and θ, the 4K filter coefficients to be determined.

For a given θ, the optimum value of C, to be denoted C^*, can be determined straightforwardly by setting the partial derivative $\partial Q(\theta,C)/\partial C$ to zero. The answer is

$$C^* = \frac{\sum_{n=1}^{N} |H_n|\,A_n}{\sum_{n=1}^{N} |H_n|^2} \tag{2-24}$$

The determination of the optimum θ, to be denoted as θ^*, cannot be done easily because of the highly nonlinear dependence of Q on θ. However, we may use numerical approximation methods to solve for θ^*. The method of Fletcher-Powell is well suited for this purpose. It is an iterative optimum seeking method, and program packages are available. In order to apply this method, we need the partial derivative of $Q(\theta,C^*)$ with respect to each of the 4K components of θ. From Eq. 2-22, we have

$$\frac{\partial Q(\theta,C^*)}{\partial \theta_i} = 2C^* \sum_{n=1}^{N} [\,|H_n|\,C^* - A_n]\, \frac{\partial |H_n|}{\partial \theta_i} \tag{2-25}$$

where θ_i is the i-th component of θ, and may denote either c_k, or d_k, or f_k, or g_k. It can be shown that

$$\frac{\partial |H_n|}{\partial c_k} = -|H_n| \; \mathbf{Re} \left[\frac{z_n^{-1}}{1 + c_k z_n^{-1} + d_k z_n^{-2}}\right]$$
$$\frac{\partial |H_n|}{\partial d_k} = -|H_n| \; \mathbf{Re} \left[\frac{z_n^{-2}}{1 + c_k z_n^{-1} + d_k z_n^{-2}}\right]$$
$$\frac{\partial |H_n|}{\partial f_k} = -|H_n| \; \mathbf{Re} \left[\frac{z_n^{-2}}{1 + c_k z_n^{-1} + d_k z_n^{-2}}\right]$$
$$\frac{\partial |H_n|}{\partial g_k} = -|H_n| \; \mathbf{Re} \left[\frac{z_n^{-2}}{1 + c_k z_n^{-1} + d_k z_n^{-2}}\right] \qquad (2\text{-}26)$$

where ***Re*** denotes the real part.

These partial derivatives can be incorporated in a computer program that searches for a minimum from a given starting value of θ, by means of the Fletcher-Powell algorithm. The search procedure is terminated when the values of θ for two successive iterations are within a prescribed limit, say 10^{-4} or 10^{-5}

The result of this procedure will be a rational transfer function that is closest to the design specification in terms of the mean squared error. However, it often turns out that the computed solution has some poles and zeros that lie outside the unit circle, since no prior restriction has been put on their location. Obviously, the unstable poles have to be relocated. The zeros that lie outside the unit circle may also have to be relocated, if a minimum phase filter is required. Suppose the H(z) obtained by the above optimization procedure has a real pole outside the unit circle, say at $z = p_1$ with $|p_1| > 1$, and suppose this pole is replaced by one located at $z = 1/p_1$. This means that the transfer function is multiplied by $(z-p_1)/(z-1/p_1)$ which, for $|z| = 1$, has a constant magnitude p_1. Thus, this correction for unstable real poles does not change the shape of the amplitude response. Furthermore, since the gain constant C is chosen to be optimum by Eq. 2-24, the relocation of this unstable pole has no other effect on the transfer function. Similarly unstable complex poles can be replaced by their reciprocals, and so can the zeros lying outside the unit circle. If the optimization program is applied again after these pole-zero inversions, with the values obtained for C^* and θ used as the starting point, then further improvements in the design are often possible. The reason for this seemingly strange behavior will become clearer in the following example taken from Steiglitz's original paper [4].

In this simple example, a second order lowpass filter is to be designed to meet the following specifications:

$A_n = 1.0, \; \lambda_n = 0, 0.01\pi, 0.02\pi, ..., 0.09\pi$

$A_n = 0.5, \; \lambda_n = 0.1\pi$

$$A_n = 0.0,\ \lambda_n = 0.11\pi, 0.12\pi, ..., 0.20\pi$$

$$A_n = 0.0,\ \lambda_n = 0.2\pi, 0.3\pi, ..., \pi$$

Thus, N = 30, and the transfer function is of the form

$$H(z) = C \frac{1 + fz^{-1} + gz^{-2}}{1 + cz^{-1} + dz^{-2}}$$

and the parameters to be determined are C and θ=(c,d,f,g). The optimization starts with an initial value for θ=(−0.25,0,0,0). The Fletcher-Powell algorithm is applied and convergence is obtained after some 90 iterations, and the value obtained for Q is 1.2611. The resulting poles are at 0.75677793 and 1.3213916, and the zeros at 0.67834430±j0.73474418. The unstable pole at 1.3213916 is very nearly the reciprocal of the stable pole. The program now inverts the unstable pole and the Fletcher-Powell optimization is applied again. Convergence is obtained after another 62 iterations, and the new poles are 0.82191163±j0.19181084. The new zeros are 0.82191163±j0.56961501. The optimum value of C, C^*=0.11733973, and the value of Q reached this time is 0.56731.

Obviously, the two poles that were so close together after the inversion, coalesced and then split into a complex conjugate pair, and this resulted in a further reduction in Q, from that achieved after the first convergence. It is clear that without the pole inversion, no further improvement was possible, because the separation of the two poles, one inside the unit circle and one outside, prevented them from becoming a double pole and then splitting, The frequency response of the final design is plotted in Figure 2.12. Table 2.1 shows the output of the computer program, listed in Appendix 2.1, for this example.

The computer program listed in Appendix 2.1, due to Steiglitz [4], consists of the following 3 steps:

1. Determination of the optimum parameters θ^* and C^* by the Fletcher-Powell algorithm, without regard to the location of the resulting poles and zeros.
2. After convergence is achieved in step 1, all poles and zeros outside the unit circle are inverted.
3. Using the new values of θ and C as the starting point, the Fletcher-Powell algorithm is applied again, and the new optimum values of θ^* and C^* obtained give the desired digital filter transfer function.

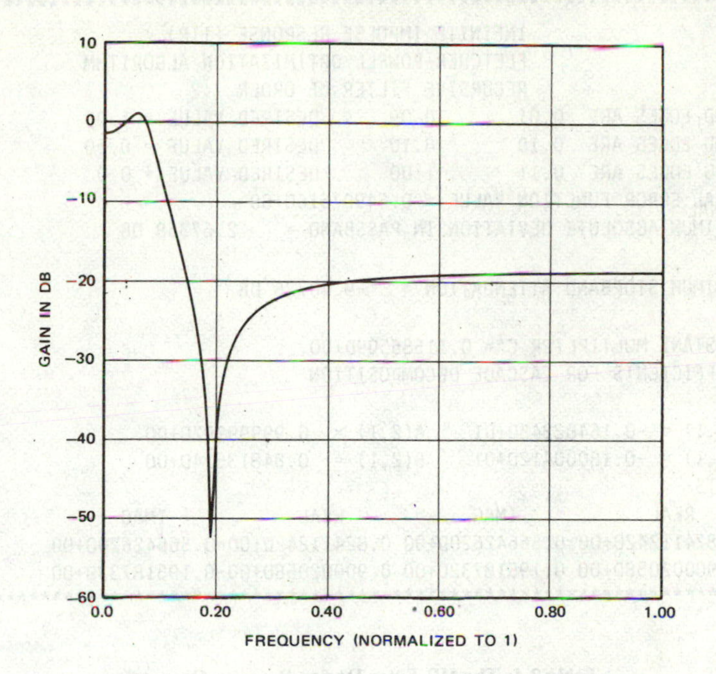

Figure 2.12 Amplitude response of filter from Table 2.1.

2.6 DESIGN OF FIR FILTERS - FOURIER SERIES METHOD

FIR filters are characterized by their finite duration impulse response, $\{h_0, h_1, \ldots, h_K\}$. Their transfer function is a polynomial in z^{-1}.

$$H(z) = \sum_{k=0}^{K} h_k z^{-k} \tag{2-27}$$

To design a filter of this type, one selects the coefficients, $\{h_k\}$, so that the transfer function has a frequency response $H(e^{j\lambda})$, $-\pi \leq \lambda \leq \pi$, that approximates the design specifications within certain tolerances.

Just as in the design of IIR filters, if the input signal is sampled at F_s, then the analog frequency f in Hz is related to the digital frequency λ through

$$\lambda = 2\pi f / F_s$$

Let $H^*(z)$ be the desired frequency response, which may be specified either in terms of real and imaginary parts, or equivalently in terms of its amplitude and

```
************************************************************************
                    INFINITE IMPULSE RESPONSE (IIR)
                    FLETCHER-POWELL OPTIMIZATION ALGORITHM
                    RECURSIVE FILTER OF ORDER   2
BAND EDGES ARE  0.01        0.09       DESIRED VALUE = 1.00
BAND EDGES ARE  0.10        0.10       DESIRED VALUE = 0.50
BAND EDGES ARE  0.11        1.00       DESIRED VALUE = 0.0
FINAL ERROR FUNCTION VALUE = 0.54907616D+00
MAXIMUM ABSOLUTE DEVIATION IN PASSBAND =    2.67358 DB

MINIMUM STOPBAND ATTENUATION =   -9.40726 DB

CONSTANT MULTIPLIER C*= 0.11586509D+00
COEFFICIENTS FOR CASCADE DECOMPOSITION

A(1,1) = -0.16482248D+01    A(2,1) =  0.99999992D+00
B(1,1) = -0.18000412D+01    B(2,1) =  0.84813514D+00
ROOTS
     REAL            IMAG           REAL            IMAG
 0.82411242D+00 0.56642620D+00 0.82411242D+00-0.56642620D+00
 0.90002058D+00 0.19518733D+00 0.90002058D+00-0.19518733D+00
************************************************************************
```

Table 2.1 The IIR Filter Design Program Output for a Simple Example.

phase. Denote the real and imaginary parts of $H^*(e^{j\lambda})$ by $R(\lambda)$ and $I(\lambda)$, respectively. That is,

$$H^*(e^{j\lambda}) = R(\lambda) + j\,I(\lambda) \tag{2-28}$$

For real impulse response, $R(\lambda)$ is an even function in λ and can be expanded into a Fourier series consisting only of cosine terms, and $I(\lambda)$ can be expanded into a Fourier series that has only sine terms.

$$R(\lambda) = a_0 + \sum_{n=1}^{\infty} a_n \cos n\lambda$$

$$I(\lambda) = \sum_{n=1}^{\infty} b_n \sin n\lambda \tag{2-29}$$

Using the elementary identities,

$$\cos n\lambda = \frac{e^{j\lambda n} + e^{-j\lambda n}}{2} = \frac{z^n + z^{-n}}{2}$$

$$\sin n\lambda = \frac{e^{j\lambda n} - e^{-j\lambda n}}{2j} = \frac{z^n - z^{-n}}{2j} \tag{2-30}$$

we may replace the sine and cosine terms of the Fourier series expansions by powers of z. Substituting Eq. 2-30 into Eq. 2-29 and then into Eq. 2-28, we have an expansion of $H^*(e^{j\lambda})$ in terms of powers of z, namely

$$H^*(e^{j\lambda}) = a_0 + \sum_{n=1}^{\infty} a_n(z^n+z^{-n})/2 + \sum_{n=1}^{\infty} b_n(z^n-z^{-n})/2$$

$$= a_0 + \sum_{n=1}^{\infty} [z^{-n}(a_n-b_n)/2 + z^n(a_n+b_n)/2] \tag{2-31}$$

Suppose the series is truncated at $n=N$, to give an approximation $H_N(z)$.

$$H_N(z) = a_0 + \sum_{n=1}^{N} [z^{-n}(a_n-b_n)/2 + z^n(a_n+b_n)/2]$$

$$= z^N[(a_N+b_N)/2 + z^{-1}(a_{N-1}+b_{N-1})/2 + \ldots + a_0 z^{-N}$$

$$+ z^{-N-1}(a_1-b_1)/2 + \ldots + z^{-2N}(a_N-b_N)/2] \tag{2-32}$$

Clearly we may pick $H(z)$ to be the polynomial in z^{-1} inside the bracket, since z^N represents simply an advance of N samples. The result is

$$H(z) = (a_N+b_N)/2 + z^{-1}(a_{N-1}+b_{N-1})/2 + \ldots + a_0 z^{-N} + z^{-N-1}(a_1-b_1)/2$$
$$+ \ldots + z^{-2N}(a_N-b_N)/2 \tag{2-33}$$

A comparison with Eq. 2-27 shows that the h_k's are related to the a_k's and b_k's through

$$h_k = (a_{N-k} + b_{N-k})/2 \qquad 0 \le k \le N-1$$
$$h_N = a_0$$
$$h_k = (a_{k-N} - b_{k-N})/2 \qquad N+1 \le k \le 2N \tag{2-34}$$

The $H(z)$ given by Eq. 2-33 approximates the desired $H^*(z)$ aside from the delay z^{-N}, which in the frequency domain corresponds to a linear phase shift and can often be ignored.

The above procedure starts with the specification of the real and imaginary parts of the desired frequency response. This is, of course, equivalent to the specification of amplitude and phase. Quite often in practice, the phase is of no particular concern and is not specified. Consequently, for simplicity one may assume a zero imaginary part $I(\lambda)$ which corresponds to a zero phase. If this is done, all b_k's are zero. From Eq. 2-33, it is seen that, in this case, the impulse response is symmetric, that is, $h_k = h_{2N-k}$. It should be noted, however, that there are many important applications where the desired filter calls for a zero real part as, for example, in the case of an ideal differentiator.

To illustrate the procedure described above, we consider two examples. In the first example, the real and imaginary parts of the desired frequency response are:

$$R(\lambda) = 1 - |\lambda|/\pi ; \quad |\lambda| \leq \pi$$
$$I(\lambda) = 0$$

Thus, $R(\lambda)$ is the same as the amplitude response. The Fourier expansion for this $R(\lambda)$ is well known,

$$R(\lambda) = 1/2 + (4/\pi^2) \sum_{n=0}^{\infty} \cos[(2n+1)\lambda]/(2n+1)^2$$

Thus, from Eq. 2-31

$$H^*(z) = 1/2 + (2/\pi^2) \sum_{n=0}^{\infty} (z^{2n+1}+z^{-2n-1})/(2n+1)^2$$

Truncating this series at n=2, a nonrecursive filter is obtained.

$$\begin{aligned} H(z) &= 2/25\pi^2 + (2/9\pi^2)z^{-2} + (2/\pi^2)z^{-4} + (1/2)z^{-5} + (2/\pi^2)z^{-6} \\ &\quad + (2/9\pi^2)z^{-8} + (2/25\pi^2)z^{-10} \\ &= 0.008106 + 0.022516z^{-2} + 0.202642z^{-4} + 0.5z^{-5} \\ &\quad + 0.202642z^{-6} + 0.022516z^{8} + 0.008106z^{-10} \end{aligned}$$

The frequency response $H(e^{j\lambda})$ which is real, is plotted in Figure 2.13.

The second example is the design of a lowpass filter with a cutoff at $\lambda=\pi/3$. Thus, $R(\lambda) = 1$ for $|\lambda| < \pi/3$ and $R(\lambda) = 0$ otherwise, $I(\lambda) = 0$. The Fourier series for $R(\lambda)$ can be found straightforwardly to be

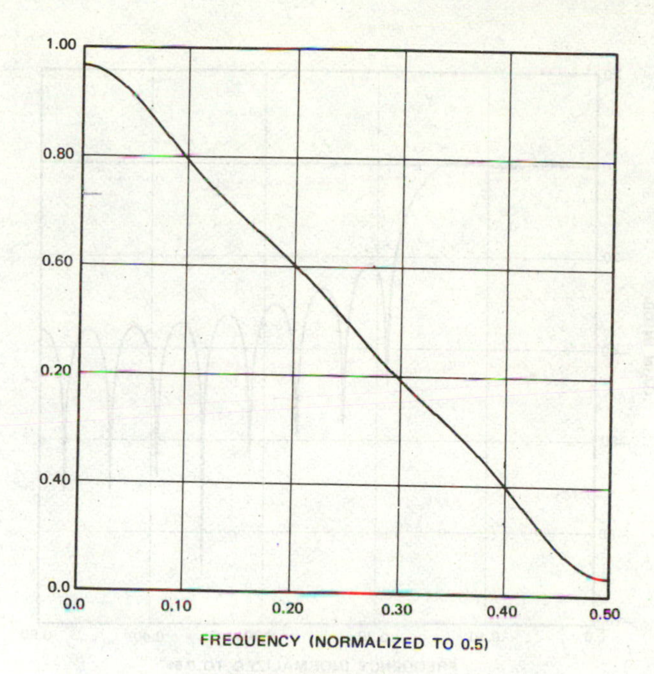

Figure 2.13 Amplitude response of nonrecursive filter designed using the Fourier series method.

$$R(\lambda) = 1/3 + (2/\pi) \sum_{k=1}^{\infty} \cos k\lambda \sin(k\pi/3)/k$$

Truncating the series to 11 terms, and proceeding as before yields

$$\begin{aligned} H(z) = \; & -0.0250604 - 0.0275664z^{-1} + 0.034458z^{-3} + 0.0393806z^{-4} \\ & - 0.0551329z^{-6} - 0.0689161z^{-7} + 0.1378322z^{-9} + \\ & 0.2756644z^{-10} + 0.3333333z^{-11} + 0.2756644z^{-12} + \\ & 0.13783222z^{-13} - 0.0689161z^{-15} - 0.0551329z^{-16} + \\ & 0.0393806z^{-18} + 0.034458z^{-19} - 0.0275664z^{-21} - \\ & 0.0252604z^{-22} \end{aligned}$$

The amplitude response of this filter is plotted in Figure 2.14. We note the familiar oscillatory behavior (Gibbs phenomenon) around $\pi/3$ due to the discontinuity in the specified $H^*(z)$ there.

Gibbs oscillations can be smoothed by using *windows* [5]. A window function $W(\lambda)$ of order M is a real function of λ of the form

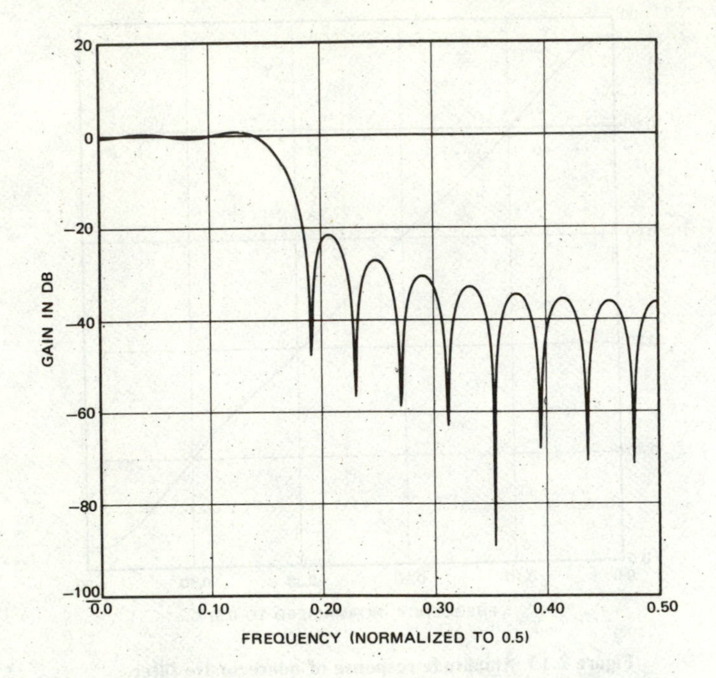

Figure 2.14 Amplitude response of 23 tap nonrecursive lowpass filter.

$$W(\lambda) = \sum_{k=0}^{M} w_k e^{j\lambda k} \tag{2-35}$$

It is essentially a narrow pulse. A commonly used window is the Hamming window function, for which the $\{w_k\}$ are given by

$$w_k = 0.54 - 0.46 \cos[\pi(k)/N]\ ,\ 0 \leq k \leq 2N \tag{2-36}$$

The $W(\lambda)$ and the $\{w_k\}$ for the Hamming window are sketched in Figure 2.15, for M=12. Suppose that the $H(e^{j\lambda})$ of our example is convolved with the window $W(\lambda)$. The convolution in the frequency domain of $H(e^{j\lambda})$ with $W(\lambda)$ is equivalent to the multiplication of each h_k by a corresponding w_k. So the impulse response of the windowed design is $\{h_k w_k\}$, where the $\{h_k\}$ are those obtained in the original design. The final transfer function is then

$$H(z) = \sum_{k=0}^{K} (w_k h_k) z^{-k} \tag{2-37}$$

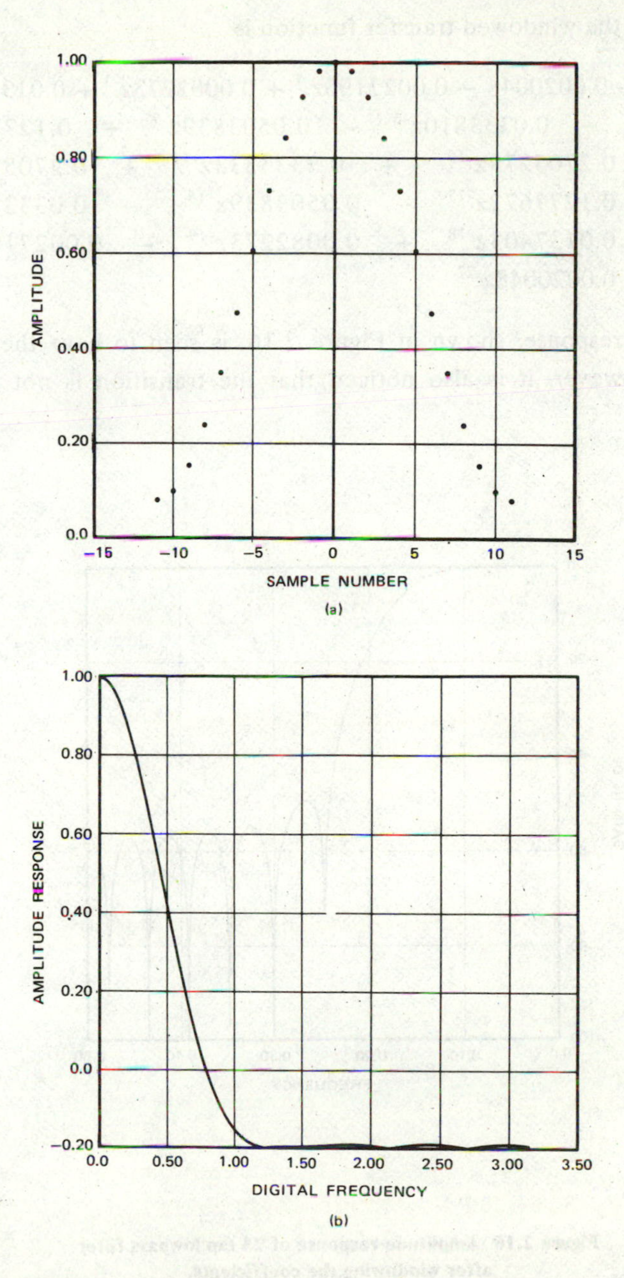

Figure 2.15 Hamming window, for (a) Eq. 2-36, M=22, (b) Eq. 2-35, M=22.

In our example the windowed transfer function is

$$H(z) = -0.0020048 - 0.0027190z^{-1} + 0.0082273z^{-3} + 0.0137403z^{-4} - 0.0333810z^{-6} - 0.0503839z^{-7} + 0.1277672z^{-9} + 0.2705279z^{-10} + 0.3333333z^{-11} + 0.2705279z^{-12} + 0.1277672z^{-13} - 0.0503839z^{-15} - 0.0333810z^{-16} + 0.0137403z^{-18} + 0.0082273z^{-19} - 0.0027190z^{-21} - 0.0020048z^{-22}$$

The frequency response, shown in Figure 2.16, is seen to have the oscillations smoothed. However, it is also noticed that the transition is not as sharp as before.

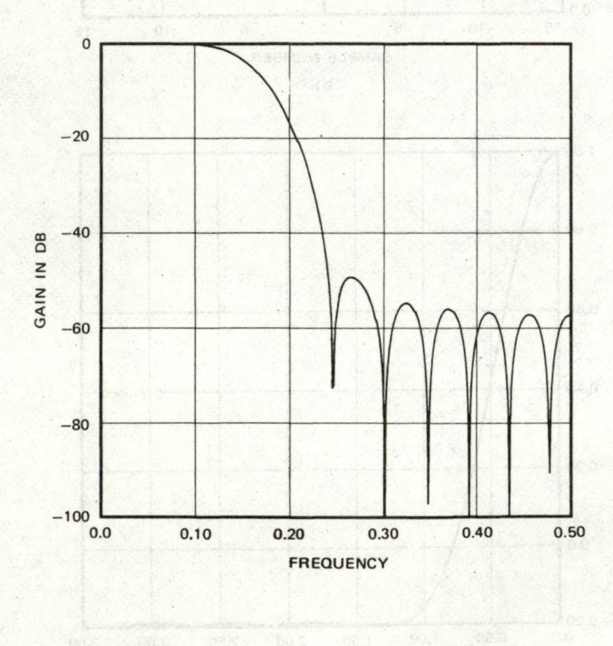

Figure 2.16 Amplitude response of 23 tap lowpass filter after windowing the coefficients.

2.7 COMPUTER AIDED DESIGN OF LINEAR PHASE FIR FILTERS

As we already know, the frequency response of a FIR filter with coefficients $\{h_k\}$ is

$$H(e^{j\lambda}) = \sum_{k=0}^{K} h_k e^{j\lambda k} \quad (2\text{-}38)$$

It can be shown that, if the phase of $H(e^{j\lambda})$ is linear in λ, the impulse response must be symmetric in the sense that

$$h_k = h_{K-k} \quad (2\text{-}39)$$

In this case, Eq. 2-38 can be written as

$$H(e^{j\lambda}) = [\sum_{n=0}^{N} a_n \cos(n\lambda)\ ]e^{-j\lambda K/2} \quad (2\text{-}40)$$

for even K, where

$$N = K/2$$
$$a_0 = h_N$$
$$a_n = 2h_{N-n}\ ,\ n = 1,2,...,N \quad (2\text{-}41)$$

For odd K, Eq. 2-38 can be rewritten as

$$H(e^{j\lambda}) = [\sum_{n=0}^{N} a_n \cos(n-1/2)\lambda\ ]e^{-j\lambda K/2} \quad (2\text{-}42)$$

where

$$N = (K+1)/2 \qquad a_n = 2h_{N-n} \quad (2\text{-}43)$$

Leaving aside for the moment the linear phase term $e^{j\lambda K/2}$ in Eqs. 2-40 and 2-42, we see that the frequency response of the filter is given by a real cosine series, the coefficients of which are simply related to the impulse response. The linear phase delay is determined by the length of the impulse response only. The problem of the design of this type of filter becomes, therefore, one of finding the values $\{h_k\}$ so that the cosine series in Eq. 2-40 or 2-42 matches the desired function of λ as closely as possible. This approach, due to McCellan et al. [6], is very useful in computer aided design of a wide class of FIR filters. The computer program listed in Appendix 2.2, after [7], enables one to automatically design multiple passband/stopband FIR filters, differentiators, and Hilbert transformers.

Figure 2.17 shows the amplitude response of a filter designed using this program. The filter has 2 passbands and 3 stopbands. The tolerance limits are given in Table 2.2 which shows the computer printout for this example. The input to the program specifies the length of the impulse response, the boundaries of the passbands and stopbands, the desired response in each band, and the relative weighting of the deviations in each band. From this information, the program produces the impulse response of a equiripple FIR filter. It also lists the deviations of the design from the desired response, for each band.

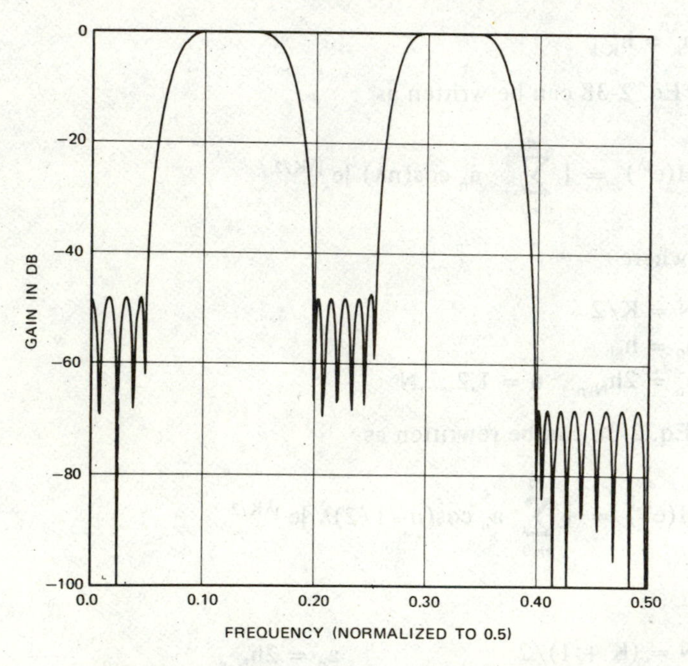

Figure 2.17 Multiband FIR filter amplitude response, filter coefficients are listed in Table 2.2

Let us consider the design of a lowpass filter discussed in the previous section. The input data for the program is:

```
23,1,2,0,0
0,0.14,0.19,0.5
1,0
1,10
```

This specifies that a 23 tap FIR filter is to be designed, with a passband from 0 to $0.14(2\pi)$ and a stopband from $0.19(2\pi)$ to $0.5(2\pi)$; that is, the input frequen-

```
**********************************************************************
                         FINITE IMPULSE RESPONSE (FIR)
                         LINEAR PHASE DIGITAL FILTER DESIGN
                         REMEZ EXCHANGE ALGORITHM
                         BANDPASS FILTER
                         FILTER LENGTH = 60
                      ***** IMPULSE RESPONSE *****
                    H(  1) = -0.13266187E-02 = H(  60)
                    H(  2) =  0.21044631E-02 = H(  59)
                    H(  3) =  0.25627436E-02 = H(  58)
                    H(  4) = -0.64310394E-02 = H(  57)
                    H(  5) =  0.29610749E-03 = H(  56)
                    H(  6) =  0.95933788E-02 = H(  55)
                    H(  7) =  0.10328032E-02 = H(  54)
                    H(  8) =  0.13788342E-02 = H(  53)
                    H(  9) =  0.73366333E-03 = H(  52)
                    H( 10) = -0.11325534E-02 = H(  51)
                    H( 11) =  0.87707490E-03 = H(  50)
                    H( 12) = -0.92005022E-02 = H(  49)
                    H( 13) =  0.73080473E-02 = H(  48)
                    H( 14) = -0.22302270E-02 = H(  47)
                    H( 15) = -0.46340838E-01 = H(  46)
                    H( 16) = -0.24636984E-02 = H(  45)
                    H( 17) =  0.27533844E-01 = H(  44)
                    H( 18) =  0.31134700E-02 = H(  43)
                    H( 19) =  0.21461532E-01 = H(  42)
                    H( 20) =  0.60995854E-02 = H(  41)
                    H( 21) =  0.15403140E-01 = H(  40)
                    H( 22) =  0.21729738E-01 = H(  39)
                    H( 23) = -0.63493103E-02 = H(  38)
                    H( 24) =  0.98729789E-01 = H(  37)
                    H( 25) = -0.16830832E-01 = H(  36)
                    H( 26) = -0.25071001E+00 = H(  35)
                    H( 27) = -0.44710323E-01 = H(  34)
                    H( 28) = -0.83216578E-02 = H(  33)
                    H( 29) = -0.11058366E+00 = H(  32)
                    H( 30) =  0.28859955E+00 = H(  31)
                        BAND  1           BAND  2          BAND  3           BAND  4
LOWER BAND EDGE     0.0            0.100000024     0.199999988      0.300000012
UPPER BAND EDGE     0.050000001    0.149999976     0.250000000      0.350000024
DESIRED VALUE       0.0            1.00000000      0.0              1.00000000
WEIGHTING           1.00000000     1.00000000      1.00000000       1.00000000
DEVIATION           0.003852384    0.003852384     0.003852384      0.003852384
DEVIATION IN DB   -48.2854004      0.066912591   -48.2854004        0.066912591
                        BAND  5
LOWER BAND EDGE     0.399999976
UPPER BAND EDGE     0.500000000
DESIRED VALUE       0.0
WEIGHTING          10.0000000
DEVIATION           0.000385238
DEVIATION IN DB   -68.2854004
**********************************************************************
```

Table 2.2 The FIR Filter Design Program Output for a 60 Tap Multiband Filter.

cy is normalized to 0.5. The third card specifies that the desired amplitude response is 1 in the passband and 0 in the stopband, and the last card specifies that the errors in the passband and stopband should be weighted by 1 and 10, respectively. For a detailed explanation of the input parameters to the program, the reader is advised to consult Appendix 2.2.

```
**************************************************************************
                      FINITE IMPULSE RESPONSE (FIR)
                      LINEAR PHASE DIGITAL FILTER DESIGN
                      REMEZ EXCHANGE ALGORITHM
                      BANDPASS FILTER
                      FILTER LENGTH =  23
                    ***** IMPULSE RESPONSE *****
                  H(  1) = -0.61297640E-02 = H(  23)
                  H(  2) =  0.14092118E-01 = H(  22)
                  H(  3) =  0.35473555E-01 = H(  21)
                  H(  4) =  0.48743814E-01 = H(  20)
                  H(  5) =  0.33617824E-01 = H(  19)
                  H(  6) = -0.11127114E-01 = H(  18)
                  H(  7) = -0.56752473E-01 = H(  17)
                  H(  8) = -0.57823956E-01 = H(  16)
                  H(  9) =  0.14385510E-01 = H(  15)
                  H( 10) =  0.14293462E+00 = H(  14)
                  H( 11) =  0.26699775E+00 = H(  13)
                  H( 12) =  0.31825352E+00 = H(  12)
                      BAND  1        BAND  2
LOWER BAND EDGE   0.0            0.189999998
UPPER BAND EDGE   0.139999986    0.500000000
DESIRED VALUE     1.00000000     0.0
WEIGHTING         1.00000000     10.0000000
DEVIATION         0.167077422    0.016707744
DEVIATION IN DB   2.92989731     -35.5416260
**************************************************************************
```

Table 2.3 The FIR Filter Design Program Output for a 23 Tap Lowpass Filter.

The output produced by the program in this case is shown in Table 2.3, and the frequency response of the filter is plotted in Figure 2.18. It is interesting to compare this design with the Fourier series design of the same filter depicted in Figures 2.14 and 2.16.

As another example, suppose the following set of input data is used

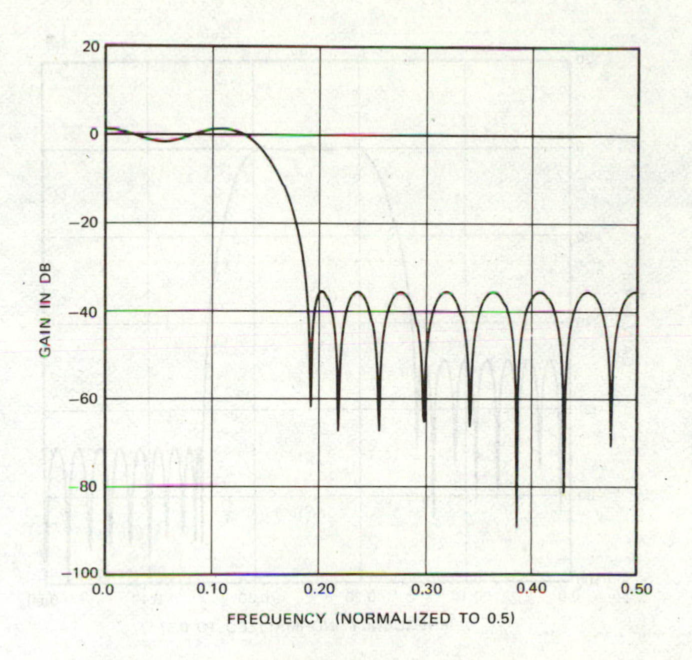

Figure 2.18 Amplitude response of 23 tap FIR filter listed in Table 2.3.

```
50,1,3,1,0
0,0.15,0.2,0.3,0.35,0.5
0,1,0
10,1,100
```

This specifies a 50 tap FIR bandpass filter. The output of the program for this case is shown in Table 2.4, and the amplitude response is plotted in Figure 2.19.

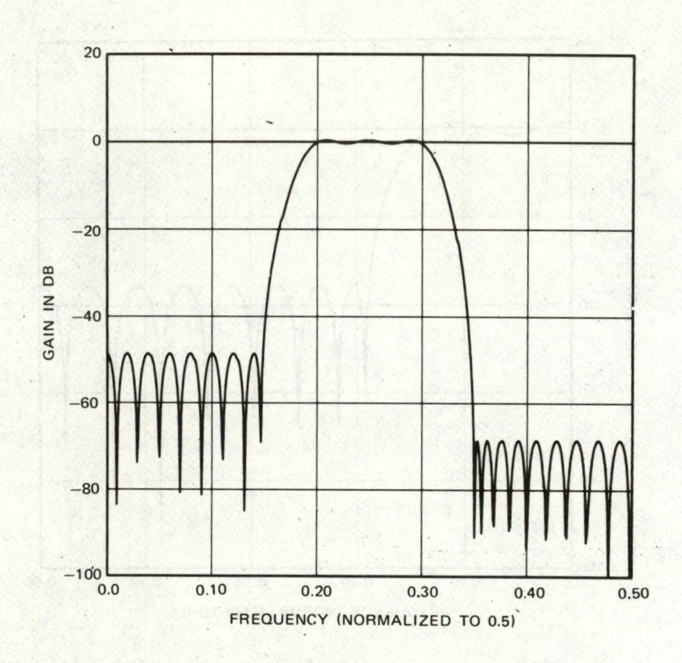

Figure 2.19 Amplitude response of 50 tap bandpass filter listed in Table 2.4.

```
**********************************************************************
                      FINITE IMPULSE RESPONSE (FIR)
                      LINEAR PHASE DIGITAL FILTER DESIGN
                      REMEZ EXCHANGE ALGORITHM
                      BANDPASS FILTER
                      FILTER LENGTH =  50
                   ***** IMPULSE RESPONSE *****
                 H(  1) =  0.15677512E-02 = H(  50)
                 H(  2) =  0.30867318E-02 = H(  49)
                 H(  3) = -0.31774966E-02 = H(  48)
                 H(  4) = -0.62076077E-02 = H(  47)
                 H(  5) =  0.74356645E-02 = H(  46)
                 H(  6) =  0.98490156E-02 = H(  45)
                 H(  7) = -0.11100803E-01 = H(  44)
                 H(  8) = -0.10112252E-01 = H(  43)
                 H(  9) =  0.89923553E-02 = H(  42)
                 H( 10) =  0.29048752E-02 = H(  41)
                 H( 11) =  0.26621167E-02 = H(  40)
                 H( 12) =  0.12020215E-01 = H(  39)
                 H( 13) = -0.20646378E-01 = H(  38)
                 H( 14) = -0.27186796E-01 = H(  37)
                 H( 15) =  0.32320708E-01 = H(  36)
                 H( 16) =  0.28301015E-01 = H(  35)
                 H( 17) = -0.20906121E-01 = H(  34)
                 H( 18) = -0.18700678E-02 = H(  33)
                 H( 19) = -0.22832230E-01 = H(  32)
                 H( 20) = -0.53931758E-01 = H(  31)
                 H( 21) =  0.90467930E-01 = H(  30)
                 H( 22) =  0.12315601E+00 = H(  29)
                 H( 23) = -0.15637612E+00 = H(  28)
                 H( 24) = -0.17732239E+00 = H(  27)
                 H( 25) =  0.19076276E+00 = H(  26)
                            BAND  1          BAND  2          BAND  3
LOWER BAND EDGE       0.0             0.199999988       0.350000024
UPPER BAND EDGE       0.149999976     0.300000012       0.500000000
DESIRED VALUE         0.0             1.00000000        0.0
WEIGHTING            10.0000000       1.00000000      100.000000
DEVIATION             0.003714367     0.037143670       0.000371437
DEVIATION IN DB     -48.6022949       0.645533919     -68.6022949
**********************************************************************
```

Table 2.4 The FIR Filter Design Program Output for a 50 Tap Bandpass Filter.

2.8 DECIMATION AND INTERPOLATION USING FIR FILTERS

In this section we discuss the design of a particular class of digital filters that are required in a growing number of practical applications. These filters are needed to achieve sampling rate conversion within a digital signal processor. The process of decreasing the sampling rate is known as *decimation* and increasing the sampling rate is referred to as *interpolation*.

The need for sampling rate conversion arises in many practical applications and is of special interest, as it has no apparent equivalent in analog signal processing. Examples of such applications are: the transmission of speech at very low bit rates using analysis and synthesis methods, extraction of very narrow bands of the spectrum of a signal for high resolution spectral analysis known as *frequency zoom* and commonly encountered in sonar and vibration signal analysis, and conversion between different digital signal code formats which inherently are derived at different sampling rates.

The problem of decimation and interpolation has received a lot of attention in recent years, and the reader is referred to references [8, 9, 10, 11] for a more advanced treatment of the subject, in particular the paper by Crochiere and Rabiner [10] which contains a comprehensive and unified treatment of this topic. In this section we only provide an introductory discussion, relying heavily on examples to illustrate the important aspects of this problem.

Let us now define more carefully the processes of decimation and interpolation. The process of decimation involves a sampling rate reduction. We start out with an input signal, $\{x_n\}$, which was derived from an analog signal x(t) by sampling it every T seconds, that is, a sampling frequency $F_S = 1/T$, as depicted in Figure 1.10. In Chapter 1, Section 1.5, we have seen that the digital spectrum of $\{x_n\}$, X(f) is periodic and consists of the spectrum of the analog signal x(t) repeated infinitely around $\pm kF_S$ multiples of the sampling frequency (k = 0, 1, 2....). We have also seen there that, if the spectrum of x(t) is not strictly bandlimited within $|f| < F_S/2$, an aliasing error will occur. Note that $2\pi f/F_s = \lambda$. Let us consider an example of such a sequence $\{x_n\}$ which has been derived by sampling at F_S, and we wish to reduce the sampling rate to $F_S/4$, that is, decimate with ratio of four. Obviously, it only makes sense to reduce the sampling rate if the information content of the signal that we wish to preserve is bandlimited to less than $F_S/8$, half the desired sampling rate, since any spectral components above this frequency will be aliased into frequencies below $F_S/8$, according to Eq. 1-56. Figure 2.20 depicts the digital spectrum of a typical sequence $\{x_n\}$ that we wish to decimate. The frequency band of interest is $[0 - f_c]$, and we have to insure that, in the decimation process, no undesirable frequency components are aliased back into it. Such a processing requirement may arise, for example, when speech is over sampled at 32 kHz. In this case, since we are interested only in a band of 0 to 4 kHz, the sampling rate can be reduced to 8 kHz. Thus, it is clear from Figure 2.20 that the first step in the decimation process has to be the filtering of the sequence $\{x_n\}$, so as to insure that the energy left above $F_S/8$ is less than a certain minimum which is application dependent. In speech, for example, we may want to keep the aliasing error at 50 dB below the peak of X(f). Once the signal is appropriately filtered, it can be decimated by simply dropping the unneeded samples. In our example, only every fourth sample is read out.

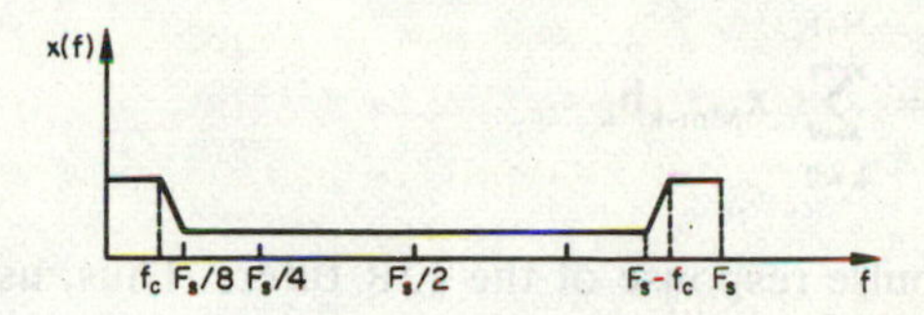

Figure 2.20 Typical spectrum of signal to be decimated.

Figure 2.21 depicts the block diagram of a general integer ratio M decimator. We see that the signal is first passed through a lowpass filter that passes undisturbed the desired information band and attenuates the band $F_S/2M$ to $F_S/2$ to prevent an excessive aliasing error. The output of the lowpass filter passes through the decimator which simply keeps every M-th point.

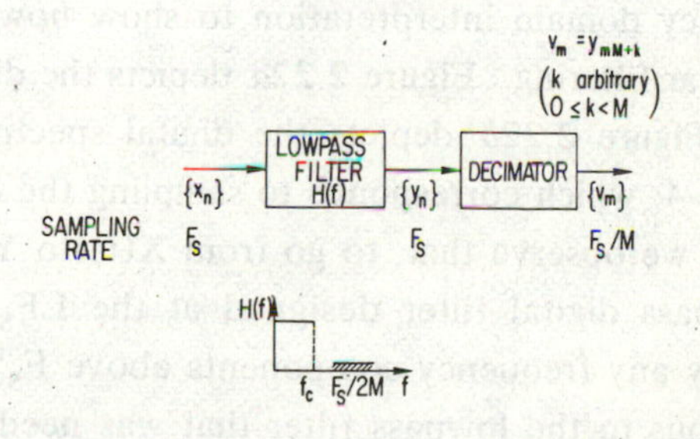

Figure 2.21 The block diagram of a M times decimator.

At this point the alert reader will realize that we do not have to compute the outputs of the lowpass filter at the rate F_S, but rather can compute only the samples at the rate F_S/M. This observation is extremely important when the question arises: should we use an IIR (recursive) or FIR (nonrecursive) filter for the lowpass filtering required. Using an IIR filter in this case has an obvious shortcoming. We cannot take advantage of the fact that we have to compute only every M-th output, since previous outputs are needed to compute the M-th output. Thus, no saving is realized. On the other side, using an FIR filter in this case implies that we can do our output computation at the rate F_S/M, and the output sequence $\{v_m\}$ is simply related to the input sequence $\{x_n\}$ by:

$$v_m = \sum_{k=0}^{N-1} x_{Mm-k} h_k \tag{2-44}$$

where $\{h_k\}$ is the impulse response of the FIR filter. Thus, using an FIR filter in the decimation process will lead to a significantly lower computation rate. Another advantage of using an FIR filter is the fact that we can easily design linear phase filters, and this is desirable in many applications. The advantage of using FIR filters in decimators is well documented in [8, 9].

The process of interpolation involves a sampling rate increase. It is the dual of decimation, so the same rule applies. We start out with a sequence $\{x_n\}$ that was derived by sampling x(t) at a sampling rate F_S^I, and we want to obtain a sequence $\{y_n\}$ that approximates as closely as possible the sequence that would have been obtained had we sampled x(t) at a rate LF_S^I. This involves inserting between any two input samples x_n and x_{n-1} an additional $L-1$ samples. We again resort to the frequency domain interpretation to show how this process can be achieved through linear filtering. Figure 2.22*a* depicts the digital spectrum of the sequence $\{x_n\}$, and Figure 2.22*b* depicts the digital spectrum that $\{y_n\}$ has to have, for the case $L=4$, which corresponds to sampling the original signal x(t) at a rate $4F_S^I$. From it we observe that, to go from X(f) to Y(f), we have to pass $\{x_n\}$ through a lowpass digital filter designed at the LF_S^I sampling rate that attenuates sufficiently any frequency components above $F_S^I/2$. This, of course, is completely analogous to the lowpass filter that was needed in the decimation process with F_S^I replaced by F_S/M. The interpolation process is depicted in the block diagram shown in Figure 2.23. As we see, the sequence $\{v_m\}$ at the input to the lowpass filter H(f) is obtained by simply inserting $L-1$ zeros between any two subsequent samples x_n and x_{n-1}. This has to be done, since H(f) is designed at the sampling rate LF_S^I and, therefore, expects input samples at this rate. Again as in the decimation process, we observe that the insertion of zeros is only conceptual, and in practice we simply treat the input sequence $\{x_n\}$ appropriately. That is, we assume that behind each x_n there are $L-1$ zero samples, when computing an output v_m. Obviously the same reasoning that led us to conclude that FIR filters are preferable in the decimation process holds here also. In the interpolation, lowpass filter using an FIR filter to compute $\{y_m\}$ implies:

$$y_m = \sum_{k=0}^{N-1} v_{m-k} h_k \tag{2-45}$$

but

$$v_{mL+i} = \begin{cases} x_m & i=0 \\ 0 & i=1,\ldots, L-1 \end{cases} \tag{2-46}$$

Substituting Eq. 2-46 into 2-45, we obtain

$$y_{mL+i} = \sum_{k=0}^{N-1} v_{mL+i-k} h_k \tag{2-47}$$

According to Eq. 2-46, v_{mL+i-k} is nonzero for a given m only when $i-k = 0$. However, since k can in general take on values larger than L, we have to take this into account. To make the discussion somewhat simpler, let us assume that the number of taps, N, is a multiple of L, $N = KL$. Then v_{mL+i-k} will take on the values x_m, x_{m-1}, ..., x_{m-K-1}, when $i-k = 0$, $i-k = L$, ..., $i-k = (K-1)L$, respectively. Thus, Eq. 2-47 becomes

$$y_{mL+i} = \sum_{j=0}^{K-1} x_{m-j}\, h_{jL+i} \tag{2-48}$$

$$i = 0, 1, \ldots, L-1$$

which implies that, although the filter has N taps, we only do $N/L = K$ multiplications per output sample, which corresponds to reducing the computation rate by the interpolation ratio. This is, of course, analogous to the reduction in computation rate obtained in the decimation process, when using an FIR filter.

Thus, we have seen that the decimation and interpolation processes both involve designing an appropriate lowpass filter to prevent aliasing, when changing

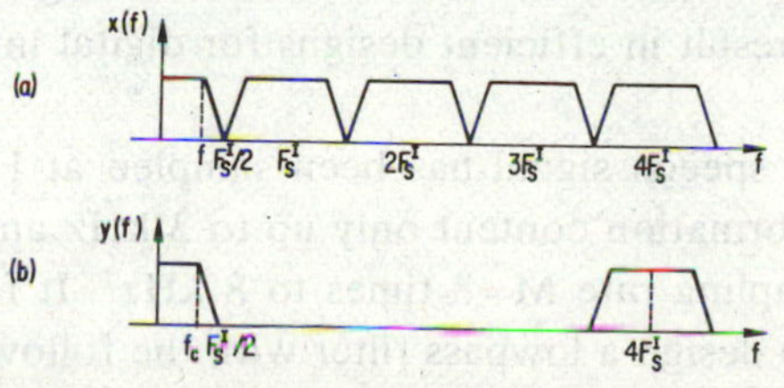

Figure 2.22 A frequency domain interpretation of the interpolation process.

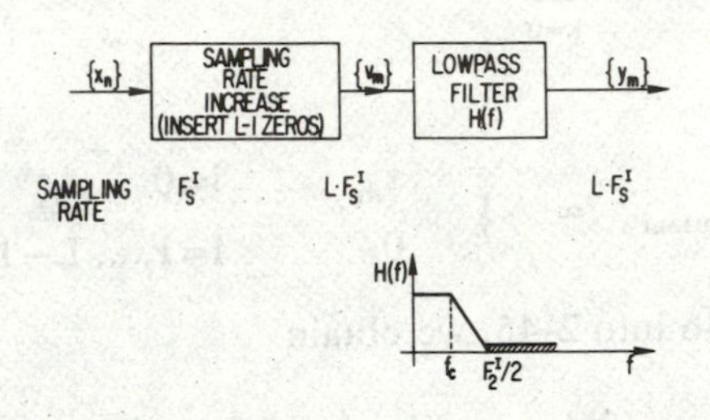

Figure 2.23 The block diagram of a L times interpolator.

sampling rates. Up to now, we have discussed only the changing of the sampling rate by an integer ratio. However, it is possible to change the sampling rate by any rational number L/M simply by interpolating up L times and then decimating M times. This involves a cascade connection of the block diagrams in Figures 2.23 and 2.21. Such an implementation will lead to using a cascade of two lowpass filters. These can be combined into one filter leading to a block diagram of a rational factor sampling rate change, as illustrated in Figure 2.24.

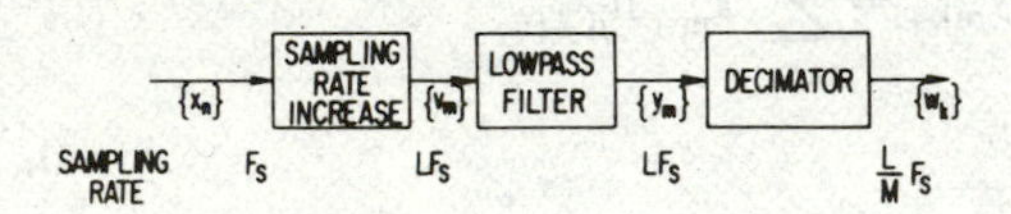

Figure 2.24 The block diagram of a rational factor sampling rate change.

We now proceed to present several examples that illustrate the decimation and interpolation processes and their usefulness in narrow band filtering. In these examples we also discuss some additional design methods and practical considerations that result in efficient designs for digital interpolators and decimators.

Suppose that a speech signal has been sampled at F_S=64 kHz, and we are interested in the information content only up to 3 kHz and would like, therefore, to decimate the sampling rate M=8 times to 8 kHz. It follows from our discussion that we have to design a lowpass filter with the following specifications:

sampling frequency (F_S): 64 kHz
passband (f_c): from 0 to 3 kHz

maximum ripple in passband (δ_1): 0.1 dB
stopband (f_s): beyond 4 kHz
minimum attenuation in stopband (δ_2): 50 dB

Using the method described in Section 2.2, we see that a 6-th order IIR elliptic filter [2, p. 142] is needed to achieve these objectives. This implies a computation rate of about 15 multiplications per output sample. On the other hand, if we design an FIR filter using the method described in Section 2.7 and the program listed in Appendix 2.2, we find that a 160 tap FIR filter will meet the required specifications. However, since the filter has symmetric coefficients and we are decimating by 8, only 10 multiplications and 20 additions are needed per output sample in this case. Thus, we have a somewhat lower computation rate, but unfortunately considerably more storage is required in the FIR case (160 words of storage versus 30 in the IIR case).

Obviously the same filter can be used, if we reverse the problem and wish to interpolate up from 8 kHz to 64 kHz sampling rate. The only difference is that now the input to the filter is at the the low rate and the output at the high rate.

We can now proceed to examine the decimation (or equivalently the interpolation) process more carefully and point out some refined design methods which lead to more efficient realizations.

In our example we stated that we wish to preserve only the information in the band 0 to 3 kHz; thus, when considering the filtering required, it suffices to filter out only those bands which will alias back into the information band. This is depicted in Figure 2.25. The dashed bands are the *care bands*, which upon decimation will alias back into our information band. It suffices, therefore, to design a multiband filter with such a frequency characteristic instead of the straightforward lowpass filter we used in our earlier discussion. Again, doing this for IIR filters will not yield any significant advantage. On the other hand, using the program in Appendix 2.2, we can design optimum FIR multiband filters.

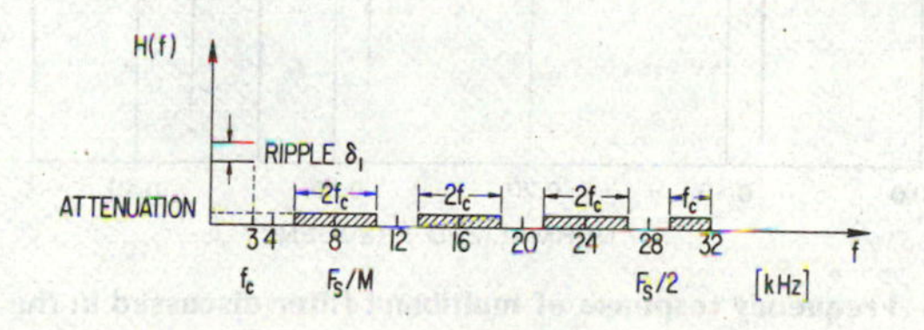

Figure 2.25 The required frequency response for a decimating (interpolating) filter, taking into account only *care regions*.

To illustrate the savings possible, we redesign our example constraining only the *care regions*. The following input cards are used as the input to the program in Appendix 2.2:

```
80,1,5,1,16
0, 0.046875,0.078125,0.171875,0.203125,0.296875,0.328125,0.421875,
0.453125,0.500000
1,0,0,0,0
1,10,10,10,10
```

The 80 tap filter obtained has characteristics exceeding the required specifications, its frequency response is plotted in Figure 2.26. Thus, we see that, in this case, constraining only the *care regions* results in a 50% saving in storage and computation rate versus the previous design (80 taps versus 160 taps). Similar savings can be expected in any decimation (interpolation) application. The only disadvantage of this method is that in the transition region f_c to $F_S/2M$ there is substantial aliasing error, and it contains unusable information, which may be undesirable in some applications.

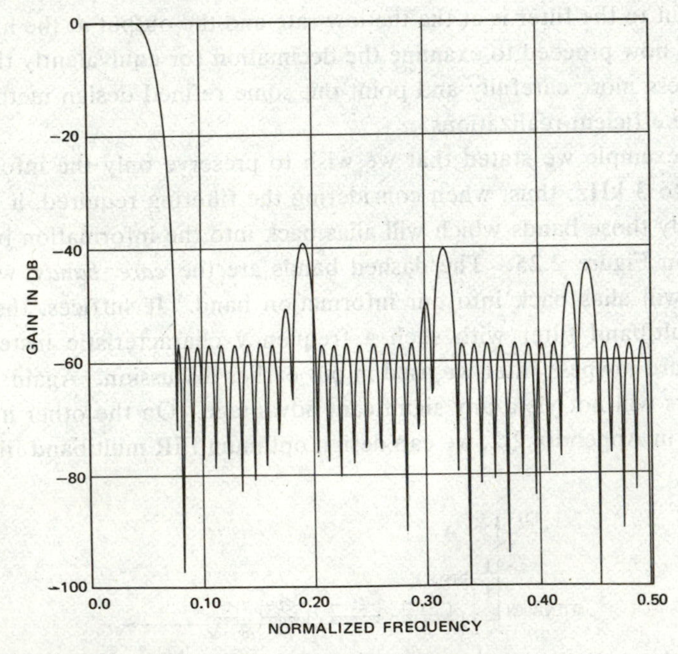

Figure 2.26 Frequency response of multiband filter discussed in the example.

The next step towards improving the efficiency of our realization of the lowpass filter required, is to do the decimation (interpolation) in several stages.

This is especially important when very high decimation ratios are required. Suppose that in the speech decimation problem, we do the decimation in two stages. In the first stage we will decimate by 4 and in the second by 2, thus obtaining an overall decimation ratio of 8 as required. Each stage can be designed following exactly the same procedure outlined above, constraining only the *care regions*. This will require two filters $H_1(f)$ and $H_2(f)$ that have frequency characteristics, as depicted in Figure 2.27. Note that we allotted each filter only half of the total ripple permitted, so that the total inband distortion will not exceed this specification. Again, we resort to the program in Appendix 2.2 with the following input cards:

```
16,1,3,1,16
0, 0.046875,0.203125,0.296875,0.453125,0.5000000
1,0,0,0
1,5,5,5

24,1,2,1,16
0, 0.1875,0.3125,0.50000
1,0
1,5
```

The filters obtained are a 16 tap filter for the first stage and a 24 tap filter for the second stage. The filter characteristics are:

```
                    BANDPASS FILTER
                  FILTER LENGTH =  16
                    BAND  1          BAND  2          BAND  3
LOWER BAND EDGE   0.0             0.203125000      0.453125000
UPPER BAND EDGE   0.046875000     0.296875000      0.500000000
DESIRED VALUE     1.00000000      0.0              0.0
WEIGHTING         1.00000000      5.00000000       5.00000000
DEVIATION         0.005486049     0.001097210      0.001097210
DEVIATION IN DB   0.095290661   -59.1941986      -59.1941986

                    BANDPASS FILTER
                  FILTER LENGTH =  24
                    BAND  1          BAND  2
LOWER BAND EDGE   0.0             0.312500000
UPPER BAND EDGE   0.187500000     0.500000000
DESIRED VALUE     1.00000000      0.0
WEIGHTING         1.00000000      5.00000000
DEVIATION         0.003898262     0.000779652
DEVIATION IN DB   0.067709863   -62.1619720
```

As we see, they exceed the required specifications._ Now let us consider the storage and computation rate required for this design. The total storage required is 40 words (16 + 24), and the computation rate is 8 multiplications per output (16/4 + 24/4). This represents a 50% saving in storage over the previous design. The computation rate is reduced by 20% from the previous design and is only half the multiplication rate required in the IIR case.

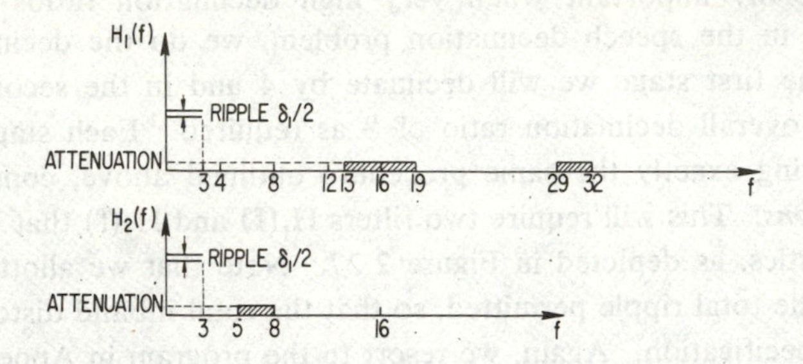

Figure 2.27 The required frequency response for the filters in a two stage design.

We now consider another example that illustrates the usefulness of FIR filters in implementing narrow band filters.

Suppose that an analog signal is sampled at 8 kHz, and we wish to examine only a 50 Hz frequency band above some center frequency f_0. This is frequently required in sonar and vibration signal processing and is commonly referred to as *frequency zoom*. To do this, we first multiply the input signal $\{x_n\}$ by $\exp[2j\pi f_0 n/F_S]$. This will shift the band of interest to the origin. Let us assume that the desired output sampling rate is 160 Hz, that is, a decimation ratio of 50. The output sequence will usually go to an FFT analyzer that may take a 1024 point transform. Since only the 50 Hz band contains usable information, only some $(1024 \times 50/80)=640$ points of the FFT are meaningful. This process is depicted in Figure 2.28.

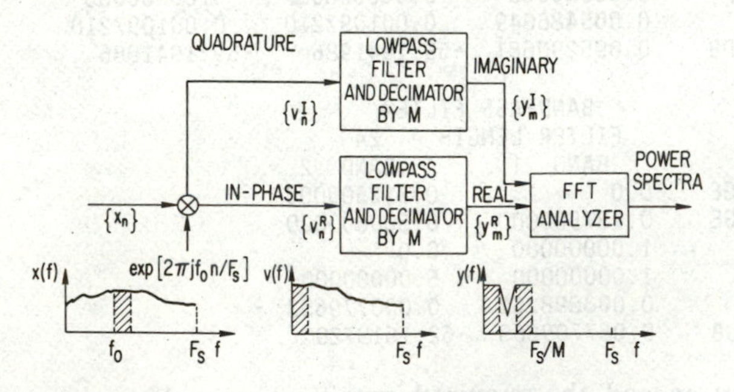

Figure 2.28 Frequency zoom filtering block diagram.

If we consider as adequate a 0.25 dB inband ripple and a 60 dB attenuation, we again could use a 6-th order elliptic filter to meet these specifications [2, pp. 119]. However, such a filter is likely to have poles very close to the unit circle and, as we shall see in Chapter 6, will require a substantial word length some 24 bits to avoid excessive round-off error. On the other hand using the multistage FIR approach, we arrive at a 2 stage design using the following input cards to the program listed in Appendix 2.2:

```
20,1,6,1,32
0,0.00625,0.99375,0.10625,0.19375,0.20625,0.29375,0.30625,
0.39375,0.40625,0.49375,0.50000
1,0,0,0,0,0
1,10,10,10,10,10
40,1,3,1,32
0,0.0625,0.1375,0.2625,0.3375,0.4625
1,0,0
1,10,10
```

The filters obtained have the characteristics listed below that obviously even exceed our specifications:

BANDPASS FILTER
FILTER LENGTH = 20

	BAND 1	BAND 2	BAND 3	BAND 4
LOWER BAND EDGE	0.0	0.993749976	0.193750024	0.293749988
UPPER BAND EDGE	0.006250001	0.106249988	0.206250012	0.306249976
DESIRED VALUE	1.00000000	0.0	0.0	0.0
WEIGHTING	1.00000000	10.0000000	10.0000000	10.0000000
DEVIATION	0.003363350	0.000336335	0.000336335	0.000336335
DEVIATION IN DB	0.058417220	-69.4645386	-69.4645386	-69.4645386

	BAND 5	BAND 6
LOWER BAND EDGE	0.393750012	0.493749976
UPPER BAND EDGE	0.406250000	0.500000000
DESIRED VALUE	0.0	0.0
WEIGHTING	10.0000000	10.0000000
DEVIATION	0.000336335	0.000336335
DEVIATION IN DB	-69.4645386	-69.4645386

BANDPASS FILTER
FILTER LENGTH = 40

	BAND 1	BAND 2	BAND 3
LOWER BAND EDGE	0.0	0.137499988	0.337499976
UPPER BAND EDGE	0.062500000	0.262499988	0.462499976
DESIRED VALUE	1.00000000	0.0	0.0
WEIGHTING	1.00000000	10.0000000	10.0000000
DEVIATION	0.005899251	0.000589925	0.000589925
DEVIATION IN DB	0.102473080	-64.5840607	-64.5840607

As we see, we need a 20 tap filter for the first stage that decimates 10 to 1 and a 40 tap filter for the second stage that decimates by 5 to 1. The total storage required is 60 words and the computation rate only 5 multiplications per output (20/20 + 40/10), which is a third of the number required should we use an IIR filter. Furthermore, when using FIR filters, we have a linear phase response and will usually require a word length of 16 bits or less, due to the fact that there is no accumulation of the round-off error.

It has been shown in [10] that most of the advantage gained by going from a single to a multistage design is realized when employing a two stage design as above. A three stage, or more, design (e.g., decimate by 2, 5, and 5) will lead to approximately the same computation rate and a somewhat lower storage requirement.

Another point worth mentioning is the possible use of so called *half band* filters. These are described in detail in [9]. Such filters are odd order FIR filters used to decimate by 2, and it can be shown that, since their frequency response is antisymmetric around $F_S/4$, they have even coefficients equal to zero. This implies that their realization requires half as many multiplications as standard FIR filters. Such a filter could be used as the first stage filter in our example in the three stage design. This will lead to a further reduction in the computation rate.

2.9 THE EFFECT OF FINITE WORD LENGTH COEFFICIENTS

In the previous sections dealing with various design methods for digital filters, we have assumed that the resulting coefficients have been computed accurately. However, in digital systems in general, and particularly in signal processors of the type we are addressing in this book, the filter coefficients can only be expressed using a finite number of bits. Furthermore, in many instances it may be advantageous to use as few bits as possible to express the filter coefficients, as long as the required performance specifications are still met. We have already examined a simple example of the effects of finite word length in Section 1.6. In this section we present a more detailed treatment of the problem, mainly through several examples that will give the reader a qualitative appreciation of the effect of finite word length coefficients on the frequency response of the digital filter.

As we have seen from the previous sections, the digital filter is characterized by a set of real numbers, namely its coefficients. Obviously, altering these numbers will alter the characteristics of the filter. By quantizing the coefficients we actually obtain a different filter from the one we have originally designed. However, since for a reasonable word length the quantized coefficients are quite close to the actual coefficients, the resulting change in the filter characteristics will usually be quite small. To assess the effect of any change in the value of a coefficient on a given filter characteristic, say the frequency response, we could differentiate $|H(e^{j\omega T})|$ with respect to that coefficient and use the value of the derivative as an indication of the sensitivity of the frequency response to changes in this particular coefficient. A second approach may be to treat the changes in the coefficient values as random statistical perturbations, and to estimate the changes that may be expected by a statistical method, or to derive bounds on these changes.

In this section we will use neither of the approaches suggested above. Instead, we choose to give the reader a qualitative appreciation of the expected effect of finite word length on the frequency response, through the presentation of three typical examples. To complement this, we have included in Appendix 2.3 a computer program that will calculate the frequency response of digital filters for a given precision (number of bits) used to represent the filter coefficients. This program can be used in conjunction with the two computer filter design programs given in Appendices 2.1 and 2.2.

Before we proceed to discuss the examples, we note two general observations. In most practical cases, the use of 16 bits (sign bit, integer bit, and 14 bits fraction) will show negligible degradation from the use of infinitely many bits, the main exception being in the case of very sharp, high order IIR filters. The second observation is that the phase linearity in linear phase FIR filters will be preserved, since it is determined only by the coefficient symmetry that $h_i = h_{N-i}$ and not by their actual values.

In the first example we consider a 24 tap lowpass FIR filter designed by using the program in Appendix 2.2. The passband is 0.0 to 0.2 and the stopband 0.3 to 0.5, in terms of normalized frequency, that is, 0.5 corresponds to half the sampling frequency. The frequency response is calculated for 16, 10 and 6 bit coefficients and plotted in Figures 2.29, 2.30, and 2.31. The response for 16 bit coefficients is within 0.01 dB of the originally designed filter. The passband ripple is ±0.05 dB, and the stopband attenuation is at least 43 dB. For 10 bit coefficients we see that the equiripple property of the filter is destroyed, the passband ripple increased to +0.14 dB and −0.9 dB, and the stopband attenuation is only 39 dB. For 6 bit coefficients, the filter only generally resembles the filter we started out with. It has a −1 dB passband ripple and only a 16 dB stopband attenuation.

The second example is a 6-th order IIR bandpass filter, designed using the program in Appendix 2.1. Its frequency response is shown for 16 and 6 bit coefficients in Figures 2.32 and 2.33. For 16 bits the passband ripple is −0.075 dB and +0.09 dB, but for 6 bits the ripple degrades to −0.78 dB and +0.05 dB and has a completely different shape. The stopband attenuation is almost the same in both cases.

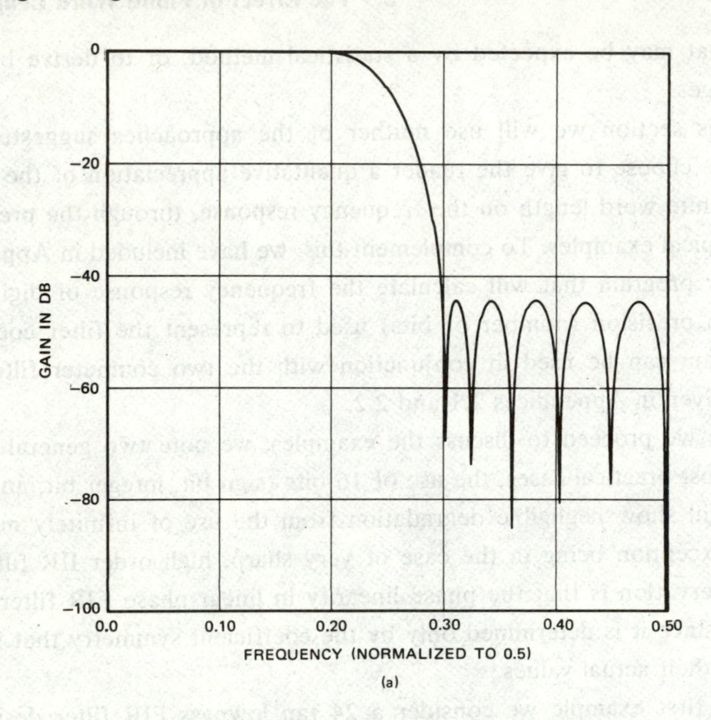

(a)

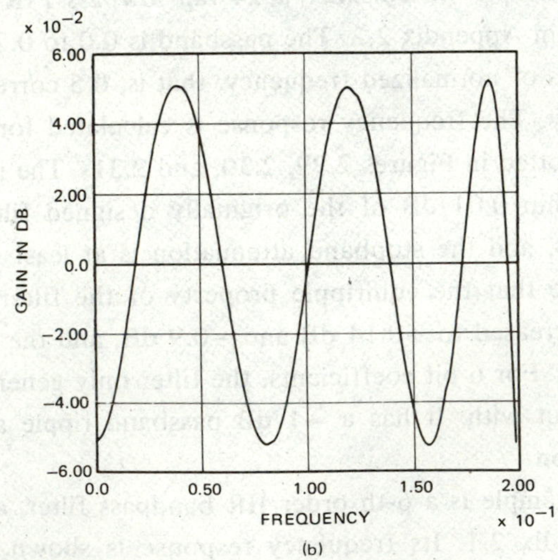

(b)

Figure 2.29 Amplitude response of 24 tap lowpass filter with 16 bit coefficients.

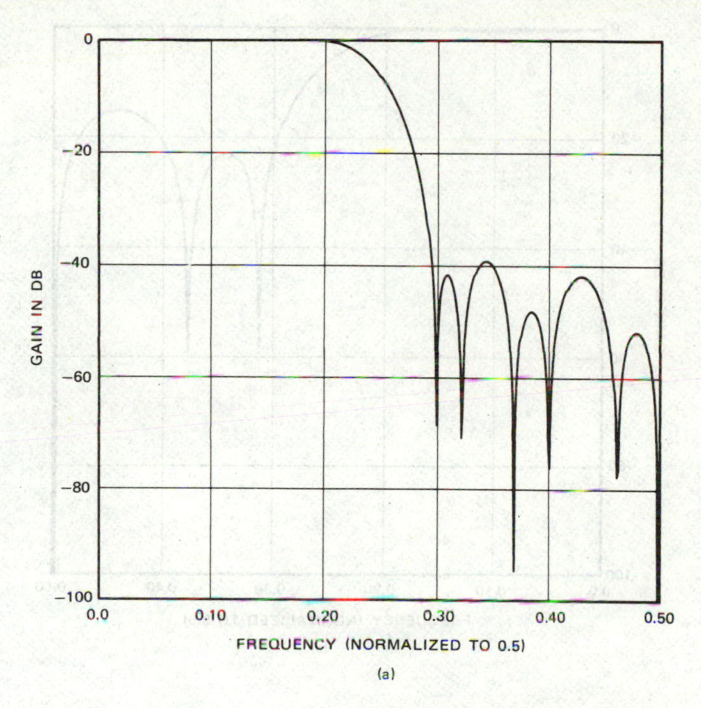

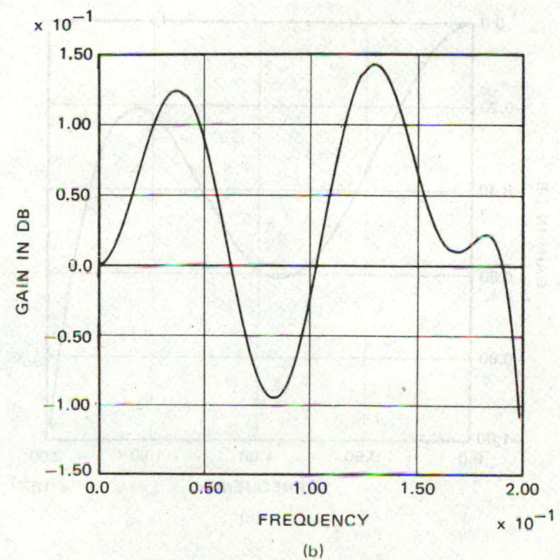

Figure 2.30 Amplitude response of 24 tap lowpass filter with 10 bit coefficients.

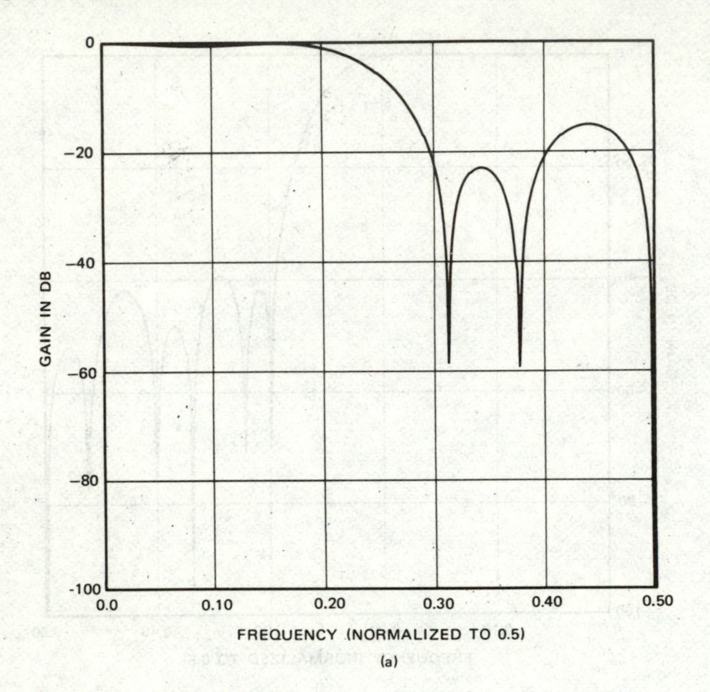

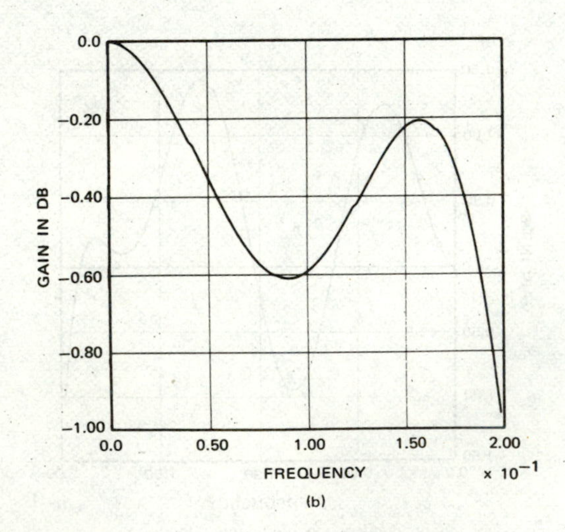

Figure 2.31 Amplitude response of 24 tap filter with 6 bit coefficients.

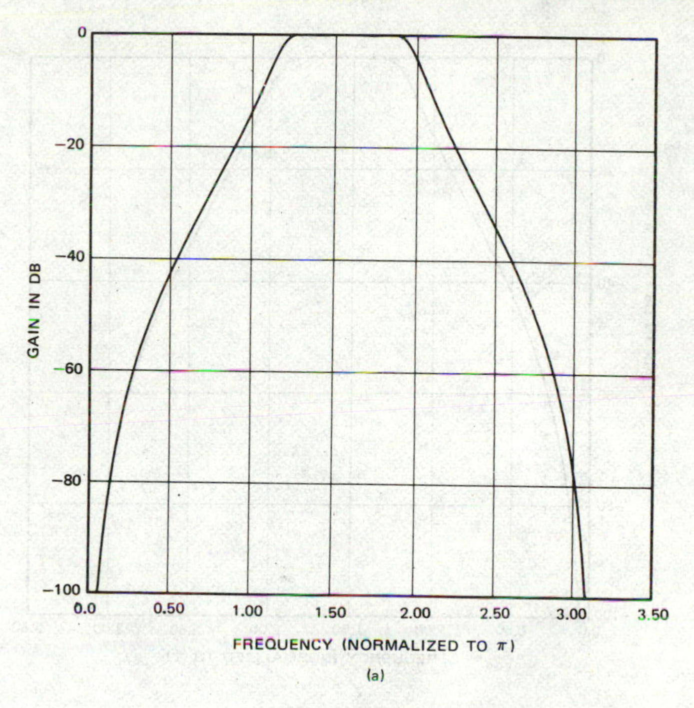

(a)

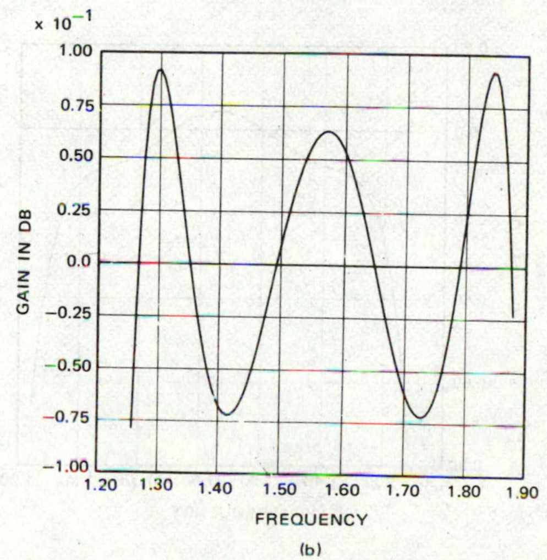

(b)

Figure 2.32 Amplitude response of 6-th order filter with 16 bit coefficients.

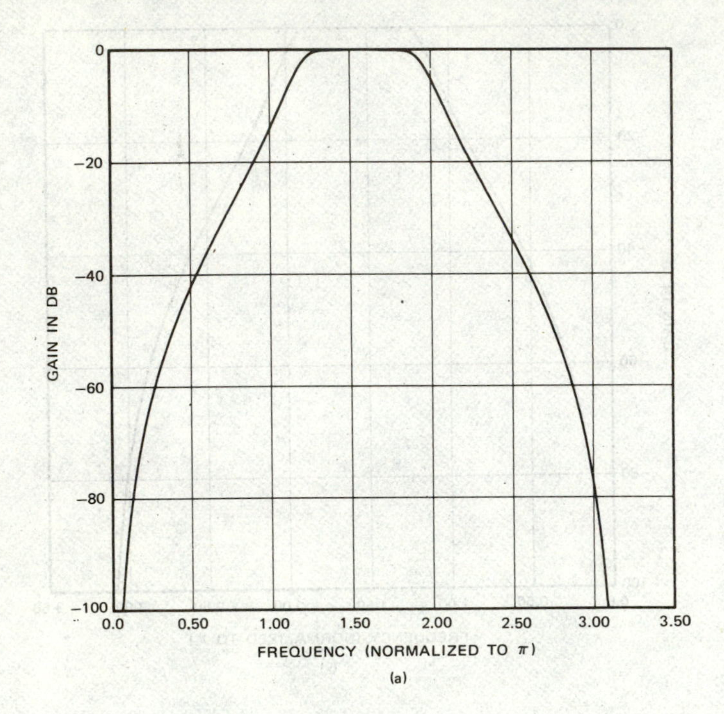

(a)

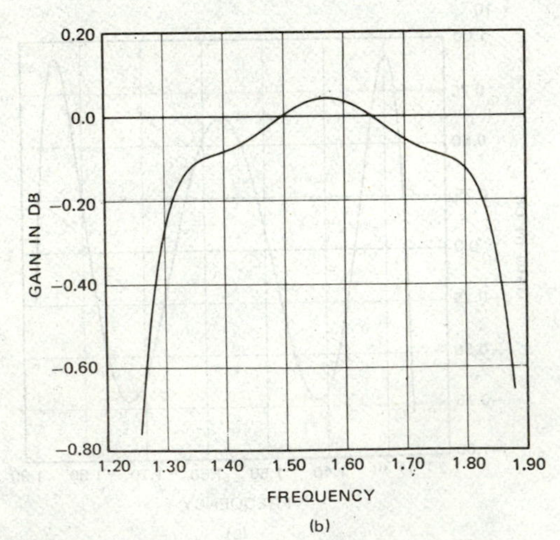

(b)

Figure 2.33 Amplitude response of 6-th order filter with 6 bit coefficients.

Finally, in the last example we consider a 69 tap multiband FIR filter. The frequency response for 16 and 10 bit coefficients is shown in Figures 2.34 and 2.35. For this case even with 16 bit coefficients, the equiripple property of the original design is not preserved, especially for the 0.4 to 0.5 band; however, the difference from the original filter is still within 3 dB for the stopbands and 0.01 dB for the passbands. For 10 bit coefficients a serious degradation is already noted, especially for the passband ripple of band 1 and the stopband attenuation of the band 0.4 to 0.5 that degraded from more than 60 dB attenuation to less than 36 dB.

These examples illustrate that the finite word length coefficients may significantly alter the frequency response of the filter from that originally designed. They also demonstrate the necessity to verify the actual filter response, especially if coefficient word lengths shorter than 16 bits are being considered. To this end the computer program in Appendix 2.3 can be used; however, the reader must be cautioned that the program assumes a direct realization of FIR filters and a cascade realization of IIR filters. The results will be different for other realizations, and suitable programs could be easily written for such realizations.

We conclude our discussion in this section with a brief mention of the design tradeoffs that appear as a result of the use of finite word length coefficients. To illustrate this point, we consider the lowpass filter discussed in the first example above. This 24 tap filter, when implemented with 6 bit coefficients, has a frequency response that could be met with a 14 tap filter using 12 bit coefficients. Obviously, the question now arises, which of the two options is preferable. As always the answer to this question is not obvious, nor clear cut. It depends on many factors and has to be answered by the designer according to the constraints of the particular system under consideration.

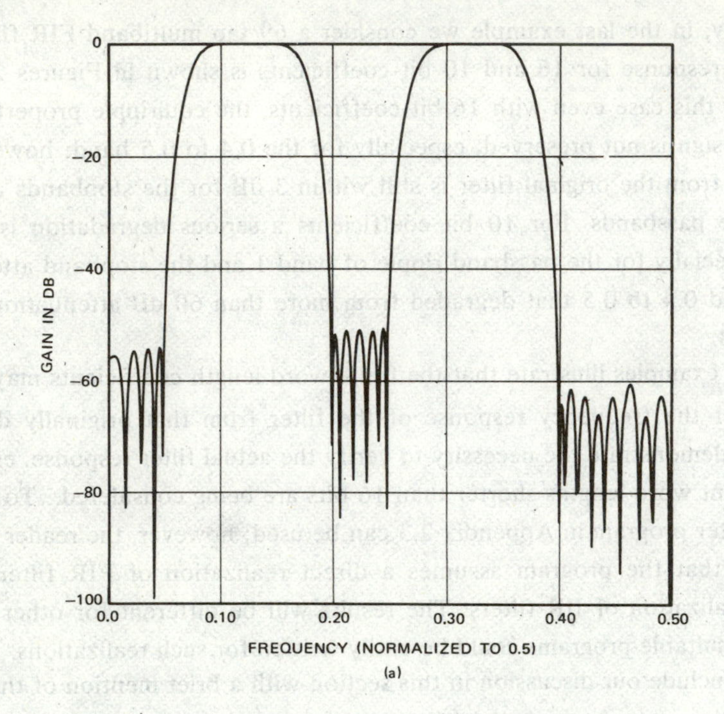

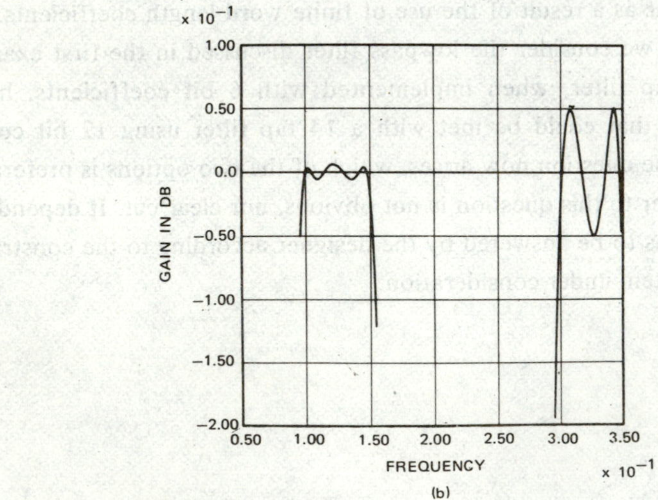

Figure 2.34 Amplitude response of 69 tap multiband filter with 16 bit coefficients.

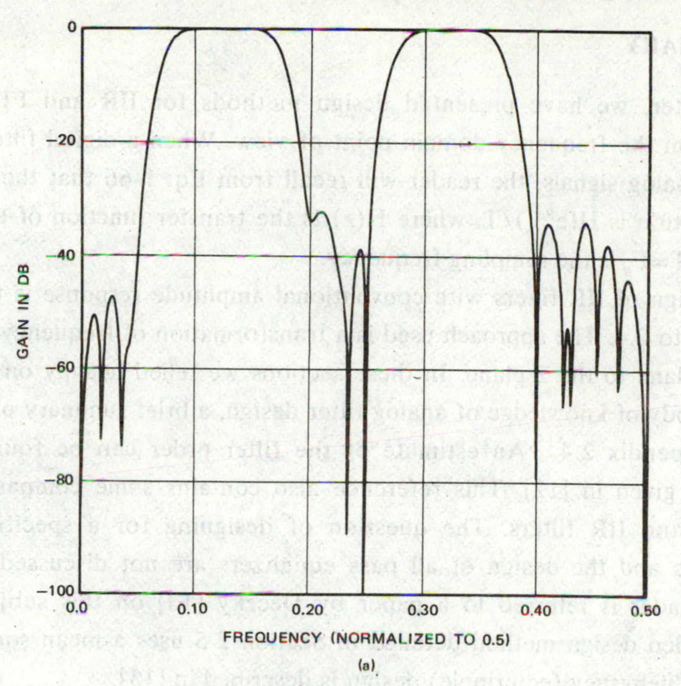

(a)

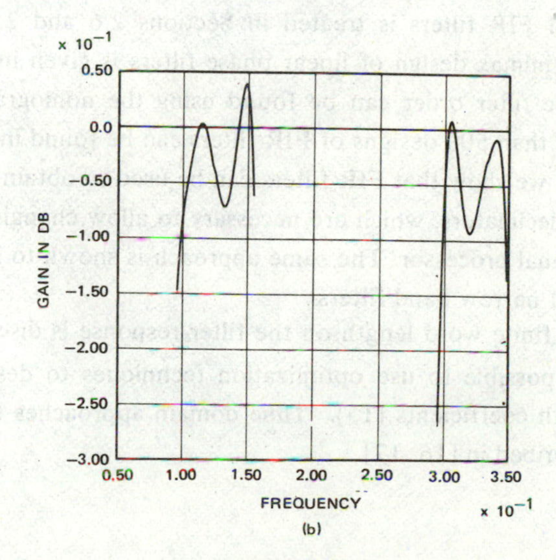

(b)

Figure 2.35 Amplitude response of 69 tap multiband filter with 10 bit coefficients.

2.10 SUMMARY

In this chapter, we have presented design methods for IIR and FIR filters, primarily from the frequency domain point of view. When a digital filter is used to process analog signals, the reader will recall from Eq. 1-66 that the effective transfer function is $H(e^{j\omega T})/T$, where $H(z)$ is the transfer function of the digital filter and $1/T=F_s$ is the sampling frequency.

The design of IIR filters with conventional amplitude response is treated in Sections 2.2 to 2.4. The approach used is a transformation of frequency variables from the s plane to the z plane. In these sections we relied heavily on the well-developed body of knowledge of analog filter design, a brief summary of which is given in Appendix 2.4. An estimate of the filter order can be found in the nomographs given in [12]. This reference also contains some comparisons between FIR and IIR filters. The question of designing for a specified phase characteristic and the design of all pass equalizers are not discussed, and the interested reader is referred to a paper by Deczky [13] on this subject. The computer aided design method detailed in Section 2.5 uses a mean square error criterion, a Chebyshev (equiripple) design is described in [13].

The design of FIR filters is treated in Sections 2.6 and 2.7. A computer program for the minimax design of linear phase filters is given in Appendix 2.2. An estimate of the filter order can be found using the nomographs in [12]. A tabulation of more than 500 designs of FIR filters can be found in [14].

In Section 2.8 we show that FIR filters can be used to obtain efficient digital interpolators and decimators, which are necessary to allow changing the sampling rate in a digital signal processor. The same approach is shown to lead to efficient implementations of narrow band filters.

The effect of finite word length on the filter response is discussed briefly in Section 2.9. It is possible to use optimization techniques to design filters with shorter word length coefficients [15]. Time domain approaches to the design of IIR filters are described in [16, 17].

EXERCISES

2.1 Design a Chebyshev filter to the specification given in Section 2.2. Sketch its amplitude response and compare it with that of the Cauer filter and the Butterworth filter.

2.2 The function $A(\lambda^2)=1/[1+a^{2n}\tan^{2n}(\lambda/2)]$ takes on the value 1 at $\lambda=0$ and the value 0 at $\lambda=\pi$. It is monotonically decreasing in the interval $(0, \pi)$. Show that, with $z=e^{j\lambda}$, it is a rational function in z, and is, therefore, a possible choice as a magnitude square function. How should n and a be chosen? Where are the poles and zeros? Compare the filter with Butterworth filters. How can the function $A(\lambda)$ be modifies so that it has equiripple response in the passband?

2.3 The impulse invariant method for the design of digital filter matches the impulse response with the samples of the impulse response of an analog filter. Let $H^a(s)$ be the analog filter, and h(t) be its impulse response. Then the desired digital filter has an impulse response $\{h_n\}$, $n\geq 0$ with $h_n=h(nT)$. Suppose it is desired to lowpass filter an analog signal sampled at $1/T=16$ KHz with a cutoff at 2 KHz.

(a) Start with an RC filter with 3 dB point at 2 KHz. Design the digital filter by the impulse invariant method. Sketch the amplitude response.

(b) Use the bilinear transformation, Eq. 2-10, to obtain a digital filter from the RC analog filter. Compare the amplitude response with that obtained in part (a).

(c) Repeat (a) and (b) but use a third order Butterworth analog filter with 3 dB at 2 KHz.

(d) Repeat (c) but with a cutoff at 4 KHz.

2.4 Prove Eq. 2-26. (*Hint*: $|H_n| = [H(z_n)H(1/z_n)]^{1/2}$).

2.5 What is the amplitude and phase response of a digital differentiator? Make necessary modifications in the approach discussed in Section 2.5 for the design of a recursive digital differentiator.

2.6 In addition to the *Hamming* windows, there are other commonly used windows. Some of these are:

(a) *Bartlett* (triangular):

$$w_k = \begin{cases} 2k/M, & 0\leq k\leq (M-1)/2 \\ 2-2k/M, & (M-1)/2\leq k\leq M \end{cases}$$

(b) *Blackman*:

$$w_k = 0.42 - 0.5\cos(2\pi k/M + 0.08\cos(4\pi k/M), 0\leq k\leq M$$

(c) *Hanning*:

$$w_k = 0.5 - 0.5\cos(2\pi k/M), \quad 0\leq k\leq M$$

(d) *Kaiser*:

$$w_k = I_0(a[M^2/4-(k-M/2)^2]^{1/2}) \,/\, I_0(aM/2), \quad 0\leq k\leq M$$

where $I_0(\bullet)$ is the zero-th order modified Bessel function, and a is a design parameter typically between 10/M and 18/M.

For each of these windows, sketch $W(\lambda)$ for $M=22$ and compare with Figure 2.16. Use these in place of the Hamming window in the lowpass filter example of Section 2.6 and compare the resulting amplitude response.

2.7 Show that a nonrecursive filter with coefficients $\{h_k\}_{k=0,K}$ has a linear phase if, and only if, $h_k=h_{K-k}$.

2.8 Design a 15 tap nonrecursive differentiator by the Fourier series method. Use two different windows and compare the result.

2.9 Use the program of Appendix 2.2 to design a 15 tap differentiator and compare with that obtained in exercise 2.8.

2.10 Use the program of Appendix 2.3 to study the effect of finite word length for the coefficients of the filter designed in exercise 2.9. Calculate for 16, 10, and 6 bits.

2.11 Use the program of Appendix 2.3 to study the effect of coefficient quantization for the Butterworth filter and the Cauer filter designed in Section 2.2. Use 10, 8, and 6 bits.

2.12 An analog signal originally sampled at 8 KHz is to be resampled at 10 KHz by using the system shown in Figure 2.24. Design the lowpass filter, assuming a minimal distortion should be introduced below 3 KHz, with the ripple in the range $0 \leq f \leq 3$ KHz to be kept below 0.1 dB. How many bits should the coefficients be quantized to?

2.13 Design a lowpass filter with the following specifications: $F_s=10$ KHz, $f_c=800$ Hz, $f_S=1$ KHz, $\delta_1=0.1$ dB, $\delta_2=60$ dB. Now repeat the design of this filter using two stage decimation and compare the computation requirement with the original design.

REFERENCES

1. M. E. Van Valkenburg, *Introduction to Modern Network Synthesis*, John Wiley & Sons Inc., New York, 1960.
2. E. Christian and E. Eisenman, *Filter Design Tables and Graphs*, John Wiley & Sons, Inc., New York, 1966.
3. R. W. Daniels, *Approximation Methods for the Design of Passive Active and Digital Filters*, McGraw-Hill, New York, 1974.
4. K. Steiglitz, *Computer-Aided Design of Recursive Digital Filters*, IEEE Trans. Audio Electroacoust., vol. AU-18, June 1970, pp. 123-129.
5. J. F. Kaiser, *Digital Filters* in *System Analysis by Digital Computer*, F. F. Kuo and J. F. Kaiser eds., John Wiley & Sons, Inc., New York, 1966.
6. J. H. McClellan and T. W. Parks, *A Unified Approach to the Design of Optimum FIR Linear Phase Digital Filters*, IEEE Trans. Circuit Theory, vol. CT-20, 1973, pp. 607-701.
7. J. H. McClellan, T. W. Parks, and L. R. Rabiner, *A Computer Program for Designing Optimum FIR Linear Phase Digital Filters*, IEEE Trans. Audio Electroacoust. vol. AU-21, December 1973, pp. 506-525.
8. R. W. Schafer and L. R. Rabiner, *A Digital Signal Processing Approach to Interpolation*, Proc. IEEE, vol. 61, June 1973, pp. 692-702.

9. M. G. Bellanger, J. L. Daguet, and G. P. Lepagnol, *Interpolation, Extrapolation, and Reduction of Computation Speed in Digital Filters*, IEEE Trans. Acoustics, Speech, and Signal Processing, vol. ASSP-22, August 1974, pp. 231-235.
10. R. E. Crochiere and L. R. Rabiner, *Optimum FIR Digital Filter Implementation for Decimation, Interpolation, and Narrow-Band Filtering*, IEEE Trans. Acoustics, Speech, and Signal Processing, vol. ASSP-23, October 1975, pp. 444-456.
11. R. R. Shivley, *On Multistage FIR Filters with Decimation*, IEEE Trans. Acoustics, Speech, and Signal Processing, vol. ASSP-23, August 1975, pp. 353-357.
12. L. R. Rabiner, J. F. Kaiser, O. Herman and M. T. Dolan, *Some Comparisons Between FIR and IIR Digital Filters*, Bell System Technical Journal, vol. 53, 1974, pp. 305-331.
13. A. G. Deczky, *Equiripple and Minimax (Chebyshev) Approximations for Recursive Digital Filters*, IEEE Trans. Audio Electroacoust., vol. ASSP-22, 1974, pp. 98-111.
14. E. Avenhaus, *On the Design of Digital Filters with Coefficients of Limited Word Length*, IEEE Trans. Audio Electroacoust., vol. AU-20, 1972, pp. 206-212.
15. R. W. Hankins, *Design Procedure for Equiripple Nonrecursive Digital Filters*, MIT-RLE Tech. Rep. 485, 1972.
16. C. S. Burrus and T. W. Parks, *Time Domain Design of Reçursive Digital Filters*, IEEE Trans. Audio Electroacoust., vol. AU-18, 1970, pp. 138-142.
17. F. Brophy and A. C. Salazar, *Recursive Digital Filter Synthesis in the Time Domain*, IEEE Trans. Acoustics, Speech and Signal Processing, vol. ASSP-22, 1974, pp. 45-55.

Appendix 2.1

A Computer Program for IIR Filter Design

This program is due to K. Steiglitz of Princeton University [4]. There are two options for input data specification: point specification and band specification. If the point specification is to be used, the first card should contain 00. The following cards give values of the frequencies constrained (the frequency is normalized to 1, that is, 1 corresponds to π in digital frequency, and to half the sampling frequency in the analog case), and the desired amplitude response at these frequencies. They are both in the F10.5 format. The last card in this set must have the frequency 1.0. The cards for the example of Section 2.5 are therefore

```
Col.  1-10  Col.  11-20
      .01         1.0
      .02         1.0
      .03         1.0
       •           •
       •           •
       •           •
      .09         1.0
      .10         0.5
      .11         0.0
      .12         0.0
       •           •
       •           •
       •           •
      .20         0.0
      .30         0.0
       •           •
       •           •
       •           •
      1.0         0.0
```

These data cards can be followed by another set of data cards, one for each filter to be designed to have the specified amplitude response. These cards have the format 2I5 followed by 2F10.5, and on it the values of N, LIMIT, EST, and EPS.

N is 4 times the number of second order sections and is limited to 32 in the program; however, this limit can be easily extended by increasing the dimensions of X, G and H. LIMIT is normally set to 1500, and may be increased if convergence is hard to achieve. EST is set to 0, and EPS can be set to 0.0001, it may be decreased if extreme accuracy is desired.

When the band specification is more convenient, it can be used and the first card should contain the number of bands in I2 format. Following that, there has to be one card for each band, containing the lower band edge, upper band edge, and the value of the desired response in that band. These cards are format free with the numbers separated by a comma. Following this data are cards, one for each filter to be designed, specifying N, LIMIT, EST and EPS, exactly as before.

As an example, suppose the following input cards are used:

```
04
0,0.1,1.0
0.2,0.3,0.0
0.4,0.6,1.0
0.8,1.0,0.0
00024030000000.000000000.00010
```

The output of the program, seen in Table 2.5, gives the final error function, the maximum deviation from the desired response (in dB), the value of the constant multiplier C^*, and the value of the coefficients for the cascade form:

$$H(z) = C^* \prod_{j=1}^{K} \frac{1 + A(1,j)z^{-1} + A(2,j)z^{-2}}{1 + B(1,j)z^{-1} + B(2,j)z^{-2}}$$

Also listed are the poles and zeros of the transfer function. Outputs also available from the program, but not shown in the tables, are a list of the desired point specification, the approximation achieved, and the error. In addition, a 201 point frequency response of the filter obtained is written on unit 10 for possible plotting.

```
************************************************************************
                         INFINITE IMPULSE RESPONSE (IIR)
                         FLETCHER-POWELL OPTIMIZATION ALGORITHM
                         RECURSIVE FILTER OF ORDER  12
 BAND EDGES ARE  0.0          0.10      DESIRED VALUE = 1.00
 BAND EDGES ARE  0.20         0.30      DESIRED VALUE = 0.0
 BAND EDGES ARE  0.40         0.60      DESIRED VALUE = 1.00
 BAND EDGES ARE  0.80         1.00      DESIRED VALUE = 0.0
 FINAL ERROR FUNCTION VALUE = 0.14498175D+01
 MAXIMUM ABSOLUTE DEVIATION IN PASSBAND =     2.49276 DB

 MINIMUM STOPBAND ATTENUATION =  -12.05523 DB

 CONSTANT MULTIPLIER C*= 0.25244430D+00
 COEFFICIENTS FOR CASCADE DECOMPOSITION

 A(1,1) =  0.53264255D+00    A(2,1) = -0.46317444D+00
 B(1,1) = -0.44271071D+00    B(2,1) =  0.59043832D+00
 A(1,2) =  0.53264255D+00    A(2,2) = -0.46317444D+00
 B(1,2) = -0.25154394D+00    B(2,2) = -0.22687963D+00
 A(1,3) =  0.53264255D+00    A(2,3) = -0.46317444D+00
 B(1,3) = -0.25154393D+00    B(2,3) = -0.22687963D+00
 A(1,4) =  0.53264255D+00    A(2,4) = -0.46317444D+00
 B(1,4) = -0.25154393D+00    B(2,4) = -0.22687963D+00
 A(1,5) = -0.11724331D+01    A(2,5) =  0.66611937D+00
 B(1,5) = -0.61800777D+00    B(2,5) =  0.21600159D+00
 A(1,6) = -0.10132787D+01    A(2,6) =  0.72480545D+00
 B(1,6) =  0.48195038D+00    B(2,6) = -0.38382065D+00
 ROOTS
       REAL             IMAG             REAL             IMAG
  0.46450118D+00                   -0.99714373D+00
  0.22135535D+00 0.73582615D+00 0.22135535D+00-0.73582615D+00
  0.46450118D+00                   -0.99714373D+00
  0.61841607D+00                   -0.36687214D+00
  0.46450118D+00                   -0.99714373D+00
  0.61841607D+00                   -0.36687214D+00
  0.46450118D+00                   -0.99714373D+00
  0.61841607D+00                   -0.36687214D+00
  0.58621654D+00 0.56786401D+00 0.58621654D+00-0.56786401D+00
  0.30900389D+00 0.34715730D+00 0.30900389D+00-0.34715730D+00
  0.50663934D+00 0.68419443D+00 0.50663934D+00-0.68419443D+00
  0.42377265D+00                   -0.90572303D+00
************************************************************************
```

Table 2.5 IIR Filter Design Program Output for Multiband Filter.

```
C     THIS IS A PROGRAM FOR DESIGNING INFINITE IMPULSE RESPONSE (IIR)
C     FILTERS USING THE FLETCHER - POWELL OPTIMIZATION ALGORITHM.
C     THIS PROGRAM HAS BEEN WRITTEN BY KEN STEIGLITZ OF PRINCETON UNIVERSITY
C     THE INPUT DATA REQUIRED FOR THE PROGRAM , AND ITS FORMAT
C     IS DESCRIBED IN APPENDIX 2.1.
C     PROCEDURE MASK THAT IS CALLED BY THE PROGRAM IS A SUBROUTINE
C     FOR HANDLING UNDERFLOW, AND SHOULD BE SUPPLIED BY THE USER
```

```
C     DEPENDING ON THE COMPUTER SYSTEM USED.
C
C......MAIN PROGRAM
      IMPLICIT REAL*8 (A-H,O-Z)
      EXTERNAL FUNCT
      DIMENSION H(624),X(32),G(32)
      COMMON/RAW/W(100),Y(100),M,ICALL
      CALL MASK
C......READ IN DATA
C     WRITE(6,51)
C  51 FORMAT('1INPUT DATA')
      READ(5,210) IBAND
  210 FORMAT(I2)
      IF(IBAND.GT.0) GOTO 211
C     THIS INPUT SECTION IS USED IF IBAND=0, THAT IS POINT SPECIFICATION
C     IS DESIRED.
      M=0
   30 M=M+1
      READ(5,21)W(M),Y(M)
   21 FORMAT(2F10.5)
C     WRITE(6,22)M,W(M),Y(M)
C  22 FORMAT(' I=',I3,' W=',D15.8,' Y=',D15.8)
      IF(W(M).LT.1.DO)GOTO30
      GOTO 299
  211 M1=0
C     THIS INPUT SECTION IS USED FOR BAND SPECIFICATION.
      DO 213 J=1,IBAND
      READ*, UPE,UNE,YDES
      M=M1
  216 M=M+1
      W(M)=UPE+(UNE-UPE)*(M-M1-1)/20.0
      Y(M)=YDES
      IF((M-M1).LT.21) GOTO 216
  213 M1=M
  299 DO 15 J=1,32
   15 X(J)=0.DO
      X(4)=.25
   99 READ(5,60)N,LIMIT,EST,EPS
   60 FORMAT(2I5,2F10.5)
C     WRITE(6,61)N,LIMIT,EST,EPS
C  61 FORMAT('1N=',I3,' LIMIT=',I5,' EST=',D15.8,' EPS=',D15.8)
      ICALL=0
   98 CALL DFMFP(FUNCT,N,X,F,G,EST,EPS,  25 ,IER,H)
      CALL ROOTS(N,X,ICALL)
      CALL INSIDE(N,X,KFLAG)
C     WRITE(6,26)IER,KFLAG,ICALL
C  26 FORMAT(' IER=',I5,' KFLAG=',I5,' ICALL=',I5)
      IF((KFLAG.NE.0.OR.IER.NE.0).AND.ICALL.LE.LIMIT)GOTO98
      ICALL=-10
      CALL FUNCT(N,X,F,G)
      GOTO99
```

```
      END
C
C
      SUBROUTINE FUNCT(N,X,F,G)
      IMPLICIT REAL*8 (A-H,O-Z)
      DIMENSION H(624),X(32),G(32),YHT(100),E(100),IGRAPH(60)
      COMPLEX*16 Z(100),NUM(100,8),DEN(100,8),Q,QBAR,ZCUR,ZCUR2
      COMMON/RAW/W(100),Y(100),M,ICALL
      DATA ISTAR,IBLANK/'*',' '/
      PI=-3.14159265358979
      K=N/4
      IF(ICALL.NE.0)GOTO0101
      DO 102 I=1,M
  102 Z(I)=CDEXP(DCMPLX(0.D0,W(I)*PI))
  101 A1=0.D0
      A2=0.D0
      DO 40 I=1,M
      ZCUR=Z(I)
      ZCUR2=ZCUR*ZCUR
      Q=DCMPLX(1.D0,0.D0)
      DO 33 J=1,K
      J4=(J-1)*4
      NUM(I,J)=1.D0+X(J4+1)*ZCUR+X(J4+2)*ZCUR2
      DEN(I,J)=1.D0+X(J4+3)*ZCUR+X(J4+4)*ZCUR2
   33 Q=Q*NUM(I,J)/DEN(I,J)
      QBAR=DCONJG(Q)
      YHT(I)=Q*QBAR
      A2=A2+YHT(I)
      YHT(I)=DSQRT(YHT(I))
   40 A1=A1+YHT(I)*Y(I)
      A=A1/A2
      F=0.D0
      DO 57 J=1,16
   57 G(J)=0.D0
      DO 42 I=1,M
      ZCUR=Z(I)
      ZCUR2=ZCUR*ZCUR
      YHT(I)=A*YHT(I)
      E(I)=YHT(I)-Y(I)
      F=F+E(I)**2
      DO 42 J=1,K
      J4=(J-1)*4
      Q=2.D0*E(I)*YHT(I)/NUM(I,J)
      G(J4+1)=G(J4+1)+Q*ZCUR
      G(J4+2)=G(J4+2)+Q*ZCUR2
      Q=-2.D0*E(I)*YHT(I)/DEN(I,J)
      G(J4+3)=G(J4+3)+Q*ZCUR
   42 G(J4+4)=G(J4+4)+Q*ZCUR2
      ICALL=ICALL+1
C     IF((ICALL/10)*10.EQ.ICALL-1)WRITE(6,25)ICALL,F
C  25 FORMAT(' CALL NO.',I4,' F=',D15.8)
```

```
      IF(ICALL.GT.0)RETURN
C......PRINT OUT
      KORDER=N/2
  350 PRINT 360,KORDER
  360 FORMAT(1H1,70(1H*)//20X,'INFINITE IMPULSE RESPONSE (IIR)'/
     120X,'FLETCHER-POWELL OPTIMIZATION ALGORITHM'/
     220X,'RECURSIVE FILTER OF ORDER',I4)
      IBC=1
  601 UNE=W(IBC)
      YDES=Y(IBC)
      IBC=IBC+1
      DO 600 I=IBC,M
      IF(Y(I).EQ.YDES) GOTO 600
      UPE=W(I-1)
      GOTO 602
  600 CONTINUE
      UPE=W(M)
  602 WRITE(6,603)UNE,UPE,YDES
  603 FORMAT('OBAND EDGES ARE ',F5.2,6X,F5.2,6X,'DESIRED VALUE =',
     1F5.2)
      IBC=I
      IF(IBC.LT.M) GOTO 601
      WRITE(6,50)F
   50 FORMAT('OFINAL ERROR FUNCTION VALUE =',D15.8)
      ATTN=0.D0
      RIPP=0.D0
      DO 501 I=1,M
      IF(Y(I).EQ.1.0) GOTO 502
      IF(Y(I).EQ.0.D0) GOTO 503
      GOTO 501
  502 IF(DABS(E(I)).GE.RIPP) RIPP=DABS(E(I))
      GOTO 501
  503 IF(DABS(E(I)).GE.ATTN) ATTN=DABS(E(I))
  501 CONTINUE
      RIPP=20.0*DLOG10(1.0+RIPP)
      ATTN=20.0*DLOG10(ATTN)
      WRITE(6,504) RIPP
  504 FORMAT('OMAXIMUM ABSOLUTE DEVIATION IN PASSBAND = ',F10.5,' DB'/)
      WRITE(6,505) ATTN
  505 FORMAT('OMINIMUM STOPBAND ATTENUATION = ',F10.5,' DB'/)
      WRITE(6,51)A
   51 FORMAT('OCONSTANT MULTIPLIER C*=',D15.8)
      WRITE(11,525) K
  525 FORMAT(I2)
      WRITE(11,526) A
  526 FORMAT(F16.11)
      WRITE(6,52)
   52 FORMAT('OCOEFFICIENTS FOR CASCADE DECOMPOSITION'/)
      DO 520 J=1,K
      J4=(J-1)*4
      J41=J4/4+1
```

```
      WRITE(11,523) X(J4+1),X(J4+2),X(J4+3),X(J4+4)
  523 FORMAT(4F16.11)
      WRITE(6,521) J41,X(J4+1),J41,X(J4+2)
  521 FORMAT('OA(1,',I1,') = ',D15.8,4X,'A(2,',I1,') = ',D15.8)
      WRITE(6,522) J41,X(J4+3),J41,X(J4+4)
  522 FORMAT('OB(1,',I1,') = ',D15.8,4X,'B(2,',I1,') = ',D15.8)
  520 CONTINUE
      ISAV=ICALL
      ICALL=-10
      CALL ROOTS(N,X,ICALL)
      ICALL=ISAV
      PRINT 366
  366 FORMAT(1HO,70(1H*)/)
C     WRITE(6,54)(G(J),J=1,N)
C  54 FORMAT('OFINAL GRADIENT ='/(' ',4D15.8))
      WRITE(6,545)
  545 FORMAT('1APPROXIMATION ACHIEVED'/)
      DO 55 I=1,M
   55 WRITE(6,56)I,W(I),Y(I),YHT(I),E(I)
   56 FORMAT(' I=',I3,' W=',D15.8,' Y=',D15.8,' YHT=',D15.8,' E=',D15.8)
      WRITE(6,59)N
   59 FORMAT('1FINAL TABLE FOR N =',I3)
      DO 60 I=1,201
      FREQ=.005DO*DFLOAT(I-1)
      ZCUR=CDEXP(DCMPLX(0.DO,FREQ*PI))
      ZCUR2=ZCUR*ZCUR
      Q=DCMPLX(1.DO,0.DO)
      PHASE=0.DO
      DO 61 J=1,K
      J4=(J-1)*4
      QBAR=1.DO+X(J4+1)*ZCUR+X(J4+2)*ZCUR2
      A1=QBAR
      A2=(0.DO,-1.DO)*QBAR
      PHASE=PHASE+DATAN2(A2,A1)
      Q=Q*QBAR
      QBAR=1.DO+X(J4+3)*ZCUR+X(J4+4)*ZCUR2
      A1=QBAR
      A2=(0.DO,-1.DO)*QBAR
      PHASE=PHASE-DATAN2(A2,A1)
   61 Q=Q/QBAR
      A1=Q*DCONJG(Q)
      A1=A*DSQRT(A1)
      PHASE=-1.DO*PHASE/PI
      LOC=(A1*60.0DO)+.5DO
      DO 100 JZAP=1,60
  100 IGRAPH(JZAP)=IBLANK
      DO 200 JFILL=1,LOC
  200 IGRAPH(JFILL)=ISTAR
C
C     THE NEXT TWO STATEMENTS CAN BE USED IF PLOTTING OF OUTPUT IS
C     DESIRED, BY REMOVING THE C DENOTATION.
```

```
C  60 WRITE(6,62)FREQ,PHASE,A1,ISTAR,(IGRAPH(LLL),LLL=1,60)
C  62 FORMAT(' W=',D15.8,' PHASE/PI=',D15.8,' YHT=',D15.8,8X,A1,60A1)
C     GAIN WILL BE GIVEN IN DB.
      A1=20.0*DLOG10(A1)
      IF(A1.LT.-100) A1=-100.0
      WRITE(10,62)FREQ,A1
   60 WRITE(8,62)FREQ,PHASE
   62 FORMAT(2F10.5)
      RETURN
      END
C
C
      SUBROUTINE INSIDE(N,X,KFLAG)
      IMPLICIT REAL*8 (A-H,O-Z)
      DIMENSION X(32)
      J=-1
      KFLAG=0
   10 J=J+2
      IF(J.GT.N)RETURN
      B=-.5D0*X(J)
      C=X(J+1)
      DISC=B*B-C
      IF(DISC.LE.0.D0)GOTO20
C......REAL ROOTS
      DISC=DSQRT(DISC)
      R1=B+DISC
      R2=B-DISC
      DR1=DABS(R1)
      DR2=DABS(R2)
      IF(DR1.LE.1.D0.AND.DR2.LE.1.D0)GOTO10
      KFLAG=1
      IF(DR1.GT.1.D0)R1=1.D0/R1
      IF(DR2.GT.1.D0)R2=1.D0/R2
      X(J)=-1.D0*(R1+R2)
      X(J+1)=R1*R2
      GOTO10
C......COMPLEX ROOTS
   20 IF(C.LE.1.D0)GOTO10
      KFLAG=1
      C=1.D0/C
      X(J+1)=C
      X(J)=X(J)*C
      GOTO10
      END
C
C
      SUBROUTINE ROOTS(N,X,ICALL)
      IMPLICIT REAL*8 (A-H,O-Z)
      DIMENSION X(32)
      IF(ICALL.EQ.-10)WRITE(6,40)
   40 FORMAT('0ROOTS'/6X,'REAL',11X,'IMAG',11X,'REAL',11X,'IMAG')
```

```
      J=-1
   10 J=J+2
      IF(J.GT.N)RETURN
      B=-.5D0*X(J)
      C=X(J+1)
      DISC=B*B-C
      IF(DISC.LE.0.D0)GOTO20
C......REAL ROOTS
      DISC=DSQRT(DISC)
      R1=B+DISC
      R2=B-DISC
      IF(ICALL.EQ.-10)WRITE(6,30)R1,R2
   30 FORMAT(' ',D15.8,15X,D15.8)
      GOTO10
C......COMPLEX ROOTS
   20 DISC=DSQRT(-1.D0*DISC)
      DISCM=-1.D0*DISC
      IF(ICALL.EQ.-10)WRITE(6,50)B,DISC,B,DISCM
   50 FORMAT(' ',4D15.8)
      GOTO10
      END
C
C     ..................................................................
C
C        SUBROUTINE DFMFP†
C
C        PURPOSE
C           TO FIND A LOCAL MINIMUM OF A FUNCTION OF SEVERAL VARIABLES
C           BY THE METHOD OF FLETCHER AND POWELL
C
C        USAGE
C           CALL DFMFP(FUNCT,N,X,F,G,EST,EPS,LIMIT,IER,H)
C
C        DESCRIPTION OF PARAMETERS
C           FUNCT  - USER-WRITTEN SUBROUTINE CONCERNING THE FUNCTION TO
C                    BE MINIMIZED. IT MUST BE OF THE FORM
C                    SUBROUTINE FUNCT(N,ARG,VAL,GRAD)
C                    AND MUST SERVE THE FOLLOWING PURPOSE
C                    FOR EACH N-DIMENSIONAL ARGUMENT VECTOR  ARG,
C                    FUNCTION VALUE AND GRADIENT VECTOR MUST BE COMPUTED
C                    AND, ON RETURN, STORED IN VAL AND GRAD RESPECTIVELY
C                    ARG,VAL AND GRAD MUST BE OF DOUBLE PRECISION.
C           N      - NUMBER OF VARIABLES
C           X      - VECTOR OF DIMENSION N CONTAINING THE INITIAL
C                    ARGUMENT WHERE THE ITERATION STARTS. ON RETURN,
C                    X HOLDS THE ARGUMENT CORRESPONDING TO THE
```

† Reprinted by permission from IBM Application Program-System/360 Scientific Subroutine Package (360A-CM-03X) Version 3. Copyright © 1968 by International Business Machines Corporation.

```
C               COMPUTED MINIMUM FUNCTION VALUE
C               DOUBLE PRECISION VECTOR.
C      F      - SINGLE VARIABLE CONTAINING THE MINIMUM FUNCTION
C               VALUE ON RETURN, I.E. F=F(X).
C               DOUBLE PRECISION VARIABLE.
C      G      - VECTOR OF DIMENSION N CONTAINING THE GRADIENT
C               VECTOR CORRESPONDING TO THE MINIMUM ON RETURN,
C               I.E. G=G(X).
C               DOUBLE PRECISION VECTOR.
C      EST    - IS AN ESTIMATE OF THE MINIMUM FUNCTION VALUE.
C               SINGLE PRECISION VARIABLE.
C      EPS    - TESTVALUE REPRESENTING THE EXPECTED ABSOLUTE ERROR.
C                A REASONABLE CHOICE IS 10**(-16), I.E.
C                SOMEWHAT GREATER THAN 10**(-D), WHERE D IS THE
C               NUMBER OF SIGNIFICANT DIGITS IN FLOATING POINT
C               REPRESENTATION.
C               SINGLE PRECISION VARIABLE.
C      LIMIT  - MAXIMUM NUMBER OF ITERATIONS.
C      IER    - ERROR PARAMETER
C               IER = 0 MEANS CONVERGENCE WAS OBTAINED
C               IER = 1 MEANS NO CONVERGENCE IN LIMIT ITERATIONS
C               IER =-1 MEANS ERRORS IN GRADIENT CALCULATION
C               IER = 2 MEANS LINEAR SEARCH TECHNIQUE INDICATES
C               IT IS LIKELY THAT THERE EXISTS NO MINIMUM.
C      H      - WORKING STORAGE OF DIMENSION N*(N+7)/2.
C               DOUBLE PRECISION ARRAY.
C
C    REMARKS
C      I) THE SUBROUTINE NAME REPLACING THE DUMMY ARGUMENT  FUNCT
C         MUST BE DECLARED AS EXTERNAL IN THE CALLING PROGRAM.
C      II) IER IS SET TO 2 IF , STEPPING IN ONE OF THE COMPUTED
C          DIRECTIONS, THE FUNCTION WILL NEVER INCREASE WITHIN
C          A TOLERABLE RANGE OF ARGUMENT.
C          IER = 2 MAY OCCUR ALSO IF THE INTERVAL WHERE F
C          INCREASES IS SMALL AND THE INITIAL ARGUMENT WAS
C          RELATIVELY FAR AWAY FROM THE MINIMUM SUCH THAT THE
C          MINIMUM WAS OVERLEAPED. THIS IS DUE TO THE SEARCH
C          TECHNIQUE WHICH DOUBLES THE STEPSIZE UNTIL A POINT
C          IS FOUND WHERE THE FUNCTION INCREASES.
C
C    SUBROUTINES AND FUNCTION SUBPROGRAMS REQUIRED
C      FUNCT
C
C    METHOD
C          THE METHOD IS DESCRIBED IN THE FOLLOWING ARTICLE
C      R. FLETCHER AND M.J.D. POWELL, A RAPID DESCENT METHOD FOR
C      MINIMIZATION,
C      COMPUTER JOURNAL VOL.6, ISS. 2, 1963, PP.163-168.
C
C    ..................................................................
C
```

```
      SUBROUTINE DFMFP(FUNCT,N,X,F,G,EST,EPS,LIMIT,IER,H)
C
C     DIMENSIONED DUMMY VARIABLES
      DIMENSION H(1),X(1),G(1)
      DOUBLE PRECISION X,F,FX,FY,OLDF,HNRM,GNRM,H,G,DX,DY,ALFA,DALFA,
     1AMBDA,T,Z,W
C
C     COMPUTE FUNCTION VALUE AND GRADIENT VECTOR FOR INITIAL ARGUMENT
      CALL FUNCT(N,X,F,G)
C
C     RESET ITERATION COUNTER AND GENERATE IDENTITY MATRIX
      IER=0
      KOUNT=0
      N2=N+N
      N3=N2+N
      N31=N3+1
    1 K=N31
      DO 4 J=1,N
      H(K)=1.DO
      NJ=N-J
      IF(NJ)5,5,2
    2 DO 3 L=1,NJ
      KL=K+L
    3 H(KL)=0.DO
    4 K=KL+1
C
C     START ITERATION LOOP
    5 KOUNT=KOUNT +1
C
C     SAVE FUNCTION VALUE, ARGUMENT VECTOR AND GRADIENT VECTOR
      OLDF=F
      DO 9 J=1,N
      K=N+J
      H(K)=G(J)
      K=K+N
      H(K)=X(J)
C
C     DETERMINE DIRECTION VECTOR H
      K=J+N3
      T=0.DO
      DO 8 L=1,N
      T=T-G(L)*H(K)
      IF(L-J)6,7,7
    6 K=K+N-L
      GO TO 8
    7 K=K+1
    8 CONTINUE
    9 H(J)=T
C
C     CHECK WHETHER FUNCTION WILL DECREASE STEPPING ALONG H.
      DY=0.DO
```

```
      HNRM=0.DO
      GNRM=0.DO
C
C     CALCULATE DIRECTIONAL DERIVATIVE AND TESTVALUES FOR DIRECTION
C     VECTOR H AND GRADIENT VECTOR G.
      DO 10 J=1,N
      HNRM=HNRM+DABS(H(J))
      GNRM=GNRM+DABS(G(J))
   10 DY=DY+H(J)*G(J)
C
C     REPEAT SEARCH IN DIRECTION OF STEEPEST DESCENT IF DIRECTIONAL
C     DERIVATIVE APPEARS TO BE POSITIVE OR ZERO.
      IF(DY)11,51,51
C
C     REPEAT SEARCH IN DIRECTION OF STEEPEST DESCENT IF DIRECTION
C     VECTOR H IS SMALL COMPARED TO GRADIENT VECTOR G.
   11 IF(HNRM/GNRM-EPS)51,51,12
C
C     SEARCH MINIMUM ALONG DIRECTION H
C
C     SEARCH ALONG H FOR POSITIVE DIRECTIONAL DERIVATIVE
   12 FY=F
      ALFA=2.DO*(EST-F)/DY
      AMBDA=1.DO
C
C     USE ESTIMATE FOR STEPSIZE ONLY IF IT IS POSITIVE AND LESS THAN
C     1. OTHERWISE TAKE 1. AS STEPSIZE
      IF(ALFA)15,15,13
   13 IF(ALFA-AMBDA)14,15,15
   14 AMBDA=ALFA
   15 ALFA=0.DO
C
C     SAVE FUNCTION AND DERIVATIVE VALUES FOR OLD ARGUMENT
   16 FX=FY
      DX=DY
C
C     STEP ARGUMENT ALONG H
      DO 17 I=1,N
   17 X(I)=X(I)+AMBDA*H(I)
C
C     COMPUTE FUNCTION VALUE AND GRADIENT FOR NEW ARGUMENT
      CALL FUNCT(N,X,F,G)
      FY=F
C
C     COMPUTE DIRECTIONAL DERIVATIVE DY FOR NEW ARGUMENT.  TERMINATE
C     SEARCH, IF DY IS POSITIVE. IF DY IS ZERO THE MINIMUM IS FOUND
      DY=0.DO
      DO 18 I=1,N
   18 DY=DY+G(I)*H(I)
      IF(DY)19,36,22
```

```
C     TERMINATE SEARCH ALSO IF THE FUNCTION VALUE INDICATES THAT
C     A MINIMUM HAS BEEN PASSED
   19 IF(FY-FX)20,22,22
C
C     REPEAT SEARCH AND DOUBLE STEPSIZE FOR FURTHER SEARCHES
   20 AMBDA=AMBDA+ALFA
      ALFA=AMBDA
C     END OF SEARCH LOOP
C
C     TERMINATE IF THE CHANGE IN ARGUMENT GETS VERY LARGE
      IF(HNRM*AMBDA-1.D10)16,16,21
C
C     LINEAR SEARCH TECHNIQUE INDICATES THAT NO MINIMUM EXISTS
   21 IER=2
      RETURN
C
C     INTERPOLATE CUBICALLY IN THE INTERVAL DEFINED BY THE SEARCH
C     ABOVE AND COMPUTE THE ARGUMENT X FOR WHICH THE INTERPOLATION
C     POLYNOMIAL IS MINIMIZED
   22 T=0.D0
   23 IF(AMBDA)24,36,24
   24 Z=3.D0*(FX-FY)/AMBDA+DX+DY
      ALFA=DMAX1(DABS(Z),DABS(DX),DABS(DY))
      DALFA=Z/ALFA
      DALFA=DALFA*DALFA-DX/ALFA*DY/ALFA
      IF(DALFA)51,25,25
   25 W=ALFA*DSQRT(DALFA)
      ALFA=DY-DX+W+W
      IF (ALFA) 250,251,250
  250 ALFA=(DY-Z+W)/ALFA
      GO TO 252
  251 ALFA=(Z+DY-W)/(Z+DX+Z+DY)
  252 ALFA=ALFA*AMBDA
      DO 26 I=1,N
   26 X(I)=X(I)+(T-ALFA)*H(I)
C
C     TERMINATE, IF THE VALUE OF THE ACTUAL FUNCTION AT X IS LESS
C     THAN THE FUNCTION VALUES AT THE INTERVAL ENDS. OTHERWISE REDUCE
C     THE INTERVAL BY CHOOSING ONE END-POINT EQUAL TO X AND REPEAT
C     THE INTERPOLATION.  WHICH END-POINT IS CHOOSEN DEPENDS ON THE
C     VALUE OF THE FUNCTION AND ITS GRADIENT AT X
C
      CALL FUNCT(N,X,F,G)
      IF(F-FX)27,27,28
   27 IF(F-FY)36,36,28
   28 DALFA=0.D0
      DO 29 I=1,N
   29 DALFA=DALFA+G(I)*H(I)
      IF(DALFA)30,33,33
   30 IF(F-FX)32,31,33
   31 IF(DX-DALFA)32,36,32
```

```
   32 FX=F
      DX=DALFA
      T=ALFA
      AMBDA=ALFA
      GO TO 23
   33 IF(FY-F)35,34,35
   34 IF(DY-DALFA)35,36,35
   35 FY=F
      DY=DALFA
      AMBDA=AMBDA-ALFA
      GO TO 22
C
C     TERMINATE, IF FUNCTION HAS NOT DECREASED DURING LAST ITERATION
   36 IF(OLDF-F+EPS)51,38,38
C
C     COMPUTE DIFFERENCE VECTORS OF ARGUMENT AND GRADIENT FROM
C     TWO CONSECUTIVE ITERATIONS
   38 DO 37 J=1,N
      K=N+J
      H(K)=G(J)-H(K)
      K=N+K
   37 H(K)=X(J)-H(K)
C
C     TEST LENGTH OF ARGUMENT DIFFERENCE VECTOR AND DIRECTION VECTOR
C     IF AT LEAST N ITERATIONS HAVE BEEN EXECUTED. TERMINATE, IF
C     BOTH ARE LESS THAN EPS
      IER=0
      IF(KOUNT-N)42,39,39
   39 T=0.D0
      Z=0.D0
      DO 40 J=1,N
      K=N+J
      W=H(K)
      K=K+N
      T=T+DABS(H(K))
   40 Z=Z+W*H(K)
      IF(HNRM-EPS)41,41,42
   41 IF(T-EPS)56,56,42
C
C     TERMINATE, IF NUMBER OF ITERATIONS WOULD EXCEED LIMIT
   42 IF(KOUNT-LIMIT)43,50,50
C
C     PREPARE UPDATING OF MATRIX H
   43 ALFA=0.D0
      DO 47 J=1,N
      K=J+N3
      W=0.D0
      DO 46 L=1,N
      KL=N+L
      W=W+H(KL)*H(K)
      IF(L-J)44,45,45
```

```
   44 K=K+N-L
      GO TO 46
   45 K=K+1
   46 CONTINUE
      K=N+J
      ALFA=ALFA+W*H(K)
   47 H(J)=W
C
C     REPEAT SEARCH IN DIRECTION OF STEEPEST DESCENT IF RESULTS
C     ARE NOT SATISFACTORY
      IF(Z*ALFA)48,1,48
C
C     UPDATE MATRIX H
   48 K=N31
      DO 49 L=1,N
      KL=N2+L
      DO 49 J=L,N
      NJ=N2+J
      H(K)=H(K)+H(KL)*H(NJ)/Z-H(L)*H(J)/ALFA
   49 K=K+1
      GO TO 5
C     END OF ITERATION LOOP
C
C     NO CONVERGENCE AFTER LIMIT ITERATIONS
   50 IER=1
      RETURN
C
C     RESTORE OLD VALUES OF FUNCTION AND ARGUMENTS
   51 DO 52 J=1,N
      K=N2+J
   52 X(J)=H(K)
      CALL FUNCT(N,X,F,G)
C
C     REPEAT SEARCH IN DIRECTION OF STEEPEST DESCENT IF DERIVATIVE
C     FAILS TO BE SUFFICIENTLY SMALL
      IF(GNRM-EPS)55,55,53
C
C     TEST FOR REPEATED FAILURE OF ITERATION
   53 IF(IER)56,54,54
   54 IER=-1
      GOTO 1
   55 IER=0
   56 RETURN
      END
```

Appendix 2.2

A Computer Program for FIR Filter Design†

This appendix contains a computer program for FIR linear phase filter design. The program with only minor modifications is after McClellan et al [3].

```
C      PROGRAM FOR THE DESIGN OF LINEAR PHASE FINITE IMPULSE
C      RESPONSE (FIR) FILTERS USING THE REMEZ EXCHANGE ALGORITHM
C      JIM MCCLELLAN, RICE UNIVERSITY, APRIL 13, 1973
C      THREE TYPES OF FILTERS ARE INCLUDED--BANDPASS FILTERS
C      DIFFERENTIATORS, AND HILBERT TRANSFORM FILTERS
C
C      THE INPUT DATA CONSISTS OF 5 CARDS
C
C      CARD 1--FILTER LENGTH, TYPE OF FILTER. 1-MULTIPLE
C      PASSBAND/STOPBAND, 2-DIFFERENTIATOR, 3-HILBERT TRANSFORM
C      FILTER.  NUMBER OF BANDS, CARD PUNCH DESIRED, AND GRID
C      DENSITY.
C
C      CARD 2--BANDEDGES, LOWER AND UPPER EDGES FOR EACH BAND
C      WITH A MAXIMUM OF 10 BANDS.
C
C      CARD 3--DESIRED FUNCTION (OR DESIRED SLOPE IF A
C      DIFFERENTIATOR) FOR EACH BAND
C
C      CARD 4--WEIGHT FUNCTION IN EACH BAND.  FOR A
C      DIFFERENTIATOR, THE WEIGHT FUNCTION IS INVERSELY
C      PROPORTIONAL TO F.
C
C      CARD 5--RIPPLE AND ATTENUATION IN PASSBAND AND STOPBAND.
C      THIS CARD IS OPTIONAL AND CAN BE USED TO SPECIFY LOWPASS
C      FILTERS DIRECTLY IN TERMS OF PASSBAND RIPPLE AND STOPBAND
C      ATTENUATION IN DB. THE FILTER LENGTH IS DETERMINED FROM
C      THE APPROXIMATION RELATIONSHIPS GIVEN IN :
C      L. R. RABINER, APPROXIMATE DESIGN RELATIONSHIPS FOR
C      LOWPASS DIGITAL FILTERS, IEEE TRANS. ON AUDIO AND
C      ELECTROACUSTICS, VOL. AU-21, NO. 5, OCTOBER 73.
```

```
C     **** WHEN THIS OPTION IS USED THE FILTER LENGTH ON
C     CARD 1 SHOULD BE SET TO 0. ******
C
C     THE FOLLOWING INPUT DATA SPECIFIES A LENGTH 32 BANDPASS
C     FILTER WITH STOPBANDS 0 TO 0.1 AND 0.425 TO 0.5, AND
C     PASSBAND FROM 0.2 TO 0.35 WITH WEIGHTING OF 10 IN THE
C     STOPBANDS AND 1 IN THE PASSBAND. THE IMPULSE RESPONSE
C     WILL BE PUNCHED AND THE GRID DENSITY IS 32.
C     SAMPLE INPUT DATA SETUP
C     32,1,3,1,32
C     0,0.1,0.2,0.35,0.425,0.5
C     0,1,0
C     10,1,10
C
C     THE FOLLOWING INPUT DATA SPECIFIES A LENGTH 32 WIDEBAND
C     DIFFERENTIATOR WITH SLOPE 1 AND WEIGHTING OF 1/H. THE
C     IMPULSE RESPONSE WILL NOT BE PUNCHED AND THE GRID
C     DENSITY IS ASSUMED TO BE 16.
C     32,2,1,0,0
C     0,0.5
C     1.0
C     1.0
C
C     THE FOLLOWING INPUT DATA SPECIFIES A LOWPASS FILTER
C     BY GIVING DIRECTLY THE RIPPLE AND ATTENUATION.
C     0,1,2,0,32
C     0,0.2,0.3,0.5
C     1,0
C     1,1
C     0.2,40
C
      COMMON DES,WT,ALPHA,IEXT,NFCNS,NGRID,PI2,AD,DEV,X,Y,GRID
      DIMENSION IEXT(66),AD(66),ALPHA(66),X(66),Y(66)
      DIMENSION H(66)
      DIMENSION DES(1045),GRID(1045),WT(1045)
      DIMENSION EDGE(20),FX(10),WTX(10),DEVIAT(10)
      DIMENSION OMEGA(50),RESPA(50)
      DOUBLE PRECISION OMEGA,SUMAR,SUMAC,RESPA,ATTN,RIPPLE,SVRIP,DELTAF
      DOUBLE PRECISION PI2,PI
      DOUBLE PRECISION AD,DEV,X,Y
      PI2=6.283185307179586
      PI=3.1415926535899793
C
C     THE PROGRAM IS SET UP FOR A MAXIMUM LENGTH OF 128, BUT
C     THIS UPPER LIMIT CAN BE CHANGED BY REDIMENSIONING THE
C     ARRAYS IEXT, AD, ALPHA, X, Y, H TO BE NFMAX/2 + 2.
C     THE ARRAYS DES, GRID, AND WT MUST DIMENSIONED
C     16(NFMAX/2 + 2).
C
      NFMAX=128
  100 CONTINUE
```

```
      JTYPE=0
C
C     PROGRAM INPUT SECTION
C
      READ *,NFILT,JTYPE,NBANDS,JPUNCH,LGRID
      IF(NFILT.GT.NFMAX.OR.NFILT.EQ.3) CALL ERROR
      IF(NBANDS.LE.0) NBANDS=1
C
C     GRID DENSITY IS ASSUMED TO BE 16 UNLESS SPECIFIED
C     OTHERWISE
C
      IF(LGRID.LE.0) LGRID=16
      JB=2*NBANDS
      READ*,(EDGE(J),J=1,JB)
      READ*,(FX(J),J=1,NBANDS)
      READ *,(WTX(J),J=1,NBANDS)
      IF(JTYPE.EQ.0) CALL ERROR
      IF(NFILT.EQ.0) GO TO 10
      NROX = 0
      GO TO 11
   10 READ *,RIPPLE,ATTN
      RIPPLE=10**(RIPPLE/20)
      ATTN=10**(-ATTN/20)
      RIPPLE=(RIPPLE-1)/(RIPPLE+1)
      RIPPLE=DLOG10(RIPPLE)
      ATTN=DLOG10(ATTN)
      SVRIP=RIPPLE
      RIPPLE=(0.005309*(RIPPLE**2)+0.07114*RIPPLE-0.4761)*ATTN
     1+(-0.00266*(RIPPLE**2)-0.5941*RIPPLE-0.4278)
      ATTN=11.01217+0.51244*SVRIP-0.51244*ATTN
      DELTAF=EDGE(3)-EDGE(2)
      NFILT=RIPPLE/DELTAF-ATTN*(DELTAF**2)+1
      NROX=1
   11 NEG=1
      IF(JTYPE.EQ.1) NEG=0
      NODD=NFILT/2
      NODD=NFILT-2*NODD
      NFCNS=NFILT/2
      IF(NODD.EQ.1.AND.NEG.EQ.0) NFCNS=NFCNS+1
C
C     SET UP THE DENSE GRID.  THE NUMBER OF POINTS IN THE GRID
C     IS (FILTER LENGTH + 1)*GRID DENSITY/2
C
      GRID(1)=EDGE(1)
      DELF=LGRID*NFCNS
      DELF=0.5/DELF
      IF(NEG.EQ.0) GO TO 135
      IF(EDGE(1).LT.DELF) GRID(1)=DELF
  135 CONTINUE
      J=1
      L=1
```

```
      LBAND=1
  140 FUP=EDGE(L+1)
  145 TEMP=GRID(J)
C
C     CALCULATE THE DESIRED MAGNITUDE RESPONSE AND THE WEIGHT
C     FUNCTION ON THE GRID
C
      DES(J)=EFF(TEMP,FX,WTX,LBAND,JTYPE)
      WT(J)=WATE(TEMP,FX,WTX,LBAND,JTYPE)
      J=J+1
      GRID(J)=TEMP+DELF
      IF(GRID(J).GT.FUP) GO TO 150
      GO TO 145
  150 GRID(J-1)=FUP
      DES(J-1)=EFF(FUP,FX,WTX,LBAND,JTYPE)
      WT(J-1)=WATE(FUP,FX,WTX,LBAND,JTYPE)
      LBAND=LBAND+1
      L=L+2
      IF(LBAND.GT.NBANDS) GO TO 160
      GRID(J)=EDGE(L)
      GO TO 140
  160 NGRID=J-1
      IF(NEG.NE.NODD) GO TO 165
      IF(GRID(NGRID).GT.(0.5-DELF)) NGRID=NGRID-1
  165 CONTINUE
C
C     SET UP A NEW APPROXIMATION PROBLEM WHICH IS EQUIVALENT
C     TO THE ORIGINAL PROBLEM
C
      IF(NEG) 170,170,180
  170 IF(NODD.EQ.1) GO TO 200
      DO 175 J=1,NGRID
      CHANGE=DCOS(PI*GRID(J))
      DES(J)=DES(J)/CHANGE
  175 WT(J)=WT(J)*CHANGE
      GO TO 200
  180 IF(NODD.EQ.1) GO TO 190
      DO 185 J=1,NGRID
      CHANGE=DSIN(PI*GRID(J))
      DES(J)=DES(J)/CHANGE
  185 WT(J)=WT(J)*CHANGE
      GO TO 200
  190 DO 195 J=1,NGRID
      CHANGE=DSIN(PI2*GRID(J))
      DES(J)=DES(J)/CHANGE
  195 WT(J)=WT(J)*CHANGE
C
C     INITIAL GUESS FOR THE EXTREMAL FREQUENCIES--EQUALLY
C     SPACED ALONG THE GRID
C
  200 TEMP=FLOAT(NGRID-1)/FLOAT(NFCNS)
```

```
      DO 210 J=1,NFCNS
  210 IEXT(J)=(J-1)*TEMP+1
      IEXT(NFCNS+1)=NGRID
      NM1=NFCNS-1
      NZ=NFCNS+1
C
C     CALL THE REMEZ EXCHANGE ALGORITHM TO DO THE APPROXIMATION
C     PROBLEM
C
      CALL REMEZ(EDGE,NBANDS)
C
C     CALCULATE THE IMPULSE RESPONSE.
C
      IF(NEG) 300,300,320
  300 IF(NODD.EQ.0) GO TO 310
      DO 305 J=1,NM1
  305 H(J)=0.5*ALPHA(NZ-J)
      H(NFCNS)=ALPHA(1)
      GO TO 350
  310 H(1)=0.25*ALPHA(NFCNS)
      DO 315 J=2,NM1
  315 H(J)=0.25*(ALPHA(NZ-J)+ALPHA(NFCNS+2-J))
      H(NFCNS)=0.5*ALPHA(1)+0.25*ALPHA(2)
      GO TO 350
  320 IF(NODD.EQ.0) GO TO 330
      H(1)=0.25*ALPHA(NFCNS)
      H(2)=0.25*ALPHA(NM1)
      DO 325 J=3,NM1
  325 H(J)=0.25*(ALPHA(NZ-J)-ALPHA(NFCNS+3-J))
      H(NFCNS)=0.5*ALPHA(1)-0.25*ALPHA(3)
      H(NZ)=0.0
      GO TO 350
  330 H(1)=0.25*ALPHA(NFCNS)
      DO 335 J=2,NM1
  335 H(J)=0.25*(ALPHA(NZ-J)-ALPHA(NFCNS+2-J))
      H(NFCNS)=0.5*ALPHA(1)-0.25*ALPHA(2)
C
C     PROGRAM OUTPUT SECTION.
C
  350 PRINT 360
  360 FORMAT(1H1, 70(1H*)//25X,'FINITE IMPULSE RESPONSE (FIR)'/
     125X,'LINEAR PHASE DIGITAL FILTER DESIGN'/
     225X,'REMEZ EXCHANGE ALGORITHM'/)
      IF(JTYPE.EQ.1) PRINT 365
  365 FORMAT(25X,'BANDPASS FILTER'/)
      IF(JTYPE.EQ.2) PRINT 370
  370 FORMAT(25X,'DIFFERENTIATOR'/)
      IF(JTYPE.EQ.3) PRINT 375
  375 FORMAT(25X,'HILBERT TRANSFORMER'/)
      PRINT 378,NFILT
  378 FORMAT(15X,'FILTER LENGTH = ',I3/)
```

```
      IF(NROX.EQ.1) PRINT 379
  379 FORMAT(15X,'FILTER LENGTH DETERMINED BY APPROXIMATION'/)
      PRINT 380
  380 FORMAT(15X,'***** IMPULSE RESPONSE *****')
      DO 381 J=1,NFCNS
      K=NFILT+1-J
      IF(NEG.EQ.0) PRINT 382,J,H(J),K
      IF(NEG.EQ.1) PRINT 383,J,H(J),K
  381 CONTINUE
  382 FORMAT(20X,'H(',I3,') = ',E15.8,' = H(',I4,')')
  383 FORMAT(20X,'H(',I3,') = ',E15.8,' = -H(',I4,')')
      IF(NEG.EQ.1.AND.NODD.EQ.1) PRINT 384,NZ
  384 FORMAT(20X,'H(',I3,') =  0.0')
      DO 450 K=1,NBANDS,4
      KUP=K+3
      IF(KUP.GT.NBANDS) KUP=NBANDS
      PRINT 385,(J,J=K,KUP)
  385 FORMAT(/24X,4('BAND',I3,8X))
      PRINT 390,(EDGE(2*J-1),J=K,KUP)
  390 FORMAT(2X,'LOWER BAND EDGE',5F15.9)
      PRINT 395,(EDGE(2*J),J=K,KUP)
  395 FORMAT(2X,'UPPER BAND EDGE',5F15.9)
      IF(JTYPE.NE.2) PRINT 400,(FX(J),J=K,KUP)
  400 FORMAT(2X,'DESIRED VALUE',2X,5F15.9)
      IF(JTYPE.EQ.2) PRINT 405,(FX(J),J=K,KUP)
  405 FORMAT(2X,'DESIRED SLOPE',2X,5F15.9)
      PRINT 410,(WTX(J),J=K,KUP)
  410 FORMAT(2X,'WEIGHTING',6X,5F15.9)
      DO 420 J=K,KUP
  420 DEVIAT(J)=DEV/WTX(J)
      PRINT 425,(DEVIAT(J),J=K,KUP)
  425 FORMAT(2X,'DEVIATION',6X,5F15.9)
      IF(JTYPE.NE.1) GO TO 450
      DO 430 J=K,KUP
      IF (FX(J).EQ.1.0) DEVIAT(J)=(1.0+DEVIAT(J))/(1.0-DEVIAT(J))
  430 DEVIAT(J)=20.0*ALOG10(DEVIAT(J))
      PRINT 435,(DEVIAT(J),J=K,KUP)
  435 FORMAT(2X,'DEVIATION IN DB',5F15.9)
  450 CONTINUE
      PRINT 455,(GRID(IEXT(J)),J=1,NZ)
  455 FORMAT(/2X,'EXTREMAL FREQUENCIES'/(2X,5F12.7))
      PRINT 460
  460 FORMAT(/1X,70(1H*)/1H1)
C       CALCULATE FREQUENCY RESPONSE
      PRINT 710
  710 FORMAT(15X,'***********  FREQUENCY  RESPONSE  **********')
      DO 610 IKA = 1,50
      OMEGA(IKA) = 0.010000000*(IKA - 1)
      SUMAR = 0.0
      SUMAC = 0.0
      DO 620 NIK = 1,NFCNS
```

```
      SUMAR=SUMAR+H(NIK)*(DCOS(PI2*OMEGA(IKA)*(NIK-1)) +
     1DCOS(PI2*OMEGA(IKA)*(NFILT-NIK)))
      SUMAC = SUMAC + H(NIK)*(DSIN(PI2*OMEGA(IKA)*(NIK-1)) +
     1DSIN(PI2*OMEGA(IKA)*(NFILT-NIK)))
  620 CONTINUE
      IF (NODD.EQ.1) SUMAR=SUMAR-H(NFCNS)*DCOS(PI2*OMEGA(IKA)*(NFCNS
     1-1))
      IF (NODD.EQ.1) SUMAC=SUMAC-H(NFCNS)*DSIN(PI2*OMEGA(IKA)*(NFCNS
     1-1))
      RESPA(IKA) = DSQRT(SUMAR**2 + SUMAC**2)
      RESPA(IKA) = 20.0*DLOG10(RESPA(IKA))
      PRINT 630, OMEGA(IKA),RESPA(IKA)
  610 CONTINUE
  630 FORMAT(15X,F5.3,8X,F9.3)
      IF(JPUNCH.NE.0) WRITE(10,730) (H(J),J=1,NFCNS)
      IF(NFILT.NE.0) GO TO 100
  730 FORMAT(5E15.8)
      RETURN
      END
      FUNCTION EFF(TEMP,FX,WTX,LBAND,JTYPE)
C
C     FUNCTION TO CALCULATE THE DESIRED MAGNITUDE RESPONSE
C     AS A FUNCTION OF FREQUENCY.
C
      DIMENSION FX(5),WTX(5)
      IF(JTYPE.EQ.2) GO TO 1
      EFF=FX(LBAND)
      RETURN
    1 EFF=FX(LBAND)*TEMP
      RETURN
      END
      FUNCTION WATE(TEMP,FX,WTX,LBAND,JTYPE)
C
C     FUNCTION TO CALCULATE THE WEIGHT FUNCTION AS A FUNCTION
C     OF FREQUENCY.
      DIMENSION FX(5),WTX(5)
      IF(JTYPE.EQ.2) GO TO 1
      WATE=WTX(LBAND)
      RETURN
    1 IF(FX(LBAND).LT.0.0001) GO TO 2
      WATE=WTX(LBAND)/TEMP
      RETURN
    2 WATE=WTX(LBAND)
      RETURN
      END
      SUBROUTINE ERROR
      PRINT 1
    1 FORMAT(' ************ ERROR IN INPUT DATA **********')
      STOP
      END
      SUBROUTINE REMEZ(EDGE,NBANDS)
```

```
C
C     THIS SUBROUTINE IMPLEMENTS THE REMEZ EXCHANGE ALGORITHM
C     FOR THE WEIGHTED CHEBYCHEV APPROXIMATION OF A CONTINUOUS
C     FUNCTION WITH A SUM OF COSINES.  INPUTS TO THE SUBROUTINE
C     ARE A DENSE GRID WHICH REPLACES THE FREQUENCY AXIS, THE
C     DESIRED FUNCTION ON THIS GRID, THE WEIGHT FUNCTION ON THE
C     GRID, THE NUMBER OF COSINES, AND AN INITIAL GUESS OF THE
C     EXTREMAL FREQUENCIES.  THE PROGRAM MINIMIZES THE CHEBYCHEV
C     ERROR BY DETERMINING THE BEST LOCATION OF THE EXTREMAL
C     FREQUENCIES (POINTS OF MAXIMUM ERROR) AND THEN CALCULATES
C     THE COEFFICIENTS OF THE BEST APPROXIMATION.
C
      COMMON DES,WT,ALPHA,IEXT,NFCNS,NGRID,PI2,AD,DEV,X,Y,GRID
      DIMENSION EDGE(20)
      DIMENSION IEXT(66),AD(66),ALPHA(66),X(66),Y(66)
      DIMENSION DES(1045),GRID(1045),WT(1045)
      DIMENSION A(66),P(65),Q(65)
      DOUBLE PRECISION PI2,DNUM,DDEN,DTEMP,A,P,Q
      DOUBLE PRECISION AD,DEV,X,Y
C
C     THE PROGRAM ALLOWS A MAXIMUM NUMBER OF ITERATIONS OF 25
C
      ITRMAX=25
      DEVL=-1.0
      NZ=NFCNS+1
      NZZ=NFCNS+2
      NITER=0
  100 CONTINUE
      IEXT(NZZ)=NGRID+1
      NITER=NITER+1
      IF(NITER.GT.ITRMAX) GO TO 400
      DO 110 J=1,NZ
      DTEMP=GRID(IEXT(J))
      DTEMP=DCOS(DTEMP*PI2)
  110 X(J)=DTEMP
      JET=(NFCNS-1)/15+1
      DO 120 J=1,NZ
  120 AD(J)=D(J,NZ,JET)
      DNUM=0.0
      DDEN=0.0
      K=1
      DO 130 J=1,NZ
      L=IEXT(J)
      DTEMP=AD(J)*DES(L)
      DNUM=DNUM+DTEMP
      DTEMP=K*AD(J)/WT(L)
      DDEN=DDEN+DTEMP
  130 K=-K
      DEV=DNUM/DDEN
      NU=1
      IF(DEV.GT.0.0) NU=-1
```

```
      DEV=-NU*DEV
      K=NU
      DO 140 J=1,NZ
      L=IEXT(J)
      DTEMP=K*DEV/WT(L)
      Y(J)=DES(L)+DTEMP
  140 K=-K
      IF(DEV.GE.DEVL) GO TO 150
      CALL OUCH
      GO TO 400
  150 DEVL=DEV
      JCHNGE=0
      K1=IEXT(1)
      KNZ=IEXT(NZ)
      KLOW=0
      NUT=-NU
      J=1
C
C     SEARCH FOR THE EXTREMAL FREQUENCIES OF THE BEST
C     APPROXIMATION
C
  200 IF(J.EQ.NZZ) YNZ=COMP
      IF(J.GE.NZZ) GO TO 300
      KUP=IEXT(J+1)
      L=IEXT(J)+1
      NUT=-NUT
      IF(J.EQ.2) Y1=COMP
      COMP=DEV
      IF(L.GE.KUP) GO TO 220
      ERR=GEE(L,NZ)
      ERR=(ERR-DES(L))*WT(L)
      DTEMP=NUT*ERR-COMP
      IF(DTEMP.LE.0.0) GO TO 220
      COMP=NUT*ERR
  210 L=L+1
      IF(L.GE.KUP) GO TO 215
      ERR=GEE(L,NZ)
      ERR=(ERR-DES(L))*WT(L)
      DTEMP=NUT*ERR-COMP
      IF(DTEMP.LE.0.0) GO TO 215
      COMP=NUT*ERR
      GO TO 210
  215 IEXT(J)=L-1
      J=J+1
      KLOW=L-1
      JCHNGE=JCHNGE+1
      GO TO 200
  220 L=L-1
  225 L=L-1
      IF(L.LE.KLOW) GO TO 250
      ERR=GEE(L,NZ)
```

```
      ERR=(ERR-DES(L))*WT(L)
      DTEMP=NUT*ERR-COMP
      IF(DTEMP.GT.0.0) GO TO 230
      IF(JCHNGE.LE.0) GO TO 225
      GO TO 260
  230 COMP=NUT*ERR
  235 L=L-1
      IF(L.LE.KLOW) GO TO 240
      ERR=GEE(L,NZ)
      ERR=(ERR-DES(L))*WT(L)
      DTEMP=NUT*ERR-COMP
      IF(DTEMP.LE.0.0) GO TO 240
      COMP=NUT*ERR
      GO TO 235
  240 KLOW=IEXT(J)
      IEXT(J)=L+1
      J=J+1
      JCHNGE=JCHNGE+1
      GO TO 200
  250 L=IEXT(J)+1
      IF(JCHNGE.GT.0) GO TO 215
  255 L=L+1
      IF(L.GE.KUP) GO TO 260
      ERR=GEE(L,NZ)
      ERR=(ERR-DES(L))*WT(L)
      DTEMP=NUT*ERR-COMP
      IF(DTEMP.LE.0.0) GO TO 255
      COMP=NUT*ERR
      GO TO 210
  260 KLOW=IEXT(J)
      J=J+1
      GO TO 200
  300 IF(J.GT.NZZ) GO TO 320
      IF(K1.GT.IEXT(1)) K1=IEXT(1)
      IF(KNZ.LT.IEXT(NZ)) KNZ=IEXT(NZ)
      NUT1=NUT
      NUT=-NU
      L=0
      KUP=K1
      COMP=YNZ*(1.00001)
      LUCK=1
  310 L=L+1
      IF(L.GE.KUP) GO TO 315
      ERR=GEE(L,NZ)
      ERR=(ERR-DES(L))*WT(L)
      DTEMP=NUT*ERR-COMP
      IF(DTEMP.LE.0.0) GO TO 310
      COMP=NUT*ERR
      J=NZZ
      GO TO 210
  315 LUCK=6
```

```
      GO TO 325
  320 IF(LUCK.GT.9) GO TO 350
      IF(COMP.GT.Y1) Y1=COMP
      K1=IEXT(NZZ)
  325 L=NGRID+1
      KLOW=KNZ
      NUT=-NUT1
      COMP=Y1*(1.00001)
  330 L=L-1
      IF(L.LE.KLOW) GO TO 340
      ERR=GEE(L,NZ)
      ERR=(ERR-DES(L))*WT(L)
      DTEMP=NUT*ERR-COMP
      IF(DTEMP.LE.0.0) GO TO 330
      J=NZZ
      COMP=NUT*ERR
      LUCK=LUCK+10
      GO TO 235
  340 IF(LUCK.EQ.6) GO TO 370
      DO 345 J=1,NFCNS
  345 IEXT(NZZ-J)=IEXT(NZ-J)
      IEXT(1)=K1
      GO TO 100
  350 KN=IEXT(NZZ)
      DO 360 J=1,NFCNS
  360 IEXT(J)=IEXT(J+1)
      IEXT(NZ)=KN
      GO TO 100
  370 IF(JCHNGE.GT.0) GO TO 100
C     CALCULATION OF THE COEFFICIENTS OF THE BEST APPROXIMATION
C     USING THE INVERSE DISCRETE FOURIER TRANSFORM
  400 CONTINUE
      NM1=NFCNS-1
      FSH=1.0E-06
      GTEMP=GRID(1)
      X(NZZ)=-2.0
      CN=2*NFCNS-1
      DELF=1.0/CN
      L=1
      KKK=0
      IF(EDGE(1).EQ.0.0.AND.EDGE(2*NBANDS).EQ.0.5) KKK=1
      IF(NFCNS.LE.3) KKK=1
      IF(KKK.EQ.1) GO TO 405
      DTEMP=DCOS(PI2*GRID(1))
      DNUM=DCOS(PI2*GRID(NGRID))
      AA=2.0/(DTEMP-DNUM)
      BB=-(DTEMP+DNUM)/(DTEMP-DNUM)
  405 CONTINUE
      DO 430 J=1,NFCNS
      FT=(J-1)*DELF
      XT=DCOS(PI2*FT)
```

```
      IF(KKK.EQ.1) GO TO 410
      XT=(XT-BB)/AA
      FT=ARCOS(XT)/PI2
  410 XE=X(L)
      IF(XT.GT.XE) GO TO 420
      IF((XE-XT).LT.FSH) GO TO 415
      L=L+1
      GO TO 410
  415 A(J)=Y(L)
      GO TO 425
  420 IF((XT-XE).LT.FSH) GO TO 415
      GRID(1)=FT
      A(J)=GEE(1,NZ)
  425 CONTINUE
      IF(L.GT.1) L=L-1
  430 CONTINUE
      GRID(1)=GTEMP
      DDEN=PI2/CN
      DO 510 J=1,NFCNS
      DTEMP=0.0
      DNUM=(J-1)*DDEN
      IF(NM1.LT.1) GO TO 505
      DO 500 K=1,NM1
  500 DTEMP=DTEMP+A(K+1)*DCOS(DNUM*K)
  505 DTEMP=2.0*DTEMP+A(1)
  510 ALPHA(J)=DTEMP
      DO 550 J=2,NFCNS
  550 ALPHA(J)=2*ALPHA(J)/CN
      ALPHA(1)=ALPHA(1)/CN
      IF(KKK.EQ.1) GO TO 545
      P(1)=2.0*ALPHA(NFCNS)*BB+ALPHA(NM1)
      P(2)=2.0*AA*ALPHA(NFCNS)
      Q(1)=ALPHA(NFCNS-2)-ALPHA(NFCNS)
      DO 540 J=2,NM1
      IF(J.LT.NM1) GO TO 515
      AA=0.5*AA
      BB=0.5*BB
  515 CONTINUE
      P(J+1)=0.0
      DO 520 K=1,J
      A(K)=P(K)
  520 P(K)=2.0*BB*A(K)
      P(2)=P(2)+A(1)*2.0*AA
      JM1=J-1
      DO 525 K=1,JM1
  525 P(K)=P(K)+Q(K)+AA*A(K+1)
      JP1=J+1
      DO 530 K=3,JP1
  530 P(K)=P(K)+AA*A(K-1)
      IF(J.EQ.NM1) GO TO 540
      DO 535 K=1,J
```

```
535 Q(K)=-A(K)
    Q(1)=Q(1)+ALPHA(NFCNS-1-J)
540 CONTINUE
    DO 543 J=1,NFCNS
543 ALPHA(J)=P(J)
545 CONTINUE
    IF(NFCNS.GT.3) RETURN
    ALPHA(NFCNS+1)=0.0
    ALPHA(NFCNS+2)=0.0
    RETURN
    END
    DOUBLE PRECISION FUNCTION D(K,N,M)
C   FUNCTION TO CALCULATE THE LAGRANGE INTERPOLATION
C   COEFFICIENTS FOR USE IN THE FUNCTION GEE.
    COMMON DES,WT,ALPHA,IEXT,NFCNS,NGRID,PI2,AD,DEV,X,Y,GRID
    DIMENSION IEXT(66),AD(66),ALPHA(66),X(66),Y(66)
    DIMENSION DES(1045),GRID(1045),WT(1045)
    DOUBLE PRECISION AD,DEV,X,Y
    DOUBLE PRECISION Q
    DOUBLE PRECISION PI2
    D=1.0
    Q=X(K)
    DO 3 L=1,M
    DO 2 J=L,N,M
    IF(J-K)1,2,1
  1 D=2.0*D*(Q-X(J))
  2 CONTINUE
  3 CONTINUE
    D=1.0/D
    RETURN
    END
    DOUBLE PRECISION FUNCTION GEE(K,N)
C   FUNCTION TO EVALUATE THE FREQUENCY RESPONSE USING THE
C   LAGRANGE INTERPOLATION FORMULA IN THE BARYCENTRIC FORM
    COMMON DES,WT,ALPHA,IEXT,NFCNS,NGRID,PI2,AD,DEV,X,Y,GRID
    DIMENSION IEXT(66),AD(66),ALPHA(66),X(66),Y(66)
    DIMENSION DES(1045),GRID(1045),WT(1045)
    DOUBLE PRECISION P,C,D,XF
    DOUBLE PRECISION PI2
    DOUBLE PRECISION AD,DEV,X,Y
    P=0.0
    XF=GRID(K)
    XF=DCOS(PI2*XF)
    D=0.0
    DO 1 J=1,N
    C=XF-X(J)
    C=AD(J)/C
    D=D+C
  1 P=P+C*Y(J)
    GEE=P/D
    RETURN
```

```
      END
      SUBROUTINE OUCH
      PRINT 1
    1 FORMAT(' ************ FAILURE TO CONVERGE **********'/
     1'OPROBABLE CAUSE IS MACHINE ROUNDING ERROR'/
     2'OTHE IMPULSE RESPONSE MAY BE CORRECT'/
     3'OCHECK WITH A FREQUENCY RESPONSE')
      RETURN
      END
```

Appendix 2.3

Frequency Response Programs

In this appendix we include two PL/1 computer programs which allow the determination of the effect that finite word length of the filter coefficients has on the amplitude response. The first program, RECRESP, determines the frequency response of IIR filters implemented in cascade form, using second order filter sections. Its output is a list of digital frequency (normalized to 2π) versus amplitude response in dB. The second program, NRECRES, determines the frequency response of FIR filters implemented in direct form.

IIR FILTER PROGRAM

```
 RECRESP:PROCEDURE OPTIONS(MAIN);
  /* THIS PROGRAM DETERMINES THE AMPLITUDE RESPONSE IN DB
     OF AN IIR FILTER IMPLEMENTED AS A CASCADED CONNECTION OF
     SECOND ORDER FILTERS. */
  /*
     THE INPUT DATA CONSISTS OF THE FOLLOWING:
     1. FIRST RECORD (CARD) CONTAINS THE FILTER ORDER
     2. SECOND RECORD CONTAINS THE GAIN CONSTANT C*
     3. SUBSEQUENT RECORDS CONTAIN THE VALUES OF
        A(1,I), A(2,I), B(1,I) AND B(2,I) FOR EACH
        SECOND ORDER SECTION (AS PER EQ. 2-44)
     4. THE LAST CARD CONTAINS THE NUMBER OF BITS, INCLUDING
        SIGN BIT AND INTEGER BIT
                                                   */
  DCL(A(10,3),C(10,2),DEL,F1,H(100),HQ(100),AX(5,3),BX(5,2),CONST)
  DECIMAL FLOAT(12);
  DCL (PI,X(800),Y(800)) DECIMAL FLOAT(12);
 QUANT:PROCEDURE(X,XQ,NQ);
  DCL (X,XQ,SX,S)  FLOAT DECIMAL(12);
 IF SIGN(X) = -1 THEN NS = 1; ELSE NS = 0 ;
  SX=ABS(X) ; IX = FLOOR(SX) ; SX = SX - IX ;
  S = 1.0000000000000 ;  XQ = 0;
  DO IQ = 1 BY 1 TO (NQ+1); S = S/2 ;
  IF SX >= S THEN DO IB=1; SX=SX-S; XQ=XQ+S;END;ELSE IB=0;
  END;
  XQ = XQ + IX ;
  IF NS=1 THEN XQ=-XQ ;
 RETURN;
 END QUANT;
  RECSEC:PROCEDURE(H,B,OM,G,PI);
```

```
 DCL(H(*),B(*),OM,G,CR1,CR2,PI,FOM)
 DECIMAL FLOAT(12);
 FOM = 2*PI*OM ;
 CR1 = SQRT( (H(1)+H(2)*COS(FOM)+H(3)*COS(2*FOM))**2 +
        (H(2)*SIN(FOM)+H(3)*SIN(2*FOM))**2 );
 CR2 = SQRT( (1.0 +B(1)*COS(FOM)+B(2)*COS(2*FOM))**2 +
         (B(1)*SIN(FOM)+B(2)*SIN(2*FOM))**2 );
 G = CR1/CR2 ;
 RETURN;
END RECSEC;
RESREC:PROCEDURE(A,C,ISEC,PI,JF,CON);
 DCL(A(*,*),C(*,*),PI,RES(1000),FR,H(3),B(2),GA,CON)
 DECIMAL FLOAT(12);
 DO IRES=1 TO 1000;
   FR = 0.000500000*(IRES-1) ;
   GA = 1.00000000; RES(IRES) = 1.0000000;
   DO IS=1 TO ISEC;
    H(*)=A(IS,*); B(*)=C(IS,*);
    CALL RECSEC(H,B,FR,GA,PI);
    RES(IRES)=RES(IRES)*GA ;
    END;
   RES(IRES)=RES(IRES)*CON ;
   IF RES(IRES) < 0.000001 THEN RES(IRES) = 0.000001 ;
   RES(IRES)=20.0*LOG10(RES(IRES));
 IF JF>0 THEN
   PUT EDIT(FR*PI*2,MAX(RES(IRES),-100.0)) (SKIP,2 F(10,5));
  END;
  RETURN;
  END RESREC;
 PI = 3.141592653589 ;
  GET EDIT(KORDER) (F(2));
  GET SKIP EDIT(CONST) (F(16,11));
  NOR = 4*KORDER ;
  DO I=1 TO NOR;
    IF MOD(I,4)=1 THEN GET SKIP EDIT(H(I)) (F(16,11)) ;
        ELSE GET EDIT(H(I)) (F(16,11));
  END;
  GET SKIP EDIT(NBI) (F(2));
     NBIT = NBI - 2 ;
  DO I=1 TO NOR;  CALL QUANT(H(I),HQ(I),NBIT) ; END;
   DO I=1 TO KORDER; J = 4*(I-1) ;
    A(I,1) = 1.000000000 ; A(I,2) = HQ(J+1) ; A(I,3) = HQ(J+2);
    C(I,1) = HQ(J+3) ; C(I,2) = HQ(J+4) ;
   END ;
   JF = 1 ;
   CALL RESREC(A,C,KORDER,PI,JF,CONST) ;
END RECRESP;
```

FIR FILTER PROGRAM

```
NRECRES:PROCEDURE OPTIONS(MAIN);
 /* THIS PROGRAM DETERMINES THE AMPLITUDE RESPONSE OF AN
    LINEAR PHASE FIR FILTER IMPLEMENTED IN DIRECT FORM  */
 /*
    THE INPUT DATA CONSISTS OF THE FOLLOWING
    1. THE FIRST CARD CONTAINS THE NUMBER OF BITS TO BE USED
       (NUMBER OF BITS AFTER THE DECIMAL POINT)
    2. THE SECOND CARD CONTAINS THE NUMBER OF TAPS OF THE
       FILTER AND THE INTEGER PART OF THE NUMBER OF TAPS/2 + 1
    3. SUBSEQUENT CARDS CONTAIN THE FILTER COEFFICIENTS
       5 PER CARD
                                                            */
DCL(H(64),F1,F2,F3,R,AT,PI,HQ(64),X(24),XC(29),HI(60),REQAT,REQRIP,
     HIQ(64),HRQ(64),HSAV(64))
   DECIMAL FLOAT(12);
  DCL NHB(64) BINARY FIXED(15);
 DCL FR(64,20) BINARY FIXED(4);
QUANT:PROCEDURE(X,XQ,NQ);
 DCL (X,XQ,SX,S)  FLOAT DECIMAL(12);
IF SIGN(X) = -1 THEN NS = 1; ELSE NS = 0 ;
 SX=ABS(X) ; IX = FLOOR(SX) ; SX = SX - IX ;
 S = 1.0000000000000 ;  XQ = 0;
 DO IQ = 1 BY 1 TO (NQ+1); S = S/2 ;
 IF SX >= S THEN DO IB=1; SX=SX-S; XQ=XQ+S;END;ELSE IB=0;
 END;
 XQ = XQ + IX ;
 IF NS=1 THEN XQ=-XQ ;
RETURN;
END QUANT;
 RESPONSE:PROCEDURE(H,NH,F1,F2,R,AT,PI);
  DCL(H(*),F1,F2,R,AT,G(500),OM,PI,FREQU) DECIMAL FLOAT(12);
  DO IRES=1 BY 1 TO 500;
    OM = 1.0*IRES*PI/500; R=0 ; AT=0 ;
    DO JRES=1 BY 1 TO NH;
      R=R+H(JRES)*COS(OM*(JRES-1));
      AT=AT+H(JRES)*SIN(OM*(JRES-1));
    END;
    G(IRES) = SQRT(R**2 + AT**2);
  END;
  NPASS=F1*1000 ; NSTOP=F2*1000 ;
  R = ABS(G(1)-1.0) ;
  DO IRES=2 BY 1 TO NPASS;
     OMX = 1.0 ;
    IF ABS(G(IRES)-OMX) > R THEN R=ABS(G(IRES)-OMX);
  END;
  AT = ABS(G(NSTOP+1));
  DO IRES=(NSTOP+2) BY 1 TO 199;
    IF ABS(G(IRES)) > AT THEN AT = ABS(G(IRES));
  END;
  R = 20.0*LOG10((1.0+R)/(1.0-R)); AT = 20.0*LOG10(AT) ;
```

```
 G  = 20.0*LOG10(G) ;
DO IRES=1 BY 1 TO 500;
     IF ABS(G(IRES)) > 100.0 THEN  G(IRES)=-100.0 ;
          PUT SKIP EDIT(0.500*(IRES-1)/500,G(IRES))
             (2 F(10,5));
END;
 RETURN;
 END RESPONSE;
  GET EDIT (NCBIT) ( F(4));
GET SKIP EDIT(NL,NCL ) (2 F(4));
  DO I=1 TO NCL;
       IF MOD(I,5)=1 THEN GET SKIP EDIT(H(I)) ( E(15,8));
              ELSE GET EDIT(H(I)) (E(15,8));
 END;
 IF NCL < NL THEN DO ;
 IF MOD(NL,2)=1 THEN KODD=0; ELSE KODD=1;
DO I=1 BY 1 TO (NL-NCL); H(NCL+I)=H(NCL-I+KODD); END;
 END;
 PI = 3.1415926536;
  NTY = 0;
DO I=1 TO NCL; CALL QUANT(H(I),HQ(I),NCBIT); END;
DO I=1 BY 1 TO (NL-NCL);HQ(NCL+I)=HQ(NCL-I+KODD); END;
 CALL RESPONSE(HQ,NL,F1,F2,R,AT,PI);
 PUT EDIT('PASSBAND RIPPLE IS',R,'DB   STOPBAND ATTENUATION IS',AT,'DB'
 ) (SKIP(4),A, 2 (F(10,4),X(4),A));
END NRECRES;
```

Appendix 2.4

Analog Lowpass Filters

In this appendix, we give a brief summary of the design of three common types of analog lowpass filters. They are, the Butterworth, the Chebyshev, and the Cauer or elliptic function filters. For a more detailed discussion the reader should consult one of the many excellent treatments of this subject [5,6].

Given a transfer function H(s), a rational function in s, its magnitude square for $s = j\omega$ is a function in ω^2. Let it be denoted by $A(\omega^2)$. Then

$$A(\omega^2) = H(s)\,H(-s)\,|_{s=j\omega} \tag{2A-1}$$

It is convenient to write $A(\omega^2)$ in the following form

$$A(\omega^2) = \frac{1}{1 + F(\omega^2)} \tag{2A-2}$$

where $F(\omega^2)$ is also a rational function of ω^2. Since $A(\omega^2) \geq 0$, we must have $F(\omega^2) \geq -1$.

To obtain H(s) from $A(\omega^2)$, one replaces ω^2 by $-s^2$, factors the numerator and denominator of $A(-s^2)$ to find its poles and zeros in the s-plane. These roots must occur in symmetry with respect to both the real axis and the imaginary axis. That is, if s_1 is a pole, then so must $-s_1$, s_1^*, and $-s_1^*$ be. To obtain a minimum phase H(s), one simply collects those factors corresponding to left half plane poles and zeros. For example, suppose $A(\omega^2) = (2+\omega^2)/(1+\omega^4)$. With $\omega^2=-s^2$, we factor $A(-s^2)$ to $(\sqrt{2}-s)(\sqrt{2}+s)/(s-\alpha)(s+\alpha)(s-\alpha^*)(s+\alpha^*)$ where $\alpha=(1+j)/\sqrt{2}$. The pole zero pattern is shown in Figure 2A.1*a*. Collecting the left half plane poles and zero, we have $H(s) = (s+\sqrt{2})/(s+\alpha)(s+\alpha^*)$. Its pole zero pattern is shown in Figure 2A.1*b*.

For H(s) to be a lowpass filter with cutoff normalized to $\omega=1$ rad/sec, its $A(\omega^2)$ must be approximately 1 in the passband, $0\leq |\omega| \leq 1$, and approximately zero in the stopband, say $|\omega| > \Omega_s$. From Eq. 2A-2, we see that the rational function $F(\omega^2)$ must be approximately zero in $0 \leq \omega^2 \leq 1$ and it must take on large values when $\omega^2 > \Omega_s^2$. The three classes of lowpass filters to be discussed presently each has a different family of $F(\omega^2)$'s.

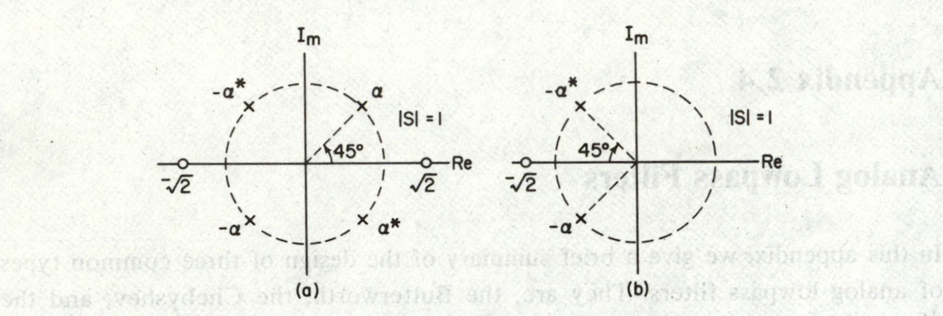

Figure 2A.1 Pole zero patterns of $A(-s^2)$ and of H(s).

(a) *Butterworth Filters*

For this class, $F(\omega^2)$ is a polynomial in ω^2 of degree n, specifically,

$$F(\omega^2) = \omega^{2n} \qquad n = 1,2,3,... \tag{2A-3}$$

We see this function is small for small ω and large for large ω, thus should produce a lowpass characteristic. With this choice of $F(\omega^2)$, the square amplitude response of the filter is

$$A(\omega^2) = \frac{1}{1 + \omega^{2n}} \tag{2A-4}$$

which is plotted in Figure 2A.2 for n=3,5,7.

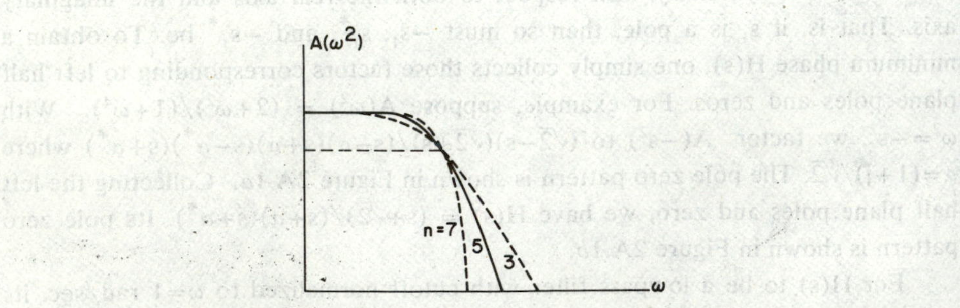

Figure 2A.2 Amplitude-square response of Butterworth lowpass filters, Eq. 2A-4.

When $\omega=1$, $A(\omega^2)=0.5$. So the 3 dB point is at $\omega=1$. $A(\omega^2)$ is monotonically decreasing. It has the maximum number of derivatives equal to zero at the origin. For this reason, Butterworth filters have *maximally flat* amplitude response. At the edge of the stopband,

$$A(\Omega_s^2) = \frac{1}{1 + \Omega_s^{2n}} \tag{2A-5}$$

This equation is used to determine the order of the filter, n, in order to satisfy the attenuation requirement in the stopband, $|\omega| > \Omega_s$.

To find H(s), we replace ω^2 in Eq. 2A-4 by $-s^2$ and factor the denominator of $A(-s^2)=1/[1+(-s^2)^n]$. The poles are given by $1+(-s^2)^n=0$, or $s^{2n}=-(-1)^n$. They are therefore the 2n-th roots of $(-1)^{n+1}$, and are located on the unit circle, equally spaced. Figure 2A.3 illustrates the location of the poles for n=4. From our previous discussion, we see that the four poles of the actual transfer function H(s) must be at $-\alpha+j\beta$, $-\alpha-j\beta$, $-\beta+j\alpha$, and $-\beta-j\alpha$, where $\alpha=\cos 22.5°$ and $\beta=\sin 22.5°$. The numerator of H(s) is, of course, 1. In general,

$$H(s) = \frac{1}{B_n(s)} \tag{2A-6}$$

where $B_n(s)$ is a polynomial of degree n with roots at $e^{j(1/2+1/2n+k/n)\pi}$, k=0,1,...,(n−1). Table 2A.1 lists $B_n(s)$ for $1 \leq n \leq 5$.

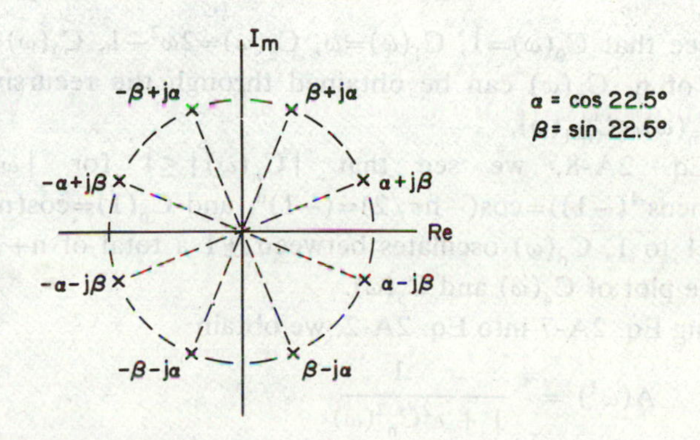

Figure 2A.3 Poles of $A(-s^2)$ for a fourth order Butterworth lowpass filter.

n	$B_n(s)$
1	$s+1$
2	$s^2+1.4142s+1$
3	$(s+1)(s^2+s+1)$
4	$(s^2+0.7654s+1)(s^2+1.8478s+1)$
5	$(s+1)(s^2+0.6180s+1)(s^2+1.6180s+1)$

Table 2A.1. The Numerator of a Butterworth Lowpass Filter of Order n

(b) *Chebyshev Filters*

The $F(\omega^2)$ of Eq. 2A-2 for this class of filter is also a polynomial in ω^2 of degree n. Specifically,

$$F(\omega^2) = \varepsilon^2 C_n^{\,2}(\omega) \tag{2A-7}$$

where $C_n(\omega)$ is the n-th degree Chebyshev cosine polynomial

$$C_n(\omega) = \begin{cases} \cos(n \cos^{-1}\omega), & |\omega| \leq 1 \\ \cosh(n \cosh^{-1}\omega), & |\omega| \geq 1 \end{cases} \tag{2A-8}$$

It is easy to see that $C_0(\omega)=1$, $C_1(\omega)=\omega$, $C_2(\omega)=2\omega^2-1$, $C_3(\omega)=4\omega^3-3\omega$. For higher values of n, $C_n(\omega)$ can be obtained through the recursive relationship $C_{n+1}(\omega)=2\omega C_n(\omega)-C_{n-1}(\omega)$.

From Eq. 2A-8, we see that $|C_n(\omega)| \leq 1$ for $|\omega| \leq 1$. Also, $C_n(-1)=\cos(n\cos^{-1}(-1))=\cos(-n\pi/2)=(-1)^n$, and $C_n(1)=\cos(n\pi/2)=1$. As ω varies from -1 to 1, $C_n(\omega)$ oscillates between ± 1 a total of n+1 times. Figure 2A.4 shows the plot of $C_4(\omega)$ and $C_5(\omega)$.

Substituting Eq. 2A-7 into Eq. 2A-2, we obtain

$$A(\omega^2) = \frac{1}{1 + \varepsilon^2 C_n^{\,2}(\omega)} \tag{2A-9}$$

In $0 \leq \omega \leq 1$, $C_n^{\,2}(\omega)$ oscillates between 0 and 1, so $A(\omega^2)$ oscillates between $1/(1+\varepsilon^2)$ and 1. Beyond $\omega=1$, $A(\omega^2)$ monotonically decreases to 0. Shown in Figure 2A.5 is a plot of $A(\omega^2)$ for n=4 and $\varepsilon^2=0.2$. Also shown is the $A(\omega^{2})$ for a fourth order Butterworth filter.

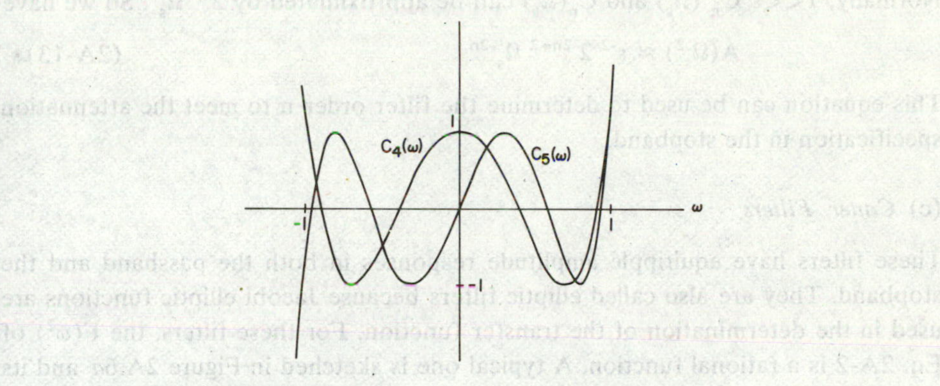

Figure 2A.4 Chebyshev polynomials of order 4 and order 5.

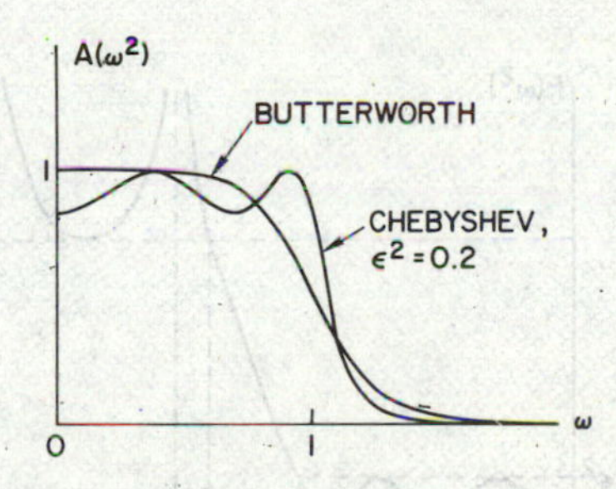

Figure A2.5 Amplitude-square response of fourth order Chebyshev and Butterworth filters.

The determination of H(s) from $A(\omega^2)$ follows our previous discussion. It can be shown that the poles of H(s) are located on an ellipse with major semiaxis of $\cosh\alpha_n$ along the $j\omega$ axis, and minor semiaxis of $\sinh\alpha_n$, where

$$\alpha_n = \frac{1}{n} \sinh^{-1} \frac{1}{\varepsilon} \tag{2A-10}$$

The n poles are at

$$p_k = \sinh\alpha_n \sin[(2k-1)\pi/2n] + j\cosh\alpha_n \cos[(2k-1)\pi/2n] \qquad k = 1,2,\ldots,n \tag{2A-11}$$

At the edge of the stopband, $\omega = \Omega_s$,

$$A(\Omega_s^{\,2}) = \frac{1}{1 + \varepsilon^2 C_n^{\,2}(\Omega_s)} \tag{2A-12}$$

Normally, $1 << \varepsilon^2 C_n^{\,2}(\Omega_s)$ and $C_n(\Omega_s)$ can be approximated by $2^{n-1}\Omega_s^{\,n}$. So we have

$$A(\Omega_s^{\,2}) \approx \varepsilon^{-2}\, 2^{-2n+2}\, \Omega_s^{\,-2n} \tag{2A-13}$$

This equation can be used to determine the filter order n to meet the attenuation specification in the stopband.

(c) *Cauer Filters*

These filters have equiripple amplitude responses in both the passband and the stopband. They are also called elliptic filters because Jacobi elliptic functions are used in the determination of the transfer function. For these filters, the $F(\omega^2)$ of Eq. 2A-2 is a rational function. A typical one is sketched in Figure 2A.6*a* and its corresponding $A(\omega^2)$ in Figure 2A.6*b*.

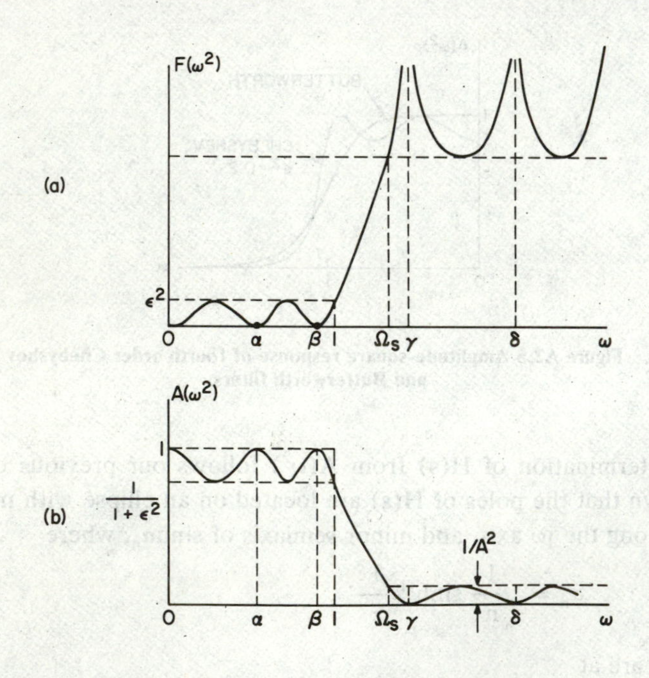

Figure 2A.6 $H(\omega^2)$ and $A(\omega^2)$ of a Cauer filter.

Unlike the two previous two classes of lowpass filters, the theory of Cauer filters is somewhat involved. A comprehensive discussion would touch upon subjects beyond the scope of this book. The computation task in the design of these filters is also not simple. Fortunately, rather complete design curves and tables are readily available [2]. These tables typically give the poles and zeros of the transfer function and the stopband edge frequency Ω_s for a given filter order, the passsband ripple and the minimum stopband attenuation. There is, however, one equation the reader may find useful in estimating the order of the filter n, to meet a given specification. It is

$$n = \frac{K(k)\, K((1-k_1^2)^{1/2})}{K(k_1)\, K((1-k^2)^{1/2})} \tag{2A-14}$$

where $k=1/\Omega_s$, $k_1=\varepsilon/(A^2-1)^{1/2})$, and K(x) is the elliptic integral given by

$$K(x) = \int_0^1 \frac{dt}{(1-t^2)^{1/2}(1-xt^2)^{1/2}} \tag{2A-15}$$

ε and A are related respectively to the ripple in the passband and the minimum attenuation in the stopband, as illustrated in Figure 2A.6.

CHAPTER 3

The Computation of the Discrete Fourier Transform and some of its Applications

3.1 INTRODUCTION

We have seen in Chapter 1 that the Discrete Fourier Transform (DFT) is the discrete-time counterpart of the familiar Fourier transform, and as such is extremely useful in many applications. The most widespread use of the DFT is in spectral analysis; however, it is also extensively applied when the frequency domain representation of the signals permits easy manipulation to achieve complex signal processing tasks. Examples include matched filtering, pulse compression, image processing, and modulation. As we saw already in Chapter 1, the straightforward computation of the DFT of a sequence of length N requires a computation rate that grows as N^2. This often represents an excessive computational burden for transforms that are sufficiently long, for example N=1024, to achieve a desirable performance level in practice, for example good resolution in spectral analysis.

In this chapter we discuss a class of algorithms to compute the DFT. They are generally known as the Fast Fourier Transform (FFT) algorithms [1,2], and their computation rate grows only as $N \log_2 N$. This represents a drastic reduction in the number of computations to be performed, for example for N = 1024 it is a slightly more than 100 fold reduction. We will discuss in some detail the FFT algorithm known as radix 2 decimation-in-time and briefly the radix 2 decimation-in-frequency algorithm. For arbitrary radix algorithms we only mention some references. The reader will find in Appendix 3.1 a PL/1 computer program for computing the FFT.

The continuing improvements in digital technology have made it possible to construct economical and cost effective *FFT boxes*. These are hardware dedicated to the computation of the FFT either as stand alone units [3], or as attachments to general purpose computers [4]. Furthermore many of the newer oscilloscopes have incorporated the capability of digital processing of the signals and contain hardware to perform the FFT algorithm [5]. In many applications, the design of dedicated FFT hardware is optimized to permit very high throughput rates to meet the high processing speed required in these applications [6]. In Chapter 5 the arithmetic unit of such an FFT processor is discussed. In the present chapter we first introduce the reader to the variety of computational structures that can be used to compute the FFT, and then discuss the various degrees of parallelism inherent in the FFT algorithm that can be exploited to obtain a wide range of performance. To this end we describe in Section 3.4 in some detail a highly parallel organization for the FFT known as the pipeline FFT.

We conclude the chapter with a brief introduction to two widely used applications of the FFT, the measurement of power spectra and performing linear convolution. The measurement of power spectra using the method of averaging modified periodograms is discussed in Section 3.5, and in Appendix 3.2 a PL/1 computer program that uses this method is given together with some examples of its use. In Section 3.6 the method of fast convolution is described. This method uses the FFT as an intermediate step in performing linear convolution to reduce the number of computations per output.

3.2 THE RADIX 2 DECIMATION-IN-TIME FFT ALGORITHM.

We have seen in Chapter 1 that the DFT of a sequence $\{x_n\}$ is given by:

$$A_p = \sum_{n=0}^{N-1} x_n W_N^{np} \qquad n,p = 0,1,2,.....(N-1) \tag{3-1}$$

where $W_N = e^{-j2\pi/N}$. $\{x_n\}$ are in general complex numbers. The resulting sequence $\{A_p\}$ is the DFT of the input sequence. In this chapter we consider only transforms of length N, which is an integral power of 2, that is, $N = 2^L$ with L an integer. The key to the development of the FFT algorithm is the fact that the sequence $\{A_p\}$ can be obtained by a suitable combination of the results of two transforms each of length N/2, and the combination requires only N multiplications. Since N is a power of 2, this decomposition process can be repeated L times, $L = \log_2 N$. Basically, we could first form N/2 transforms of length 2, combine them to get N/4 transforms of length 4, combine these to get N/8 transforms of length 8, and so on until finally after L such steps, the transform of length N is obtained. Since in taking transforms of length 2, $W_2 = e^{-j2\pi/2} = -1$, only addition and subtraction are needed, the computation of the DFT using this approach requires multiplication only in the steps of combining half length transforms to form full length transforms. As a result, the total number of multiplications is $N \times L = N \log_2 N$. We now proceed to show how this is done.

Let us divide the original input sequence $\{x_n\}$ into two parts as follows;

$$x_n^e = x_{2n} \qquad n = 0,1,\ldots,(N/2-1)$$
$$x_n^o = x_{2n+1} \qquad n = 0,1,\ldots,(N/2-1) \tag{3-2}$$

That is, $\{x_n^e\}$ contains the numbers in $\{x_n\}$ with even indices and $\{x_n^o\}$ contains the numbers in $\{x_n\}$ with odd indices, and each subsequence is of length N/2. We also note that W_N^2 is equal to $W_{N/2}$ due to the simple identity

$$W_N^2 = e^{-j2(2\pi/N)} = e^{-j(2\pi/(N/2))} = W_{N/2} \tag{3-3}$$

and therefore we can rewrite Eq. 3-1 as;

$$A_p = \sum_{n=0}^{N-1} x_n W_N^{np}$$

$$= \sum_{n=0}^{N/2-1} (x_{2n} W_N^{2np} + x_{2n+1} W_N^{2np+p})$$

$$= \sum_{n=0}^{N/2-1} x_n^e W_{N/2}^{np} + W_N^p \sum_{n=0}^{N/2-1} x_n^o W_{N/2}^{np}$$

$$p = 0,1,2\ldots,N-1 \tag{3-4}$$

By replacing p by p + N/2 in Eq. 3-4, we arrive at

$$A_{p+N/2} = \sum_{n=0}^{N/2-1} x_n^e W_{N/2}^{n(p+N/2)} + W_N^{p+N/2} \sum_{n=0}^{N/2-1} x_n^o W_{N/2}^{n(p+N/2)}$$

$$= \sum_{n=0}^{N/2-1} x_n^e W_{N/2}^{np} - W_N^p \sum_{n=0}^{N/2-1} x_n^o W_{N/2}^{np} \tag{3-4a}$$

where we have used the identities

$$W_{N/2}^{n(p+N/2)} = W_{N/2}^{np} \times W_{N/2}^{n(N/2)} = W_{N/2}^{np} \times e^{-j2\pi} = W_{N/2}^{np}$$

and

$$W_N^{p+N/2} = W_N^p \times W_N^{N/2} = W_N^p \times e^{-j\pi} = -W_N^p$$

The first summation in Eqs. 3-4 and 3-4a is recognized as the transform of the subsequence $\{x_n^e\}$, and the second summation is simply the DFT of the odd subsequence $\{x_n^o\}$. Let us denote these by $\{A_p^e\}$ and $\{A_p^o\}$ respectively. That is

$$A_p^e = \sum_{n=0}^{N/2-1} x_n^e W_{N/2}^{np}$$

$$A_p^o = \sum_{n=0}^{N/2-1} x_n^o W_{N/2}^{np}$$

$$p = 0,1,\ldots,N/2-1 \qquad (3\text{-}4b)$$

We now recall from Chapter 1, Eq. 1-80, that these are periodic sequences, and therefore $A_p^e = A_{p+N/2}^e$ and $A_p^o = A_{p+N/2}^o$. Thus we can combine Eqs. 3-4, 3-4a and 3-4b to obtain

$$A_p = A_p^e + W_N^p A_p^o$$

$$A_{p+N/2} = A_p^e - W_N^p A_p^o$$

$$p = 0,1,\ldots,(N/2-1) \qquad (3\text{-}5)$$

Eq. 3-5 constitutes the heart of the FFT radix 2 decimation-in-time algorithm and is commonly referred to as the *butterfly*, and the factor W_N^p is known as the twiddle factor. Eq. 3-5 also suggests that A_p and $A_{p+N/2}$ should be computed together by first forming the product $W_N^p A_p^o$ and then adding and subtracting the result from A_p^e. This is carried out for each $p = 0,1,\ldots,N/2-1$. Thus a total of $N/2$ multiplications and N additions are needed for the calculation of $\{A_p\}_{p=0,1,\ldots,N-1}$ from the two half length transforms $\{A_p^e\}$ and $\{A_p^o\}$.

This decomposition step is repeated L times, as illustrated in Figure 3.1, until we reach the step in that the computation of a 2 point transform is required, which as we recall, requires only an addition and subtraction. Therefore it is clear that a grand total of $L \times N/2 = (N/2)\log_2 N$ complex multiplications and $N\log_2 N$ complex additions are needed to compute an N point transform. This is illustrated in Figure 3.1.

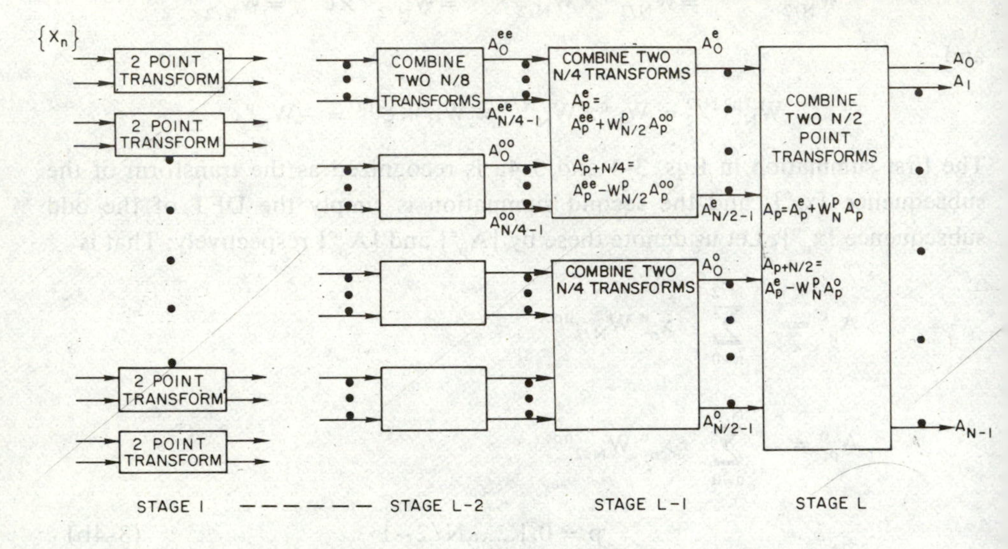

Figure 3.1 Performing the DFT by successive decomposition of a full length transform into two half length transforms.

Let us now go through a simple example for N = 8 to illustrate this FFT algorithm. The example also serves to introduce several additional concepts as well as to present some details which arise in the FFT computation regarding data shuffling and storage requirements.

Consider an 8 point input sequence $x_0,...,x_7$ and we wish to compute its DFT, $A_0,...,A_7$. Following our earlier development as well as the block diagram of Figure 3.1, we first divide the input sequence x_n into its even and odd parts $x_n{}^e$ and $x_n{}^o$:

$$
\begin{aligned}
x_0{}^e &= x_0 & \qquad x_0{}^o &= x_1 \\
x_1{}^e &= x_2 & \qquad x_1{}^o &= x_3 \\
x_2{}^e &= x_4 & \qquad x_2{}^o &= x_5 \\
x_3{}^e &= x_6 & \qquad x_3{}^o &= x_7
\end{aligned}
$$

Let $A_p{}^e$ and $A_p{}^o$ denote the 4 point transforms of $x_n{}^e$ and $x_n{}^o$ respectively. A_p is obtained via Eq. 3-5. This computation is depicted in Figure 3.2 using a special flow graph notation. The flow graph notation has been used in the past in the FFT literature to represent pictorially the algorithm. The following conventions are used to draw the flow graph of the FFT. A heavy dot • represents an add subtract operation, with the result of the addition appearing in the upper outgoing branch and the result of the subtraction appearing in the lower outgoing branch

An arrow on a branch represents a multiplication by the constant written above the arrow. In the absence of an arrow the constant is taken to be 1.

We can now decompose each 4 point DFT into a computation of two 2 point DFTs by assigning the even and odd indices in x_n^e to x_n^{ee}, x_n^{eo}, and x_n^o to x_n^{oe} and x_n^{oo}:

$$x_0^{ee} = x_0^e = x_0 \qquad x_0^{eo} = x_1^e = x_2$$
$$x_1^{ee} = x_2^e = x_4 \qquad x_1^{eo} = x_3^e = x_6$$

$$x_0^{oe} = x_0^o = x_1 \qquad x_0^{oo} = x_1^o = x_3$$
$$x_1^{oe} = x_2^o = x_5 \qquad x_1^{oo} = x_3^o = x_7$$

To calculate the 2 point DFT's, we note that the twiddle factor is $W_2 = e^{-j\pi} = -1$. Therefore no multiplication is needed. For example, A_0^{ee} and A_1^{ee} are the DFT of x_0^{ee} and x_1^{ee}, and according to the definition of the DFT (Eq. 3-1), are obtained as follows:

$$A_0^{ee} = x_0^{ee} + x_1^{ee}$$

$$A_1^{ee} = x_0^{ee} - x_1^{ee} \qquad (3\text{-}6)$$

The computation of the whole 8 point FFT is shown in the flow graph given in Figure 3.3.

Several important features of the FFT algorithm should be noted from this figure. First the input sequence is not addressed in its natural order by the FFT algorithm, but rather in a shuffled manner. The order in which the input sequence numbers are addressed is easily determined. The procedure to determine this is known as bit reversal, since it involves converting the indices of the input sequence into binary numbers, reversing them, and reading the numbers out according to the bit reversed indices. The following example illustrates the bit reversal technique for the 8 point DFT under discussion.

Read in	Binary	Reverse	Read out
0	000	000	0
1	001	100	4
2	010	010	2
3	011	110	6
4	100	001	1
5	101	101	5
6	110	011	3
7	111	111	7

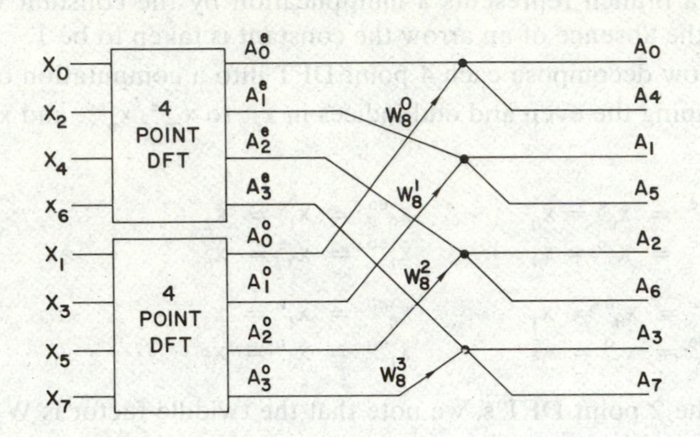

Figure 3.2 An 8 point DFT can be obtained through a suitable combination of two 4 point DFTs.

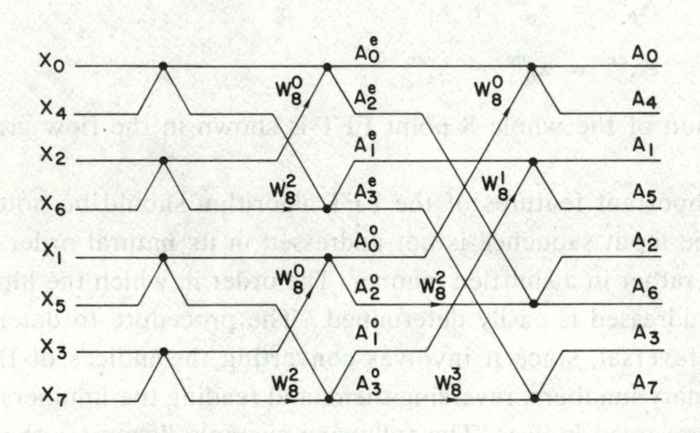

Figure 3.3 A flow graph pictorial representation of an 8 point radix 2 decimation-in-time FFT.

We see the last column is indeed the order in which the input is addressed in Figure 3.3. When the input sequence is shuffled the output sequence is obtained in natural order (A_0,A_1,A_2,A_3) and their butterfly pairs (A_4,A_5,A_6,A_7). The order in which the data is addressed and written back at each computational stage of the FFT is easily determined from the flow graph. The control section of the FFT processor has to generate the actual addresses from which data is directed into and from the arithmetic unit.

It is also evident from Figure 3.3 that if the data is fed in sequentially the algorithm cannot start the computations until at least $N/2 + 1$ input values are available. In this example, the delay is 5.

The FFT algorithm is executed in $L = \log_2 N$ stages, and at each stage $N/2$ *butterflies* are performed. Furthermore, the computations can be done *in-place*, if the input sequence is permitted to be erased, since the results of each *butterfly* can be written into the locations that provided the input to the *butterfly*. This implies that in addition to a small amount of intermediate storage in the arithmetic unit, only N memory locations are needed to perform an N point FFT. Since in general the input data, as well as the intermediate results and the output are complex numbers, the memory locations referred to above should be regarded as containing complex (double word) values. Furthermore, even if the input data is real, the same observation will still be valid since the intermediate results will still be complex.

We now consider in some detail the control section of the FFT that generates the actual addresses required to direct the data flow into and from the arithmetic unit. We assume that it is desired to do the computation *in-place*, that is the input sequence will be erased and replaced by its DFT sequence. First we have to shuffle the input sequence in *bit reversed* order, so as to insert correctly indexed values into a naturally ordered address space. In our example depicted in Figure 3.3, we call the 8 memory locations M0 to M7. Then the following assignment of input sequence to memory location has to be performed:

$$M0 \leftarrow x_0 \;;\; M1 \leftarrow x_4 \;;\; M2 \leftarrow x_2 \;;\; M3 \leftarrow x_6$$

$$M4 \leftarrow x_1 \;;\; M5 \leftarrow x_5 \;;\; M6 \leftarrow x_3 \;;\; M7 \leftarrow x_7$$

The flowchart in Figure 3.4 defines an algorithm for performing this exchange in place. Initially x_0 to x_{N-1} are stored in M0 to M(N-1) respectively, and upon termination of the algorithm the required shuffling is completed.

The next step is to perform the first stage of the DFT, in which numbers $2^0 = 1$ apart, starting from location zero (M0), are used as inputs to the *butterfly*, which performs a 2 point DFT and therefore requires no multiplications, this corresponds to twiddle factor $W_8^{\,0}$. The results are written back into the locations the inputs came from. In our example we input to the *butterfly* (M0 , M1), (M2 , M3) ..... (M6 , M7). Upon completion of the first stage we can start the second stage of the algorithm. At this step, starting from location zero, numbers $2^1 = 2$ locations apart are used as an input to the *butterfly* and the twiddle factors are N/4 apart, that is, $W_N^{\,0}, W_N^{\,N/4}$. In our example this implies that we input to the *butterfly* (M0, M2; $W_8^{\,0}$), (M1, M3; $W_8^{\,2}$), (M4, M6; $W_8^{\,0}$) and (M5, M7; $W_8^{\,2}$).

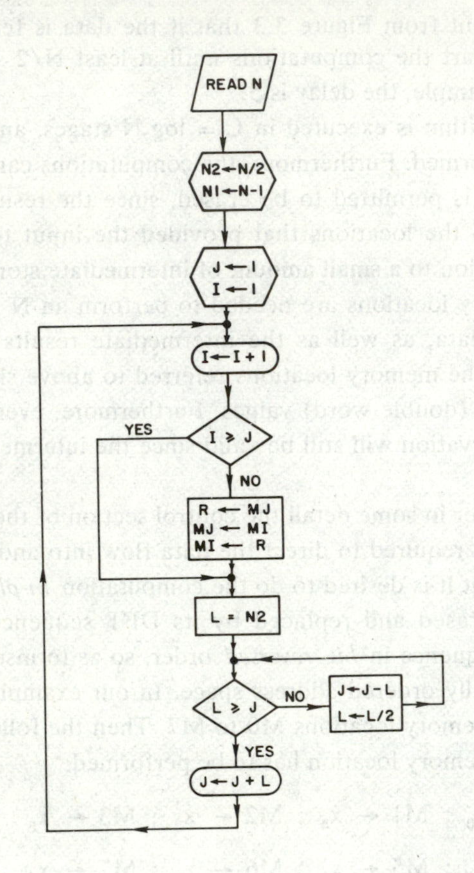

Figure 3.4 An algorithm for performing in-place bit reversal shuffling.

The subsequent stages proceed in a similar fashion. In general, at stage J, we take numbers 2^{J-1} locations apart, and use 2^{J-1} distinct twiddle factors:

$$W_N^{\,0}, W_N^{\,N/2^J}, W_N^{\,2N/2^J} \ldots\ldots W_N^{\,N(2^{J-1}-1)/2^J}$$

Thus in the last stage in our example, with J = 3 and N = 8, we should take (M0 , M4 ; $W_8^{\,0}$), (M1 , M5 ; $W_8^{\,1}$), (M2 , M6 ; $W_8^{\,2}$) and (M3 , M7 ; $W_8^{\,3}$), which is indeed correct as can be seen from the flow graph in Figure 3.3.

In Appendix 3.1 a PL/1 computer program is given for performing the radix 2 decimation-in-time algorithm. Additional programs for a variety of radices and arbitrary transform length, and in different programming languages are abundant in the literature [7,8,9,10,11].

3.3 THE RADIX 2 DECIMATION-IN-FREQUENCY FFT

In this section we describe briefly another widely used version of the FFT algorithm, known as the decimation-in-frequency algorithm. In this case the input sequence is taken in natural order and the output DFT sequence is obtained in the bit reversed order. The two algorithms are computationally equivalent, both requiring $(N/2)\log_2 N$ complex multiplications and $N \log_2 N$ complex additions.

To obtain the decimation-in-frequency algorithm we divide the input sequence into two parts:

$$x_n^1 = x_n \qquad n = 0,1,......N/2-1$$
$$x_n^2 = x_{n+N/2} \qquad n = 0,1,.......N/2-1 \tag{3-7}$$

Then the N point DFT can be written as;

$$A_p = \sum_{n=0}^{N-1} x_n W_N^{np}$$

$$= \sum_{n=0}^{N/2-1} x_n W_N^{np} + \sum_{n=N/2}^{N-1} x_n W_N^{np}$$

$$= \sum_{n=0}^{N/2-1} x_n^1 W_N^{np} + \sum_{n=0}^{N/2-1} x_n^2 W_N^{np} W_N^{pN/2} \tag{3-8}$$

We now *decimate* the frequency samples into the odd and even parts, by setting p = 2k when it is even and p = 2k + 1 when it is odd. Thus

$$A_{2k} = \sum_{n=0}^{N/2-1} (x_n^1 W_{N/2}^{nk} + x_n^2 W_{N/2}^{nk} W_N^{kN})$$

$$A_{2k+1} = \sum_{n=0}^{N/2-1} (x_n^1 W_{N/2}^{nk} + x_n^2 W_{N/2}^{nk} W_N^{kN+N/2}) W_N^n \tag{3-9}$$

where we used Eq. 3-3 to set $W_N^2 = W_{N/2}$. Furthermore, since $W_N^{kN} = 1$ and $W_N^{N/2} = -1$ we can rewrite Eq. 3-9 as

$$A_{2k} = \sum_{n=0}^{N/2-1} (x_n^{\ 1} + x_n^{\ 2}) W_{N/2}^{\ nk}$$

$$A_{2k+1} = \sum_{n=0}^{N/2-1} (x_n^{\ 1} - x_n^{\ 2}) W_N^{\ n} W_{N/2}^{\ nk} \qquad (3\text{-}10)$$

$\{A_{2k}\}$ is recognized as the N/2 point DFT of the sequence $\{x_n^{\ 1} + x_n^{\ 2}\}$, and $\{A_{2k+1}\}$ is recognized as the transform of $\{(x_n^{\ 1} - x_n^{\ 2})W_N^{\ n}\}$. Thus we have again reduced the computation of an N point DFT to the computation of two N/2 point DFTs, the two subsequences obtained from the original sequence by two additions and one multiplication. Therefore the *butterfly* computation in the decimation-in-frequency is of the form

$$a_n = x_n^{\ 1} + x_n^{\ 2}$$

$$b_n = (x_n^{\ 1} - x_n^{\ 2}) W_N^{\ n} \qquad (3\text{-}11)$$

Figure 3.5 illustrates in flow graph notation the computation of a 16 point DFT by the decimation-in-frequency algorithm described above. As we see from the figure at stage J numbers 2^{L-J} locations apart are presented to the *butterfly*, and the twiddle factors have 2^{L-J} distinct values;

$$W_N^{\ 0}, W_N^{\ N/2^{L-J+1}}, \ldots\ldots, W_N^{\ (2^{L-J}-1)N/2^{L-J+1}}$$

The reader may benefit from trying to work out for himself the example depicted in Figure 3.5 according to the rules given above, and possibly write a computer program to perform the FFT using the decimation-in-frequency algorithm. Additional insight into the control of this algorithm can also be gained from Appendix 5.2 which describes the implementation of an FFT using it.

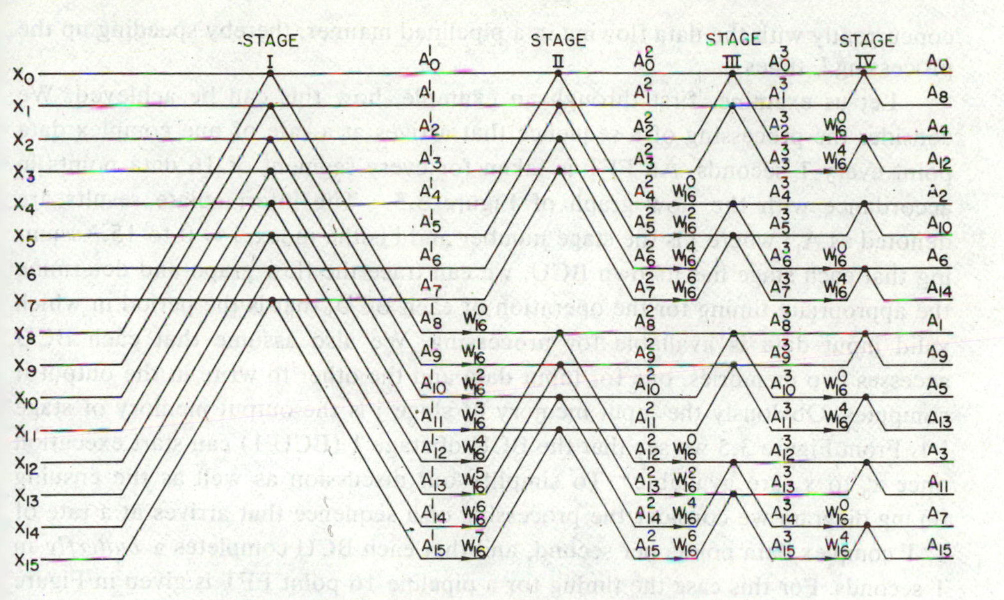

Figure 3.5 A 16 point radix 2 decimation-in-frequency FFT.

3.4 THE PIPELINE FFT.

In the previous two sections we have seen that the DFT can be obtained by performing repeatedly the basic butterfly computations. In many applications, most notably those involving the processing of wideband radar signals, very high processing speeds are required [12]. This leads to a necessity to perform many *butterflies* in a limited time period, which in many cases is beyond the capabilities of a single arithmetic unit. Thus several computational elements operating in parallel are needed to perform the required arithmetic. In this section we describe an organization for the FFT algorithm that is particularly suited to parallel processing. It is known as the pipeline FFT and was first proposed in [13]. We will refer to an arithmetic unit capable of performing the *butterfly* operation as a basic computation unit (BCU). In Chapter 5 an efficient hardware implementation for the BCU of such a pipeline FFT processor is discussed.

As we see from the flow diagrams describing the FFT algorithm, Figures 3.3 and 3.5, there are $L = \log_2 N$ stages in the computation. At each stage N numbers are computed which are then used by the next stage as inputs. This structure suggests a hardware architecture consisting of a BCU for each stage and N storage locations for each stage. With suitable timing, the various FFT stages can operate concurrently with the data flowing in a pipelined manner, thereby speeding up the processing L times.

Let us examine, first through an example, how this can be achieved. We consider the processing of a sequence that arrives at a rate of one complex data point every T seconds. An FFT is taken for every segment of 16 data points in accordance with the flow graph of Figure 3.5. The intermediate results are denoted as A_i^j, where j is the stage number and i is the index, i = 0 to 15. Assuming that each stage has its own BCU, we can trace the flow graph and determine the appropriate timing for the operation of each BCU, that is the period in which valid input data is available for processing. We also assume that each BCU accesses two memories, one for input data and the other to write in the output it computes. Obviously the input memory of stage j is the output memory of stage j-1. From Figure 3.5 we see that the BCU of stage 1 (BCU 1) can start execution once x_0 to x_8 are available. To simplify our discussion as well as the ensuing timing diagram we consider the processing of a sequence that arrives at a rate of 1/T complex data points per second, and that each BCU completes a *butterfly* in T seconds. For this case the timing for a pipeline 16 point FFT is given in Figure 3.6. From it we see that BCU 1 operates, after x_0 to x_8 arrived, for 8T seconds during which the 8 *butterflies* of stage 1 are executed. After that BCU 1 is idle for 8T seconds while waiting for the next batch of input data ($x_{16},\ldots,x_{31}$). The computation resumes when x_{24} arrives, and BCU 1 is again active for 8T seconds, and so on. BCU 2 is ready to start computing as soon as BCU 1 has written out A_0^1 to A_4^1, therefore a delay of 6T between the starting of BCU 1 and BCU 2 is necessary. After BCU 2 starts it is also active for 8T seconds, idle for the next 8T seconds and so on. Similarly in the following stages, BCU J starts its operation delayed by $(N/2^J + 2)$T seconds relative to BCU 1, is active for 8T and then idle for the next 8T seconds. Finally a bit reversal shuffling has to be done at the end to read out correctly the output DFT sequence. As we see in this configuration the BCUs, once initialization is completed, are active only half the time, making this system only 50% efficient in terms of BCU utilization. As we will see in Section 3.5 in a practical case of real time processing with a 2 to 1 overlap in data blocks this system can be 100% efficient. A 2 to 1 overlap in data blocks implies that x_{19} to x_{27} are equal to x_9 to x_{18}, and therefore BCU 1 upon completing the first 8 *butterflies* has already valid input data to continue processing. This effect will ripple on to other stages, causing them to be continuously busy. In other cases in which the input data blocks are not overlapped, suitable input buffering can be provided to allow the system to be utilized at full efficiency.

Figure 3.7 depicts the block diagram of the first three stages of an N point FFT. It can also be regarded as a complete 8 point FFT. The extension to the

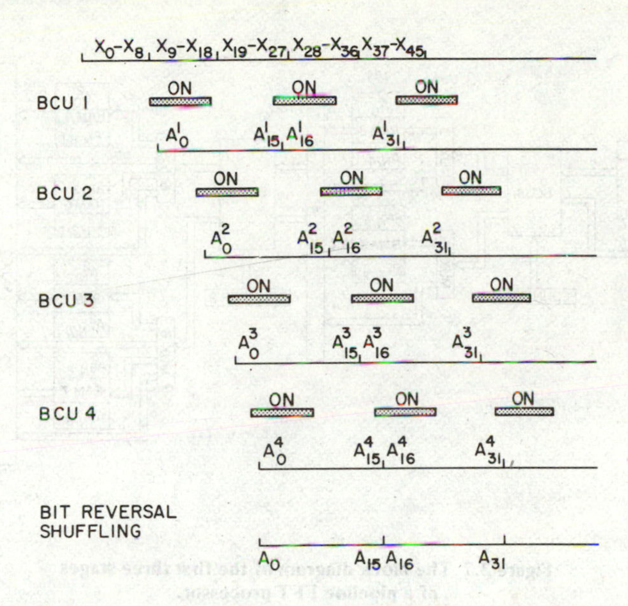

Figure 3.6 The timing for the operation of the BCUs in a pipeline FFT.

following stages is straightforward. In the first stage the multiplexer MUX 1 routes blocks of N/2 data points alternately into the memory and directly into BCU 1. BCU 1 receives from counter 1 the value k, the power of the twiddle factor; the counter simply counts down from N/2 in decrements of 1. The two outputs of BCU 1 A_i^1 and $A^1_{i+N/2}$ are routed through multiplexers MUX 2 and MUX 3, so that N/4 outputs are sent alternately to RAM 2 and RAM 3 respectively, and directly into MUX 4 that permits the sharing of BCU 2 to first execute the upper N/4 *butterflies* whose inputs come from MUX 2 and RAM 2, and than the lower N/4 *butterflies* whose inputs come from RAM 3 and MUX 3. BCU 2 receives the values k from counter 2, that counts down from N/2 in decrements of 2. The third stage operates in a similar manner as the previous two. The output of BCU 2 is split now into 4 data streams each consisting of N/8 values, and BCU3 is shared by the 4 different port pairs. The timing for activating the various units in the block diagram can be deduced directly from Figure 3.6. The memory addressing for reading in values into the BCUs and writing them out is simply sequential.

The pipeline FFT described in this section is only one example of the many possible parallel processing schemes that are inherent in the FFT algorithm. In the

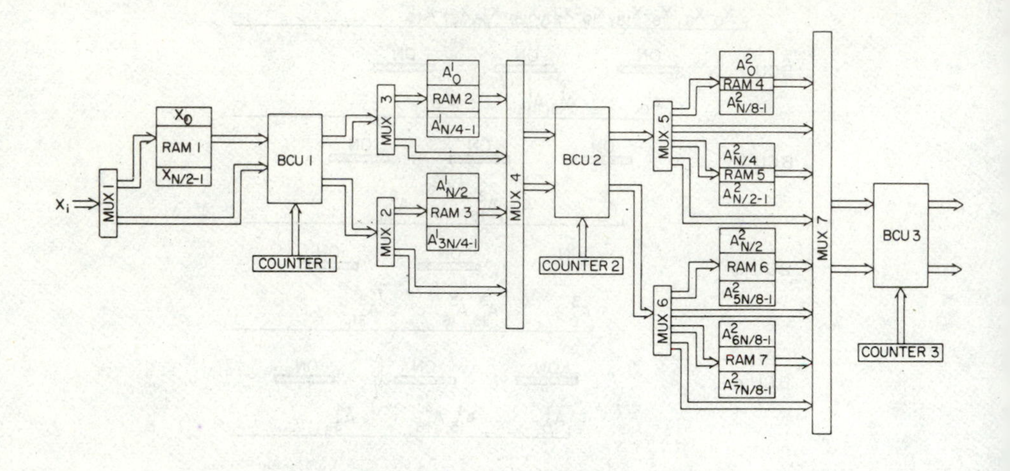

Figure 3.7 The block diagram of the first three stages of a pipeline FFT processor.

most extreme case $(N/2)\log_2 N$ BCUs operating concurrently can be used to obtain a complete DFT sequence in only one time period T once all data is available. Furthermore the pipelining technique can be extended to the design of the BCU itself, thus further reducing the throughput time. An excellent treatment of the various degrees of parallelism in the computation of the FFT can be found in [14].

3.5 THE MEASUREMENT OF POWER SPECTRA

In this section we discuss one of the most important applications of the FFT, the measurement of power spectra. We will not present here a thorough nor detailed treatment of this subject, as this topic is suitable for a complete book [15]. Furthermore, it requires the use of many concepts from statistical estimation theory that are beyond the scope of this book. Instead we present in an ad-hoc fashion a method for the measurement of power spectra originally proposed by Welch [16], that is computationally efficient and yields *good* estimates of the power spectra.

Given an input sequence $\{x_i\}$ whose power spectra we wish to measure and whose length N is a large number, say 32,000 or more, it is impractical to take its DFT directly. In addition, this fine frequency domain partition may not even be necessary to realize the resolution and accuracy requirements. The method to be discussed here computes the DFT of relatively short data blocks and then averages their results.

The input sequence $\{x_i\}$ of length N is partitioned into segments of length J each, with their starting points M units apart. For our discussion we will assume that J is chosen to be an integral power of 2, and that the input segments overlap, that is, $M < J$. A particularly *good* estimate is obtained for very large N, when the overlap is chosen to be half, that is $M = J/2$. It has been shown that under certain conditions, this choice actually minimizes the variance of the resulting estimate. For this case the sequence will be divided into K segments, $K = 2N/J$. This is depicted in Figure 3.8, where each segment is denoted as x_i^k and is defined by:

$$x_i^k = x_{k+(k-1)J/2} \qquad i = 0,1,\ldots.J-1 \quad k = 1,2,\ldots.K \tag{3-12}$$

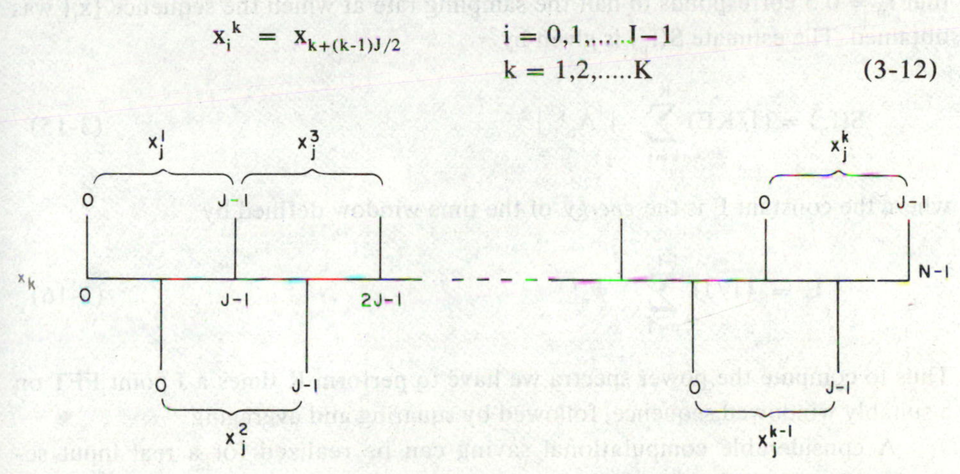

Figure 3.8 An N point sequence divided into overlapped short segments of length J each.

To avoid discontinuities due to the Gibbs phenomena when taking the transform of these finite length records, it is advisable to multiply the input sequence by a time window w_i. A commonly used window for this purpose is the triangular window, often referred to in the statistics literature as the Parzen window. It is defined by

$$w_i = 1 - \frac{|\, i - (J-1)/2 \,|}{|\, (J+1)/2 \,|} \qquad i = 0,1,\ldots.J-1 \tag{3-13}$$

Another widely used window is the Hamming window, mentioned in Chapter 2 Eq. 2-36 in connection with filter design.

Using such a time window, we compute, for k=1,2, ...,K, the DFT of the windowed input segment $\{x_n^k w_n\}$. That is we compute

$$A_p^k = (1/J) \sum_{n=0}^{J-1} x_n^k w_n e^{-jnp(2\pi/J)}$$

$$p = 0,1,....J-1 \qquad k = 1,2,....K \tag{3-14}$$

These are called the modified periodograms. We denote by $S(f_p)$ the estimates of the power spectrum at the normalized frequency $f_p = pJ$, $p = 0,1,....,J/2$. Note that $f_p = 0.5$ corresponds to half the sampling rate at which the sequence $\{x_i\}$ was obtained. The estimate $S(f_p)$ is given by

$$S(f_p) = (J/KE) \sum_{k=1}^{K} | A_p^k |^2 \tag{3-15}$$

where the constant E is the *energy* of the time window defined by

$$E = (1/J) \sum_{n=0}^{J-1} w_n^2 \tag{3-16}$$

Thus to compute the power spectra we have to perform K times a J point FFT on a suitably windowed sequence, followed by squaring and averaging.

A considerable computational saving can be realized for a real input sequence, due to the fact that it is possible to transform two J point real sequences in one J point FFT [10]. We define a complex sequence $\{y_n^k\}$ as

$$y_n^k = x_n^k w_n + j x_n^{k+1} w_n$$

$$n = 0,1,.....J-1 \qquad k = 1,2,....K/2 \tag{3-17}$$

for even K. We now compute the DFT of $\{y_n^k\}$ and denote the resulting sequence as $\{B_p^k\}$, that is,

$$B_p^k = (1/J) \sum_{n=0}^{J-1} y_n^k e^{-jnp(2\pi/J)} \tag{3-18}$$

than it is possible to show by simple algebraic manipulation that $S(f_p)$ is given by [16]:

$$S(f_p) = (J/2KE) \sum_{k=1}^{K/2} (|B_p^k|^2 + |B_{J-p}^k|^2) \tag{3-19}$$

Thus in this case only K/2 FFTs have to be computed to obtain the power spectra, which is a considerable computational saving over the direct method of Eq. 3-15.

A PL/1 computer program for power spectrum estimation is given in Appendix 3.1.

3.6 FAST CONVOLUTION

In many applications of digital signal processing the need arises to perform a linear convolution of two sequences. Often, one sequence is a short finite sequence, say the impulse response of a filter, and the other is a comparatively long data sequence. Actually the data sequence can be viewed as an infinitely long sequence that comes in continuously at a given rate. This is the case in real time digital processing of speech, radar, and many other signals. Let us denote the short sequence by $\{h_n\}_{n=0}^{N-1}$, the data sequence by $\{x_n\}$, and the result of their linear convolution by $\{y_n\}$, than y_n is given by:

$$y_n = \sum_{k=0}^{N-1} h_k x_{n-k} \tag{3-20}$$

$$n = \ldots -1, 0, 1, \ldots$$

A direct computation of y_n through Eq. 3-20 will require performing N multiplications and $N-1$ additions for each output sample y_n. Since N can be a rather large number, in many cases this turns out to be an excessive computation rate. We now proceed to show how this convolution can be performed using the FFT algorithm, which will lead to significant computational savings.

In Chapter 1, Section 1.7 we discussed the circular convolution of two periodic sequences, and we saw that the DFT of the output sequence, in that case is equal to the term by term product of the DFT's of the two sequences that are convolved. This property can be used to compute y_n the result of the linear convolution.

Let us partition the incoming data sequence x_n into segments of length L each, such that if x_n^k are the elements of the k-th segment, they are obtained from the original sequence by the relation:

$$x_n^k = \begin{cases} x_n & kL \le n \le (k+1)L-1 \\ 0 & \text{otherwise} \end{cases} \tag{3-21}$$

$$k = \ldots -1, 0, 1, \ldots$$

This is depicted in Figure 3.9, from which we also note that x_n, the original sequence can be viewed as a summation over k of the nonoverlapping segments $\{x_n^k\}$. Let $\{q_n^k\}$ be the convolution of $\{h_n\}$ with the k-th data segment $\{x_n^k\}$. Thus Eq. 3-20 can be rewritten as:

$$y_n = \sum_{k=-\infty}^{\infty} \sum_{j=0}^{N-1} x_{n-j}^k h_j = \sum_{k=-\infty}^{\infty} q_n \tag{3-22}$$

The sequences $\{q_n^k\}$ are of length $L+N-1$ each. Also, two successive ones, say $\{q_n^k\}$ and $\{q_n^{k+1}\}$ overlap by $N-1$ points.

We will illustrate this through a simple example. Let $N=4$ and $L=9$, than $\{q_n^0\}$ is obtained by convolving x_0^0, x_1^0,...x_8^0, with h_0, h_1, h_2, h_3 and has the terms q_0^0, q_1^0, q_2^0,...q_8^0, q_9^0, q_{10}^0, q_{11}^0 (that is, $9+4-1=12$ terms). Similarly $\{q_n^1\}$ is obtained by convolving x_9^1, x_{10}^1,...,x_{17}^1 with h_0, h_1, h_2, h_3 and has the terms q_9^1, q_{10}^1, q_{11}^1,...q_{17}^1, q_{18}^1, q_{19}^1, q_{20}^1. Thus we see that the partial result sequences q^0 and q^1 overlap for $4-1=3$ values of n. In our case the overlap occurs at $n=9,10,11$. To obtain the correct result for y_n we have to add the values of the sequences $\{q^k\}$ at these points. In our example

$$y_{10} = x_{10} h_0 + x_9 h_1 + x_8 h_2 + x_7 h_3$$

but we obtain y_{10} through Eq. 3-22 by;

$$y_{10} = q_{10}^0 + q_{10}^1$$

from the definition of q_n^k and the segments $\{x^k\}$ it follows that

$$\begin{aligned} q_{10}^0 &= x_{10}^0 h_0 + x_9^0 h_1 + x_8^0 h_2 + x_7^0 h_3 \\ &= 0 \times h_0 + 0 \times h_1 + x_8 h_2 + x_7 h_3 = x_8 h_2 + x_7 h_3 \end{aligned}$$

and

$$\begin{aligned} q_{10}^1 &= x_{10}^1 h_0 + x_9^1 h_1 + x_8^1 h_2 + x_7^1 h_3 \\ &= x_{10} h_0 + x_9 h_1 + 0 \times h_2 + 0 \times h_3 = x_{10} h_0 + x_9 h_1 \end{aligned}$$

that will indeed give the correct result for y_{10}. This example is also illustrated in Figure 3.10.

This method of sectioning the input data sequences into nonoverlapping segments, and obtaining the linear convolution by adding the overlapping partial

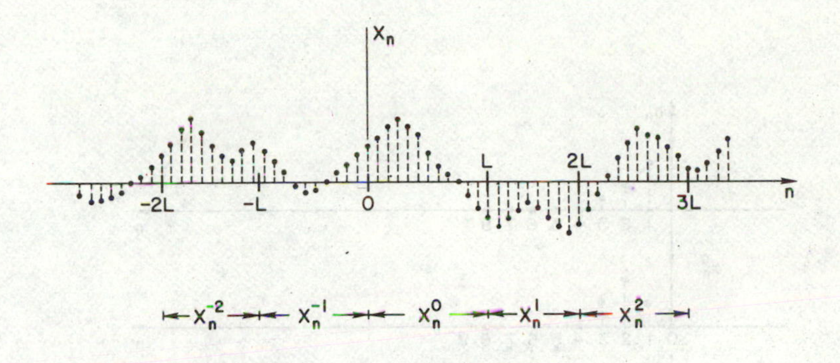

Figure 3.9 Partitioning the input sequence into nonoverlapping segments of length L each.

result sequences is commonly referred to as the *overlap and add* method (naturally!). Another method that is commonly used is the *overlap and save* method in which the input data sequence is partitioned into segments of length L overlapping by N−1 points, and in this case the first N−1 points of the partial result sequence are incorrect and have to be substituted by the last N−1 points of the previous partial result sequence.

Thus, regardless of the method used for combining the partial results, we have reduced the problem of performing the linear convolution of Eq. 3-20 to performing a series of convolutions of two finite sequences to obtain partial result sequences $\{q^k\}$ that are afterwards suitably combined to obtain y_n. We recall from our discussion in Chapter 1 that the DFT of $\{q_n{}^k\}$, $\{Q_p{}^k\}$, can be obtained by multiplying term by term the DFT's of $\{h_n\}$, $\{H_p\}$ and of $\{x_n{}^k\}$, $\{X_p{}^k\}$. Therefore to obtain $\{q_n{}^k\}$ we first do a N+L−1 point DFT on the data segment $\{x_n{}^k\}$, multiply the resulting DFT $\{X_p{}^k\}$ by the (N+L−1 point) DFT of the sequence $\{h_n\}$. Note that the DFT of $\{h_n\}$ can be precomputed and only the values of $\{H_p\}$ need to be stored. and then an inverse DFT (IDFT) on $\{Q_p{}^k\}$ will yield the values of $\{q_n{}^k\}$. The following equations define the sequence of operations just mentioned:

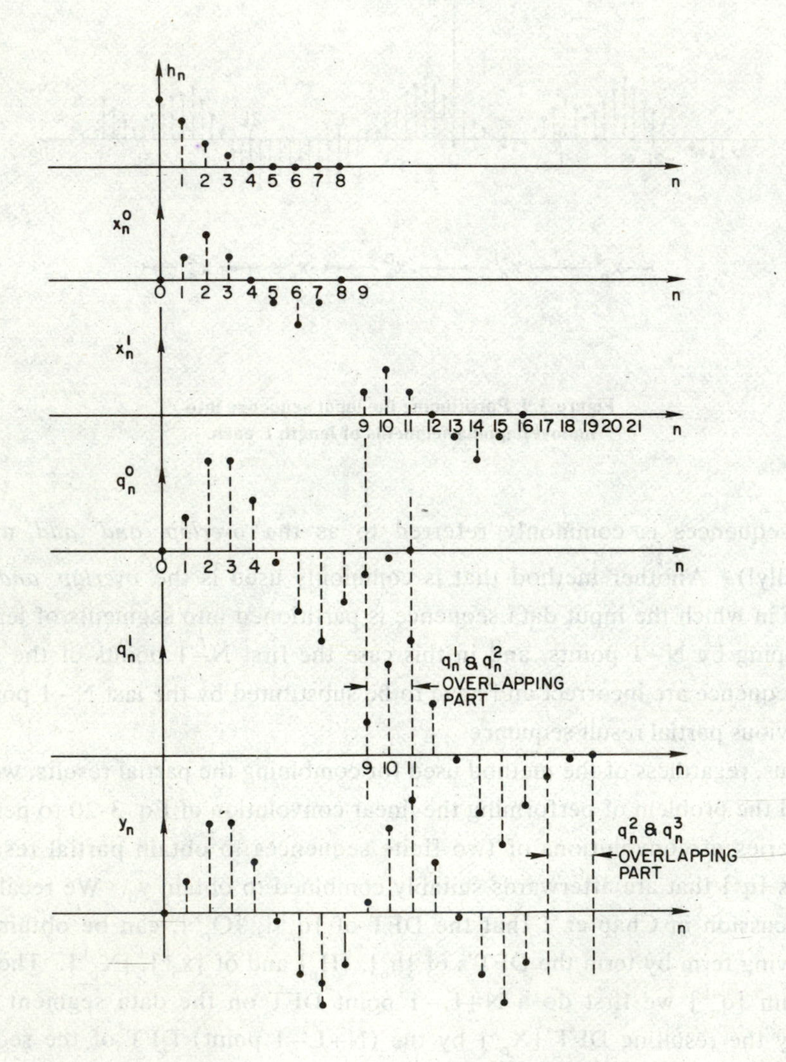

Figure 3.10 An illustration of the overlap and add method for a simple example.

$$H_p = \sum_{n=0}^{N-1} h_n W^{np} \tag{3-23}$$

where $W = e^{-2j\pi/(N+L-1)}$

$$X_p{}^k = \sum_{n=0}^{L-1} x_{n+kL}{}^k W^{np} \tag{3-24}$$

$$Q_p{}^k = H_p X_p{}^k \tag{3-25}$$

$$q_{n+kL}{}^k = 1/(N+L-1) \sum_{p=0}^{N+L-1} Q_p{}^k W^{np} \tag{3-26}$$

This procedure is known as the *fast convolution* method. At this point the reader must wonder where the term fast originated, since at least at first glance this procedure seems more complicated than the direct convolution. However this is not the case, and the name fast convolution will seem indeed justified as soon as we count the number of multiplications required per output sample. Let us assume for the moment that our data is complex, (for the case of real data we will compute simultaneously the DFT of two consecutive segments, using the method described in the last part of this section), then we saw earlier that to do $N+L-1$ point DFT (or IDFT) using the FFT algorithm $(N+L-1)/2 \log_2(N+L-1)$ complex multiplications are required. Therefore to obtain the partial result sequence $\{q_n{}^k\}$ through fast convolution $(2x(N+L-1)/2) \log_2 N + (N+L-1)$ complex multiplications will be required. Since we obtain L valid output points y_n, the number of complex multiplications per output sample is $(N+L-1)(1+\log_2 N)/L$ instead of N in a direct computation. To illustrate the reduction in the number of multiplications let us consider two examples. First $N=32$ and $L=481$, then the direct convolution requires 31 multiplications per output sample and the fast convolution 6.43, that is, a 5 fold reduction. Next let $N=128$ and $L=1921$, then instead of 128 multiplications we require only 8.56 multiplications per output sample in the fast convolution, a 16 fold reduction. For very large N it is easy to see that instead of doing N multiplication we do $(1+\log_2 N)$ which is indeed a drastic reduction, and therefore the name fast convolution is fully justified.

We conclude this section with a brief discussion of a method of performing a DFT on real sequences. In most practical cases the data sequence $\{x_n\}$ is real and doing an N point DFT on it will still require about $(N/2)\log_2 N$ complex multiplication, that is, the same as if it was a complex sequence. We will show now that using the properties of the DFT it is possible to do an N point DFT to transform two N point real sequence.

Let $\{x_n^1\}$ and $\{x_n^2\}$ be two N point real sequences and denote by $\{X_p^1\}$ and $\{X_p^2\}$ the DFT's of $\{x_n^1\}$ and $\{x_n^2\}$ respectively. Then as was shown in Chapter 1, due to the fact that the sequences are real, the resulting DFT has the symmetry property:

$$X_p^1 = (X_{N-p}^1)^*$$

$$X_p^2 = (X_{N-p}^2)^* \tag{3-27}$$

where * denotes complex conjugate.

We now define a complex sequence a_n as

$$a_n = x_n^1 + jx_n^2 \tag{3-28}$$

$$n=0,1,...N-1$$

From the linearity property of the DFT it follows that the DFT of a_n, A_p satisfies:

$$A_p = X_p^1 + jX_p^2 \tag{3-29}$$

Substituting the relations given in Eq. 3-27 we obtain:

$$A_{N-p} = X_{N-p}^1 + jX_{N-p}^2 = X_p^{1*} + jX_p^{2*} \tag{3-30}$$

and taking complex conjugate of both sides of Eq. 3-30 yields:

$$A_{N-p}^* = X_p^1 - jX_p^2 \tag{3-31}$$

Combining Eqs. 3-29 and 3-31 it is possible to obtain X_p^1 and X_p^2 in terms of A_p and A_p^*

$$X_p^1 = (A_p + A_{N-p}^*)/2$$

$$X_p^2 = (A_p - A_{N-p}^*)/2j \tag{3-32}$$

Therefore, to obtain the DFT of the two N point real sequences $\{x_n^1\}$ and $\{x_n^2\}$ we can do a N point FFT on the complex sequence $\{a_n\}$ defined above (Eq. 3-28) and than use Eq. 3-32 to obtain the two transforms $\{X_p^1\}$ and $\{X_p^2\}$. This will require only an extra of N additions, and we saved almost half the multiplications over the straightforward transforming of each sequence independently.

3.7 SUMMARY

In this chapter we have discussed some of the more common variants of the FFT algorithm. In Section 3.2 the radix 2 decimation-in-time algorithm was described in detail, and in Section 3.3 the radix 2 decimation-in-frequency algorithm was presented. In Section 3.4 the pipeline FFT was introduced as an example of the various degrees of parallelism inherent in the FFT algorithm. Such parallelism can be exploited to speed up the processing at the expense of additional hardware. The next two sections deal with some of the more common applications of the FFT. In Section 3.5 the use of the FFT in power spectra measurement was detailed to acquaint the reader with one of the more important uses of the FFT and in Section 3.6 the use of the FFT to achieve fast convolution is described.

At this point we want to caution the reader that this is only an introductory treatment of the many aspects and refinements of the FFT algorithm and its applications. The reader will benefit from consulting the references mentioned to gain a deeper understanding of these issues. We have thus far ignored the effect of finite register length, and the need for scaling to avoid overflow. These topics are of considerable importance and should be given careful consideration in any application. We address some of these problems in Chapter 6, and for a more detailed treatment the reader is advised to consult references [17,18,19] to obtain an appreciation and perspective of the problems involved.

EXERCISES

3.1 Develop the radix 2 decimation-in-time algorithm for N=16 in detail showing that the computations can be done in-place. Follow the same procedure as was used for the example in Section 3.2. Draw the flow graph for this case.

3.2 Develop the radix 2 decimation-in-frequency algorithm for N=16 in detail and draw its flow graph. Compare with Figure 3.5.

3.3 Show that if the input is given (or addressed) in bit reversed order, then the twiddle factors are addressed in normal order, that is, W_N^{rm}, are required for the butterflies with r fixed in each stage and m varying in ascending order. Use as an example N=16.

3.4 Show that the twiddle factors can be computed recursively as follows.

$$W_N^{rm} = W_N^r \; W_N^{r(m-1)}$$

Do this calculation on a computer, once using full precision and once using only 8 bits, (for example, in PL/1 use once FLOAT DECIMAL (16) and once BINARY FIXED (8,7), and compare the results obtained. You will note appreciable error accumulation. Show that this can be partly remedied by resetting the result to a correct value at intermediate points, for example any time rm=N/4 set the result to −j. Run the modified recursion and note the improvement.

3.5 Prove that the DFT of the sequence $\{x_n\}$, $\{X_p\}$ can be computed by passing the sequence

$\{x_n W_N^{n^2/2}\}=\{v_n\}$ through a linear FIR filter with an impulse response $h_n=W_N^{-n^2/2}$ and dividing the result by $W_N^{p^2/2}$, that is,

$$X_p = W^{p^2/2} \sum_{n=0}^{N-1} v_n h_{p-n}$$

(*Hint*: use the identity $np=[n^2+p^2-(p-n)^2]/2$.)
Note: this method of computing the DFT is used quite often in radar and is known as the *chirp z-transform*. The name *chirp* is derived from the fact that the sequence $\{h_n\}$ can be thought of as having linearly increasing frequency, and in radar such signals are called *chirp signals*.

3.6 Show that an N point DFT can be computed by decomposing it into an M point DFT and an L point DFT when $N=M\times L$, as follows:

Let $n=M\,k+m$ and $p=L\,r+q$, then

$$X_p = X_{Lr+q} = \sum_{m=0}^{M-1} V_{q,m} W_M^{mr}$$

where

$$V_{q,m} = W_N^{qm} \sum_{k=0}^{L-1} x_{Mk+m} W_L^{qk}$$

$$r, m = 0, 1, \ldots, M-1$$
$$q, k = 0, 1, \ldots, L-1$$

This is the generalization of the radix 2 algorithms and allows derivation of any radix algorithm.

3.7 Using the Equations you proved in Exercise 3.6, derive Eq. 3-5 by taking M to be 2.

3.8 Derive a radix 4 algorithm using the results of Exercise 3.6, by letting M=4. Draw a flow graph of this algorithm for N=16. Compare the number of multiplications with the number for the radix 2 algorithm, not counting multiplications by ± 1 or $\pm j$.

3.9 Prove Eq. 3-19, using Eqs. 3-15 to 3-18.

3.10 Develop an example for the overlap and save method for doing fast convolution using N=4 and L=9. Follow the procedure used for the example in Section 3.6

3.11 Compare the number of multiplications for the case N=64 and L=48, when doing direct convolution versus fast convolution using overlap and add.

3.12 We wish to do a convolution to compute an FIR filter having N=71 taps. Choose the segment length L, which will result in the lowest number of multiplications per output sample when doing fast convolution.

3.13 Let $\{x_n\}_{n=0,N-1}$ and $\{h_n\}_{n=0,N-1}$ be periodic sequences and $\{y_n\}_{n=0,N-1}$ the result of their periodic convolution. Show that if we are interested only in every other output, that is, in y_{2n} only, it is given by

$$y_{2n} = \frac{2}{N} \sum_{p=0}^{N/2-1} Q_p W_{N/2}^{pn}$$

where

$$Q_p = X_p H_p + X_{p+N/2} H_{p+N/2} \qquad p=0, 1, \ldots, N/2-1$$

and $\{X_p\}$, $\{H_p\}$ are the DFT of $\{x_n\}$ and $\{h_n\}$ respectively.
Note: this result implies that if we wish to decimate the output by two, we can compute only a half size inverse transform, when performing the convolution in the frequency domain (fast). This result can be extended to cover convolution of nonperiodic sequences using overlap and add or overlap and save, and used to implement long FIR filters that decimate.

REFERENCES

1. J. W. Cooley and J. W. Tukey, *An Algorithm for the Machine Calculation of Complex Fourier Series*, Mathematics of Computation, vol. 19, no. 90, 1965, pp. 297-301.
2. G-AE Subcommittee on Measurement Concepts, *What is the Fast Fourier Transform*, IEEE Trans. Audio Electroacoust., vol. AU-15, June 1967, pp. 45-55.
3. OMNIFEROUS™ FFT ANALYZER, Federal Scientific Corp. 615 West 131 St., New York, N.Y. 10027
4. J. R. Fisher, *Architecture and Application of the SPS-41 and SPS-81 Programmable Digital Signal Processors*, EASCON 74 Record, pp. 674 -678, IEEE Publication 74CH0883-1 AES.
5. NI2001 Programmable Calculating Oscilloscope, Norland Instruments, Norland Drive, Fort Atkinson, Wisconsin 53538.
6. J. K. Hartt and L. Sheats, *Application of Pipeline Fast Fourier Transform Technology in Radar Signal and Data Processing*, EASCON 71 Record, pp. 216-221, IEEE New York.
7. M. J. Corinthios, *A Fast Fourier Transform for High Speed Signal Processing*, IEEE Trans. Comput., vol. C-20, no. 8, Aug. 1971, pp. 843-846.
8. R. C. Singleton, *An Algorithm for Computing the Mixed Radix Fast Fourier Transform*, IEEE Trans. Audio Electroacoust., vol. AU-17, June 1969, pp. 93-103.
9. R. C. Singleton, *A Method for Computing the Fast Fourier Transform with Auxiliary Memory and Limited High Speed Storage*, IEEE Trans. Audio Electroacoust., vol. AU-15, June 1967, pp. 91-97.
10. J. W. Cooley, P. A. W. Lewis and P. D. Welch, *The Fast Fourier Transform Algorithm: Programming Considerations in the Calculation of Sine, Cosine and Laplace Transforms*, J. Sound Vibration, vol. 12, July 1970, pp. 315-337.
11. E. O. Bringham, *The Fast Fourier Transform*, Prentice-Hall Inc., 1974, Engelwood Cliffs, New Jersey.
12. L. W. Martinson and R. J. Smith, *Digital Matched Filtering with Pipelined Floating Point Fast Fourier Transform*, IEEE Trans. Acoustics, Speech and Signal Processing, vol. ASSP-23, April 1975, pp. 222-234.
13. H. L. Gorginsky and G. A. Works, *A Pipeline Fast Fourier Transform*, IEEE Trans. Comput., vol. C-19, Nov. 1970, pp. 1015-1019.
14. B. Gold and T. Bially, *Parallelism in Fast Fourier Transform Hardware*, IEEE Trans. Audio Electroacoust., vol. AU-21, February 1973, pp. 5-16.
15. G. M. Jenkins and D. G. Watts, *Spectral Analysis and Its Application*, Holden Day Inc., 1968, San Francisco.

16. P. D. Welch, *The Use of Fast Fourier Transform for the Estimation of Power Spectra: A Method Based on Time Averaging over Short, Modified Periodograms*, IEEE Trans. Audio Electroacoust., vol. AU-15, June 1967, pp. 70-73.
17. T. Kaneko and B. Liu, *Accumulation of Round-Off Error in Fast Fourier Transform*, J. of the Association for Computing Machinery, vol. 17, October 1970, pp. 637-654.
18. P. D. Welch, *A Fixed-Point Fast Fourier Transform Error Analysis*, IEEE Trans. Audio Electroacoust., vol. AU-17, June 1969, pp. 151-157.
19. C. J. Weinstein, *Round-off Noise in Floating Point Fast Fourier Transform Computation*, IEEE Trans. Audio Electroacoust., vol. AU-17, Sept. 1969, pp. 209-215.

Appendix 3.1

An FFT Program

In this appendix the listing for a PL/1 program to compute the DFT of a complex sequence of N points, X(*), using the radix 2 decimation-in-time algorithm. The calculation is done in place, and the resulting DFT is returned in X(*). The vector SINT is a sine table for determining the twiddle factors. INV is a character, if INV = '1' the inverse transform is performed, otherwise the direct transform is performed.

```
DFT:PROCEDURE(X,INV,N,SINT);
 /*THIS PROCEDURE COMPUTES THE DFT OF THE COMPLEX INPUT SEQUENCE X.
   X(0:N) - GIVEN COMPLEX INPUT SEQUENCE, RESULTING COMPLEX OUTPUT
        SEQUENCE.
   N - NUMBER OF ELEMENTS IN INPUT ARRAY - 1.
   SINT(0:(N+1)/4- GIVEN INPUT SINE TABLE, QUARTER PERIOD.
   INV = '0' FORWARD TRANSFORM IS COMPUTED, INV = '1' - INVERSE
   TRANSFORM IS COMPUTED.  */
DECLARE
   (X(*),R,R1,S)
   COMPLEX FLOAT DECIMAL(16),
   (SINT(*),S1,S2)
   FLOAT DECIMAL(16),
   (N,I,J,K,L,M,NT,NH,NQ,IR,IT)
   BINARY FIXED(15),
   (ID,ND) BINARY FIXED(31),
   INV CHAR(1);
NT = N;
NQ = NT/4 + 1;
NH = NQ+NQ;              /*REORDER INITIAL VECTOR X*/
J = 0;               /*USING THE BIT REVERSAL TECHNIQUE*/
   DO I =0 TO NT-2;
   IF J > I THEN DO;
      R = X(J); X(J) = X(I); X(I) = R;
      END;
   K = NH;
      DO WHILE (J >= K);
      J = J-K;
      K = K/2;
      END;
   J = J+K;
   END;
ID = 1;          /*CALCULATE THE DISCRETE FOURIER TRANSFORM*/
 IR = 1; M = 0; IT = 0;
```

```
TRAN: I = ID;
  ID = ID+ID;
     DO J =0 TO I-1;
     S2 = SINT(M);
     IF INV ¬= '1' THEN S2 = -S2;
     S1 = SINT(NQ-M);
     IF J >= IR THEN DO;
        M = M - IT; S1 = -S1;
        END;
     ELSE M  = M+IT;
        DO ND=J TO NT BY ID;
        K  = ND;
        L  = K+I;
        S = CPLX(S1,S2);
        R1 = X(L);         /*BUTTERFLY COMPUTATION*/
        R = S*R1;
        R1 = X(K);
        X(K) = R1 + R;
        X(L) = R1 - R;
        END;
     END;
  IR = I;
  IT = NQ/I;
  IF I <  NH THEN GO TO TRAN;
 RETURN ;
END DFT;
```

Appendix 3.2

A Power Spectra Computation Program

In this appendix a PL/1 program for the estimation of power spectra is given. The procedure should be used in conjunction with the FFT program in Appendix 3.1. Figure 3A.1 shows a plot of a typical output of tne program, it is the power spectra of white noise passed through a bandpass filter.

```
SPECTRA:PROCEDURE(Y,NY,S,NSP,DELT,NT);
DCL(Y(*),S(*),SS1,DELT,XR(0:511),XI(0:511),SINT(0:129),W(0:511),S1,S2)
  FLOAT DECIMAL(16);
DCL INV CHARACTER(1);
DCL X(0:511) COMPLEX DECIMAL FLOAT(16);
/*THIS PROCEDURE COMPUTES AN ESTIMATE OF THE POWER SPECTRA OF THE
 SEQUENCE Y(I) USING THE TIME AVERAGING OF MODIFIED PERIODOGRAMS
AFTER P. WELCH |16|.  */
/* NY IS LENGTH OF Y, NSP IS NUMBER OF SEGMENTS (EVEN), NSP*NT/2=NY
  FOR 50  OVERLAP. */
SS1=0; S=0;
DO J=0 BY 1 TO (NT-1); W(J)=1-ABS(J-(NT-1)/2)/((NT+1)/2);
  SS1=SS1+W(J)**2; END;
SS1=SS1/NT;
INV='0'; NFT=NT-1;
  S1 = 6.283818530718E+00/NT;
   N1 = (NT - 1)/4 + 1;
   DO J = 0 TO N1;
    S2 = FLOAT(J)*S1; SINT(J) = SIN(S2);
   END;
DO ISP=0 BY 2 TO (NSP-2);
 DO J=0 BY 1 TO NT-1 ;
 XR(J)=Y(1+J+ISP*NT/2); XR(J)=XR(J)*W(J);
 XI(J)=Y(1+J+(ISP+1)*NT/2); XI(J)=XI(J)*W(J);
 END;
  X = CPLX(XR,XI);
 CALL DFT(X,INV,NFT,SINT);
DO J=0 BY 1 TO (NT-1);
 S(J+1) = S(J+1) + ABS(X(J))**2 + ABS(X(NT-J))**2;
END;
END;
S=S*(NT/(2*SS1*NSP));
/* NORMALIZING THE POWER SPECTRUM */
 SS1=S(1); DO J=2 BY 1 TO NT; IF S(J)>SS1 THEN SS1=S(J);END;
  PUT EDIT('NORMALIZING FACTOR OF SPECTRA',SS1) (SKIP,A, F(15,8));
```

```
S = S/SS1 ;
DO J=1 TO NT; IF S(J) > 0.0000000001 THEN S(J) = 10.0*LOG10(S(J));
     ELSE S(J) = -100.0000 ; END;
RETURN;
END SPECTRA;
```

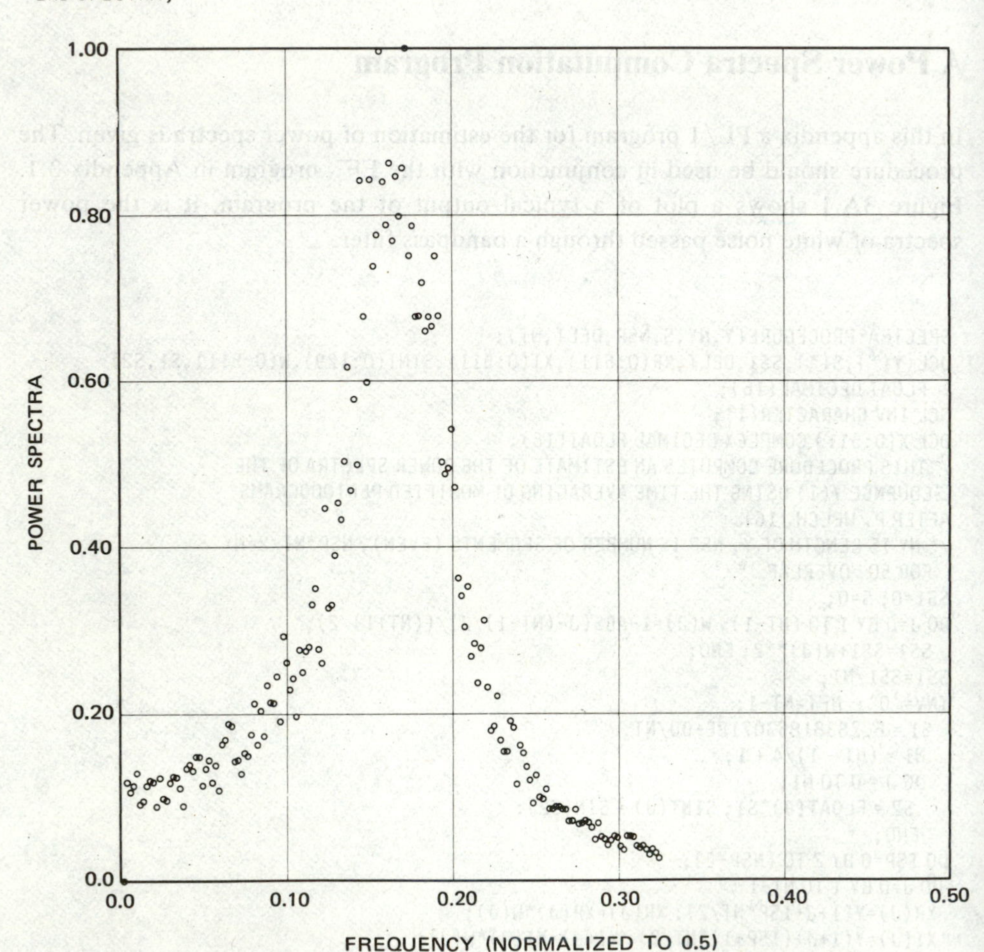

Figure 3A.1 Power spectra of white gaussian noise passed a bandpass filter.

CHAPTER 4

The Implementation of General Purpose Digital Signal Processors

4.1 INTRODUCTION

This chapter is devoted to the hardware implementation of digital signal processors. Although in principle all digital signal processing algorithms can be implemented by programming general purpose digital computers. It has been realized that this solution is not cost effective in many applications. The reason for this is the mismatch between a general purpose computer architecture and most signal processing algorithms that require a very large number of repeated arithmetic operations of a relatively simple nature and a low number of input/output operations. Furthermore, if real time operation is desirable, even a modest signal processing need would require a computational rate found only in very large computers, yet on the other hand, only a small fraction of the sophisticated operating system such systems offer would be used. Due to these factors, considerable effort has been invested in the design and implementation of general purpose and specialized digital signal processors. The subject of special purpose signal processors is treated in Chapter 5. In this chapter, we provide the reader with an introduction to the implementation of general purpose digital signal processors.

The first two sections are intended to provide a minimal background and review of the basics of binary arithmetic and of digital hardware, especially as relating to the implementation of digital signal processors. Section 4.4 presents some general design considerations for signal processors setting the framework for the more specific discussion of processor architectures to be discussed in Section 4.5. To illustrate the various design considerations we introduce a hypothetical processor called DISP and analyze its basic characteristics. Section 4.6 contains some further detail on the organization of the arithmetic unit. It is obvious that in a single chapter it is impossible to cover all the important details or even provide enough information that will enable the reader to design a processor completely on his own. The material we have included is intended to provide the reader with some insight into the problems arising in the implementation of such processors and some of the more frequently used solutions. Readers who are familiar with the material discussed in the next three sections should proceed directly to Section 4.4.

4.2 BINARY ARITHMETIC

In this section we discuss the representation of numbers in binary notation, then proceed to the methods that are used to perform the basic arithmetic operations on numbers expressed in binary notation.

We focus mainly on the sign-magnitude and 2's complement representations, the most widely used binary notations, which have many properties that make them especially suitable for use in digital processors. We also discuss the canonical signed digit code, which is used later on to obtain *efficient* mechanization of multiplication.

The basic digital computing elements can distinguish only between two states of the input and produce only one of two states as the output. To use these elements to perform arithmetic operations on numbers, the numbers have to be expressed in a unique way by a combination of elements each having only two possible values. The binary representation of numbers in which each number is expressed by a unique string of ones and zeros, called *bits*, is therefore a suitable representation of numbers inside a digital computing element. Basically, the binary representation of numbers is completely analogous to the perhaps more accustomed decimal representation, and differs in principle mostly by the way in which signed numbers are handled. In the decimal number representation we have ten digits, 0 to 9. The symbols + and −denote positive and negative numbers respectively, and the decimal point is used to separate between the integer and the fractional portion of a number. However, since in digital hardware we have only two bits available and nothing else, the same bits have to be used together with well defined conventions to also fill the place of the three symbols; +, ., and −. The various binary notations used differ mainly in the conventions adopted to handle the representation of negative numbers.

Let us begin with the binary representation of an unsigned integer, which is common to all binary notations. A B bit binary number $(x^{B-1}, x^{B-2}, \ldots x^{0})$, where $x^{j} = 0$ or 1 are the bits, is interpreted as

$$x = \sum_{j=0}^{B-1} x^j 2^j . \qquad (4\text{-}1)$$

$$x^j = 0 \text{ or } 1$$

Thus the binary number 11010 represents the decimal number 26, since

$$0\times2^0 + 1\times2^1 + 0\times2^2 + 1\times2^3 + 1\times2^4 = 2 + 8 + 16 = 26$$

This, of course, is completely analogous to the decimal number system where the number base is 10, that is, $26 = 2\times10^1 + 6\times10^0$, or $2763 = 2\times10^3 + 7\times10^2 + 6\times10^1 + 3\times10^0$. Using Eq. 4-1 it is easy to verify that the largest integer that can be represented by a B bit binary number is 2^B-1. Thus 10, 16, or 32 bits can represent all integers up to and including 1023, 65,435, or 4,150,998,095 respectively.

The modification of the representation described above to represent fractional numbers is straightforward. It is done by assuming a fixed scaling by a negative power of 2 of all numbers. We illustrate this with a simple example. Suppose 16 bits are used to represent decimal numbers, and all numbers scaled by $2^{-5}=0.03125$, then the largest number we can now represent is $(2^{16}-1)2^{-5} = 2047$, and the smallest fraction is 0.03125. This is equivalent to assuming a *binary point* in the fifth position to the left of the rightmost or the *least significant bit* (LSB) and computing the fractional part as in Eq. 4-1 only using negative powers of 2. Thus in this case the binary number 0101100010110110 represents $0\times2^{10} + 1\times2^9 + 0\times2^8 + 1\times2^7 + 1\times2^6 + 0\times2^5 + 0\times2^4 + 0\times2^3 + 1\times2^2 + 0\times2^1 + 1\times2^0 + 1\times2^{-1} + 0\times2^{-2} + 1\times2^{-3} + 1\times2^{-4} + 0\times2^{-5}$ or the decimal number 709.6875.

In digital processors (as well as in computers) it is often assumed that we are dealing only with fractions, that is, all numbers are less than or equal 1 in magnitude. This is done by assuming that a B bit number has its binary point at the left of the most significant bit (MSB). In other words, all numbers are assumed to be scaled by 2^{-B}, which makes the largest binary number 11111...1, $(2^B-1)/2^{-B}=1-2^{-B}$ which is less than 1. Obviously the range of the numbers is not changed and only a scaling of their range into numbers less than 1 is performed.

We now proceed to discuss the representation of signed numbers.

Sign – Magnitude Notation

In the sign-magnitude notation, numbers that have the same absolute value are represented by the same binary number, but an extra bit called the sign bit is added to the left of the MSB to distinguish positive from negative numbers, for example

$$+0.828125 = 0110101$$
$$-0.828125 = 1110101$$

Thus a B bit binary number in sign-magnitude notation with an assumed binary point between the sign bit and the MSB lies in the range $-(1-2^{-B+1}) \leq x \leq (1-2^{-B+1})$, and there are as many negative numbers as there are positive numbers. Note that zero can be represented as either +0 or −0 which seems redundant, but maybe convenient in some operations.

2's Complement Notation

In 2's complement notation positive numbers are expressed exactly as in the sign-magnitude notation, where the first bit is taken to denote the sign. A negative number is represented by taking its absolute value and subtracting it from 2, and since the absolute value is less than 1, this will result in the first bit always being one. We illustrate this by a simple example. The 2's complement representation of −0.375 is obtained by 2 − 0.375 = 1.625 whose binary representation is 1101. A more convenient interpretation of the 2's complement notation is that in this notation the first bit (sign bit) is assumed to multiply −1. Therefore if x is represented by a B bit binary number $(x^0, x^1, \ldots x^{B-1})$ in 2's complement notation, the decimal value of x is given by:

$$x = -x^0 + \sum_{j=1}^{B-1} x^j 2^{-j} \qquad (4\text{-}2)$$

Thus according to Eq. 4-2 the 2's complement binary number 1101 represents the decimal number

$$-1 + 1\times 2^{-1} + 0\times 2^{-2} + 1\times 2^{-3} = -1 + 0.5 + 0.125 = -0.375$$

It is easy to verify, using Eq. 4-2, that the largest number expressable in 2's complement notation is $1-2^{-B+1}$ (01111...1) and the smallest is −1 (1000...0). Furthermore, here zero has only one representation (0000...0), and there is one more negative number (−1) than there are positive numbers.

The Canonical Signed Digit Code (CSDC)

This code is different from most binary representations, in that it allows each bit to have a sign. This code has some interesting properties that make it useful in the mechanization of multiplication. Given an integer L in the range $-2^B+1 \leq L \leq 2^B-1$ than it can be represented in a generalized B bit binary code of signed bits β^j such that

$$L = \sum_{j=0}^{B-1} \beta^j 2^j \tag{4-3}$$

$$\beta^j = 0, +1 \text{ or } -1$$

(for example 15 is represented by 1000−1 and −15 by −10001.) Let T_L denote the number of nonzero bits in this representation of the integer L, that is,

$$T_L = \sum_{j=0}^{B-1} |\beta^j| \tag{4-4}$$

It has been proved [1] that for any such integer L there exists a unique representation of the form given in Eq. 4-3 in which no two consecutive bits are nonzero, that is

$$\beta^j \times \beta^{j-1} = 0\,, \qquad 0 \leq j \leq B-1 \tag{4-5}$$

and furthermore, such a representation has the least number of nonzero bits. A representation of this form satisfying Eq. 4-5 is called the canonical signed digit code (CSDC). Thus a representation of L in the CSDC requires a minimum number of nonzero bits β^j. It has also been shown in [1] that a number L represented in the CSDC will have on the average T_L nonzero digits, where

$$T_L = B/3 + [1-(-2^{-B})]/9 \tag{4-6}$$

So for moderately large B the average number of nonzero bits is approximately B/3 as opposed to B/2 in the usual (nonsigned) binary representations.

The conversion from the 2's complement code to the CSDC is straightforward, it is based on the identity:

$$2^{k+n+1} - 2^k = 2^{k+n} + 2^{k+n-1} + 2^{k+n-2} + \dots + 2^k \tag{4-7}$$

which indicates that a string of 1's in the 2's complement notation will be replaced by a 1 followed by zeros followed by a −1. For example, the 2's complement number 011111 (31) in the CSDC is 10000−1 (32−1 = 31). To convert an arbitrary 2's complement number, one first groups together the strings of consecutive 1's that are separated by one or more 0's. A string of two or more 1's is converted into the CSDC according to Eq. 4-7. Isolated single 1's are left unchanged. After this conversion, the entire new bit string obtained is examined. Any pair (-1 1) is now changed to (0 -1), and any triple (0 1 1) is now changed to (1 0 -1). This process may have to be repeated a few times until no two adjacent nonzero digits remain. The reader should verify that by using this method the 2's complement number 110101101101 is 100-10-100-10-101 in CSDC. Obviously anything said so far about the representation of integers applies equally to fractions by simply assuming the existence of a binary point in the appropriate location. This code has been used to obtain an efficient mechanization of multiplication in digital signal processors, and the interested reader may find a detailed description in [2].

We now proceed to discuss how arithmetic operations are performed on binary numbers. We will discuss only arithmetic operations on numbers represented in 2's complement notation, since this is the most commonly used notation and as we will see in the next section a variety of integrated circuits are readily available to perform these operations.

2's Complement Addition and Subtraction

The basic computation element to perform addition and subtraction in any binary arithmetic is the two bit full adder (FA). It accepts three binary inputs X, Y, and C_{in} that are the *addend* bit, the *augend* bit, and the *carry* in. Two output bits are produced, the result Z and the carry out C_{out}. The FA is illustrated in Figure 4.1*a* together with the truth table defining the outputs for all possible input combinations. To add two positive or unsigned B bit numbers $(x^0, x^1, ..., x^{B-1})$ and $(y^0, y^1, ..., y^{B-1})$, we may connect B FAs as shown in Figure 4.1*b*.

To add two numbers in 2's complement notation the addend is added to the augend (including the sign bit) and the carry in into the LSB is set to zero (C_0=0). To subtract two numbers in 2's complement notation the 1's complement of the *subtrahend* is first obtained by complementing each bit that is, setting every 0 to 1 and vice versa. The result is added to the *minuend* and the carry in into the LSB is set to one (C_0=1). Thus with the addition of some simple logic, the adder shown in Figure 4.1 will be capable of performing subtraction as well.

Clearly when adding or subtracting two arbitrary numbers the result may be larger than 1, which is an overflow and will cause an incorrect result. We illustrate this with a simple example; 0.5625 + 0.6875 = 1.25. When this computation is carried out in 2's complement notation we obtain

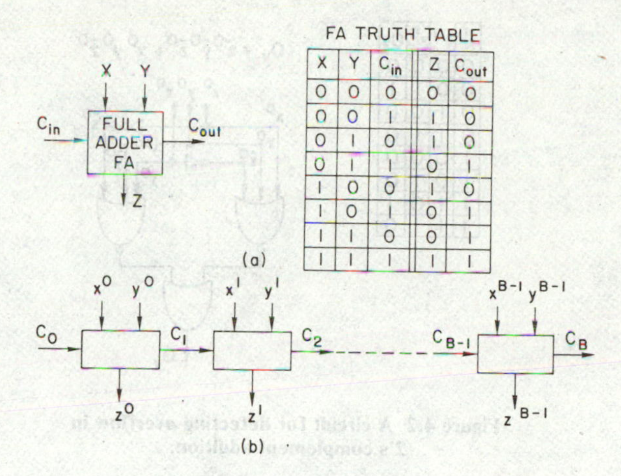

X	Y	C_{in}	Z	C_{out}
0	0	0	0	0
0	0	1	1	0
0	1	0	1	0
0	1	1	0	1
1	0	0	1	0
1	0	1	0	1
1	1	0	0	1
1	1	1	1	1

Figure 4.1 The full adder, (a) a single stage and its truth table, (b) a B bit adder made up by using B full adders.

$$010010 + 0100110 = 101000 \rightarrow -0.75$$

which of course is incorrect. Fortunately it is very easy to detect an overflow. We note in the example that although we added two positive numbers the result obtained is interpreted as negative, and it is easy to verify that the opposite is also true. Thus to detect an overflow in 2's complement arithmetic it is enough to monitor the sign bits of the addend and augend and compare them with the sign bit of the result. Figure 4.2 gives the truth table for detecting an overflow ($Q_v=1$ in case of overflow) and the logic circuit that implements it. In it x^0 and y^0 denote the addend and the augend sign bits and z^0 is the result sign bit. The strategy to handle an overflow is either to stop operation and alert the operator (programmer) that an overflow has occurred, or to force the result to the highest possible (negative or positive) value that corresponds to clipping and is an undesirable nonlinear effect. Another possibility is to make the register holding the result larger (that is, having an extra significant bit to the left) and upon detecting an overflow rescale the result and give an indication that a rescaling has occurred.

Let us now examine what happens when while adding several numbers an intermediate result is larger than 1, even though the final result is less than 1. As we see in 2's complement arithmetic an intermediate overflow does not affect the final result. We illustrate this through a simple example. Suppose we want to add the numbers $0.875 + 0.40625 + (-0.34375) = 0.9375$ in 2's complement notation the following occurs:

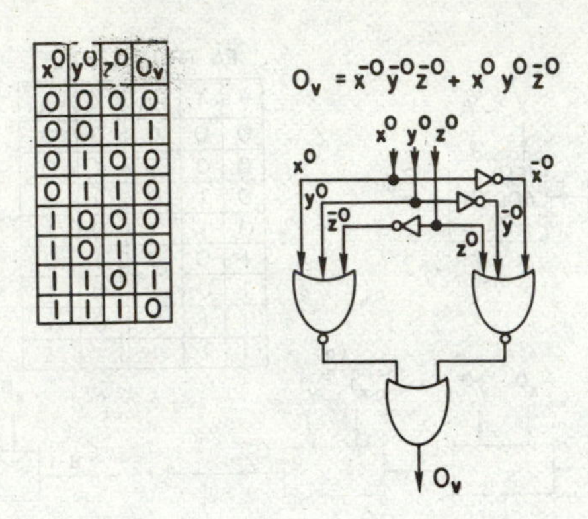

Figure 4.2 A circuit for detecting overflow in 2's complement addition.

0.875 →	011100		
+	+		
0.40625 →	001101		
1.28125	101001 →	−0.71875	incorrect intermediate result due to overflow
+	+		
−0.34375 →	110101		
0.9375	011110 →	0.9375	correct final result

Although an intermediate result has overflowed, since the final result is less than 1 in magnitude, it is still correct. It is this important property of the 2's complement notation that makes it very useful in handling arithmetic operations. Since the commutative property of addition is preserved, the addition can be performed in any order.

2's Complement Multiplication

Let us now consider the multiplication of two numbers x and y represented in a B bit 2's complement notation. If $z = x \times y$, then by using Eq. 4-2 we obtain:

$$z = (-x^0 + \sum_{j=1}^{B-1} x^j 2^{-j})\, y = -x^0 y + \sum_{j=1}^{B-1} (x^j y) 2^{-j} \qquad (4\text{-}8)$$

thus to multiply the two numbers we can proceed as follows. We first multiply y by the bits of x, starting with least significant bit x^{B-1}. Since x^j is either zero or 1 this means that the result is either zero or y, add y x^j to the previous result, the first time at bit x^{B-1} we simply add yx^{B-1} to 0 and shift the sum one bit to the right which corresponds to multiplying by 2^{-1}. (Note that an arithmetic shift of a 2's complement number requires shifting in the sign bit, for example $-0.5 \rightarrow 1.100$ and $-0.25 = -0.5 \times 2^{-1} \rightarrow 1.1100$.) Upon reaching x^0 a subtraction has to be performed instead of an addition and z is obtained. A conceptual block diagram of such a multiplier is shown in Figure 4.3. It follows, therefore, that multiplying two B bit numbers will require B^2 AND operations (multiplying a single bit by a single bit), B−1 additions of B bit numbers and B−1 shifts. The result z will be a 2B−1 bit number. We illustrate this procedure with a simple example in which $x = -0.4375 \rightarrow 110010$ and $y = -0.40625 \rightarrow 110011$. These steps are illustrated below.

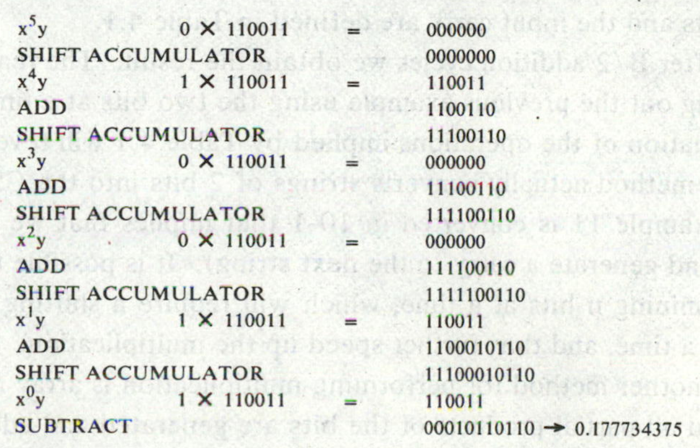

x^5y	0 × 110011	=	000000
SHIFT ACCUMULATOR			0000000
x^4y	1 × 110011	=	110011
ADD			1100110
SHIFT ACCUMULATOR			11100110
x^3y	0 × 110011	=	000000
ADD			11100110
SHIFT ACCUMULATOR			111100110
x^2y	0 × 110011	=	000000
ADD			111100110
SHIFT ACCUMULATOR			1111100110
x^1y	1 × 110011	=	110011
ADD			1100010110
SHIFT ACCUMULATOR			11100010110
x^0y	1 × 110011	=	110011
SUBTRACT			00010110110 → 0.177734375

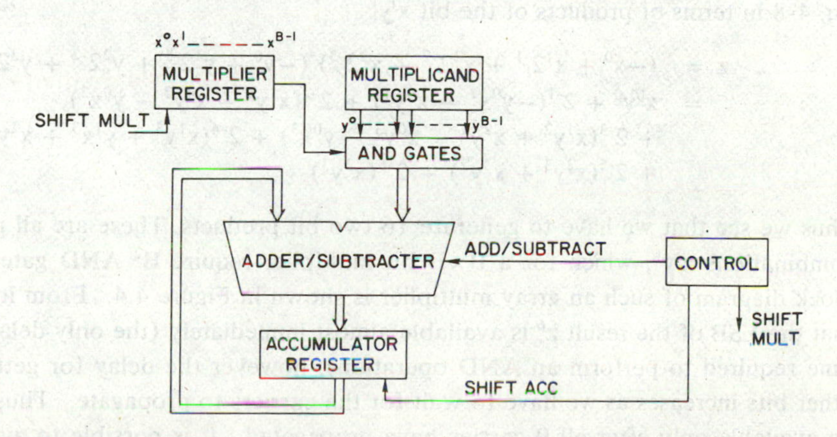

Figure 4.3 The block diagram of a serial multiplier.

A more careful examination of the procedure outlined will reveal that we could speed up the multiplication by skipping across zeros, that is any time $x^j=0$ we simply shift the accumulator and continue. Although this may be appealing, this results in a variable multiplication time, which leads to complications in the control circuitry.

Let us now consider another multiplication method, in which we examine two bits of the multiplier at a time. In this case we assume that the accumulator can be shifted 2 bits to the right at once, and that the register holding the multiplicand can be shifted left (multiplied by 2). Then if we define C_i as a carry into the pair (x^i, x^{i-1}), starting with the LSB pair (x^B, x^{B-1}) and $C_B=0$, and proceeding from right to left the operations performed for each possible combination of the two bits and the input carry are defined in Table 4.1.

After B/2 addition cycles we obtain the result. The reader may benefit from working out the previous example using the two bits at a time method. A careful examination of the operations implied by Table 4.1 will reveal that this two bit at a time method actually converts strings of 2 bits into the CSDC discussed earlier (for example 11 is converted in 10-1 that implies that we subtract the multiplicand and generate a carry to the next string). It is possible to extend this method to examining n bits at a time, which will require a shifting capability of up to n bits at a time, and thus further speed up the multiplication.

Another method for performing multiplication is array multiplication. In this method all partial products of the bits are generated and added up. We illustrate this by considering a simple example of a 4×4 bit multiplier. First we rewrite Eq. 4-8 in terms of products of the bit $x^j y^k$

$$\begin{aligned} z = & \ (-x^0 + x^1 2^{-1} + x^2 2^{-2} + x^3 2^{-3})(-y^0 + y^1 2^{-1} + y^2 2^{-2} + y^3 2^{-3}) \\ = & \ x^0 y^0 + 2^{-1}(-y^0 x^1 - x^0 y^1) + 2^{-2}(x^1 y^1 - x^0 y^2 - y^0 x^2) \\ & + 2^{-3}(x^1 y^2 + x^2 y^1 - x^0 y^3 - y^0 x^3) + 2^{-4}(x^1 y^3 + y^1 x^3 + x^2 y^2) \\ & + 2^{-5}(x^2 y^3 + x^3 y^2) + 2^{-6}(x^3 y^3) \end{aligned}$$

Thus we see that we have to generate 16 two bit products. These are all possible combinations $x^j y^k$, which for a B×B bit multiplier require B^2 AND gates. The block diagram of such an array multiplier is shown in Figure 4.4. From it we see that the LSB of the result z^6 is available almost immediately (the only delay is the time required to perform an AND operation), however the delay for getting the other bits increases as we have to wait for the carriers to propagate. Thus z^0 will be available only after all 9 carries have propagated. It is possible to avoid this delay by using carry look-ahead circuits, which further increases the amount of hardware required. This will be discussed in the next section.

C_i	x^i	x^{i-1}	C_{i-2}	Operation
0	0	0	0	NOP
0	0	1	1	SUBTRACT 2× MULTIPLICAND
0	1	0	0	ADD MULTIPLICAND
0	1	1	1	SUBTRACT MULTIPLICAND
1	0	0	0	ADD MULTIPLICAND
1	0	1	1	SUBTRACT MULTIPLICAND
1	1	0	0	ADD 2× MULTIPLICAND
1	1	1	1	NOP

Note: 1. After each operation the accumulator is shifted 2 bits to the right except for the following end condition; for an even number of bits in the multiplier after the operation for pair (x^0, x^1) the accumulator is shifted only one bit right, for an odd number of bits the last pair considered is $(0, x^0)$ and the accumulator is not shifted.
2. NOP designates no operation.

Table 4.1 Multiplication of 2's Complement Binary Numbers Two Bits at a Time.

Some remarks are in order before we end this section. The intent of this section is to acquaint the reader who is unfamiliar with the concepts discussed with a basic understanding of the principles of performing the binary arithmetic operations required in digital signal processing algorithms. The subjects discussed are mainly addition and multiplication techniques that will aid the reader in following the subsequent sections dealing with hardware implementation. For a more advanced and detailed treatment of this topic the interested reader is advised to consult [3].

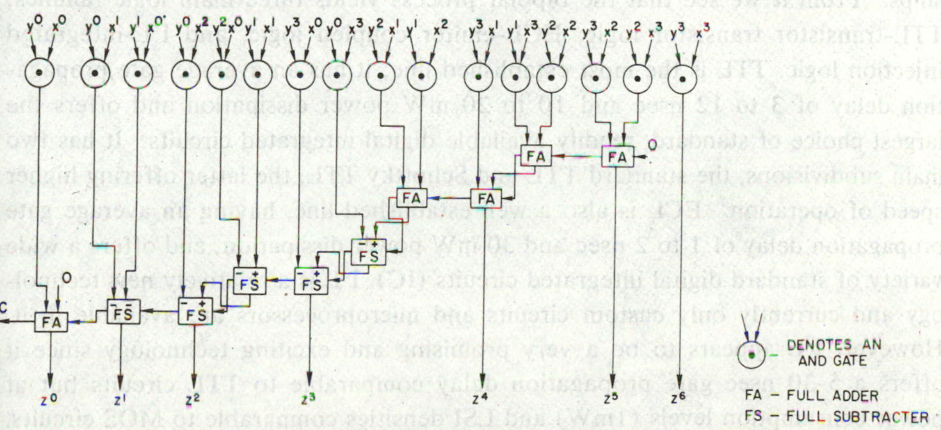

Figure 4.4 A 4 bit array multiplier.

4.3 AN OVERVIEW OF DIGITAL HARDWARE

This section is intended to provide the reader who has little prior exposure to digital hardware with an overview of the subject so as to facilitate the understanding of this and the next chapter dealing with the hardware implementation of digital signal processors. In the first part of this section we discuss the major semiconductor technologies and their corresponding logic families with an emphasis on a comparison of their relative merits, especially as related to their possible use in digital signal processors. In the second part of this section we describe the main functional units that are commercially available, and will be used in the ensuing sections as the basic building blocks to implement digital signal processors.

Many device technologies are used to manufacture digital integrated circuits and they yield a large number of logic families, each having its own merits and limitations. A closer look reveals that there are two main technologies, better known by their generic names, bipolar and MOS (metal oxide semiconductors), which yield devices with distinctly different characteristics. Generally speaking bipolar devices are faster, consume more power, and are less amenable to large scale integration (LSI), that is, the total number of devices per integrated circuit is less than their MOS counterparts. The names bipolar and MOS are derived from the physical structure of the transistor used in these technologies. Depending on the manner in which the transistors are interconnected to form gates, flip-flops, and so on, or the details of the process used to fabricate them a large number of logic families exist with bipolar and MOS devices.

Figure 4.5 lists the major logic families currently available and their relationships. From it we see that the bipolar process yields three main logic families: TTL-transistor transistor logic, ECL-emiter coupled logic, and I^2L-integrated injection logic. TTL is the most established line; it has an average gate propagation delay of 3 to 12 nsec and 10 to 20 mW power dissipation and offers the largest choice of standard, readily available digital integrated circuits. It has two main subdivisions, the standard TTL and Schottky TTL, the latter offering higher speed of operation. ECL is also a well established line, having an average gate propagation delay of 1 to 2 nsec and 30 mW power dissipation, and offers a wide variety of standard digital integrated circuits (IC). I^2L is a relatively new technology and currently only custom circuits and microprocessors are available in it. However, I^2L appears to be a very promising and exciting technology since it offers a 5-30 nsec gate propagation delay comparable to TTL circuits but at power consumption levels (1mW) and LSI densities comparable to MOS circuits. Therefore it will undoubtedly gain prominence in the near future.

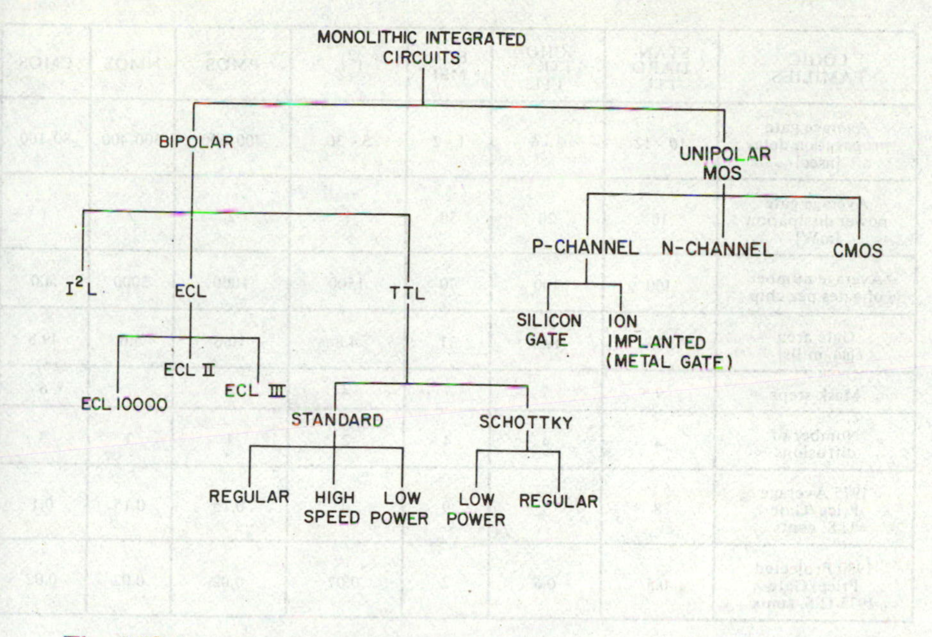

The MOS process, which requires less diffusion steps and is therefore simpler and cheaper, yields three main logic families; P-channel, N-channel and complementary MOS (CMOS). Generally MOS circuits have a gate propagation delay of 100 to 600 nsec and a power dissipation of 1 to 2 mW. Due to their low power dissipation more than a 1000 gates can be on one LSI IC. The largest choice of standard logic circuits is available in CMOS and the other two are used mainly for memories, microprocessors and custom integrated circuits.

Table 4.2 lists the main characteristics of the major logic families, and thus permits a comparison between them. The numbers in the table were compiled from references [4,5,6] and are only intended to serve as a general guideline for comparison purposes. In light of the rapid developments and advances in this area they may be somewhat inaccurate. From Table 4.2 it is quite obvious that I^2L holds great promise of significantly better overall performance in applications where its speed is adequate. Another important observation is that the choice of a particular technology can be a difficult task as the boundaries in performance are overlapping and a clear cut decision is not obvious.

The same basic technologies are used to manufacture semiconductor memories. Two main types of memories are currently used, read only memory (ROM) and random access memory (RAM). As in many instances the acronyms are confusing. The ROM is actually also a random access memory, that is,, we can read out consecutively any two locations, and the RAM acronym is used for

LOGIC FAMILIES	STANDARD TTL	SCHOTTKY TTL	ECL MSI	I^2L	PMOS	NMOS	CMOS
Average gate propagation delay [nsec]	10 - 12	3 - 6	1 - 2	5 - 30	400-600	300-400	80-100
Average gate power dissipation [mW]	10	20	30	1	2	2	1
Average number of gates per chip	100	100	70	1500	1000	2000	500
Gate area [sq. mills]	52.8	19.9	31	4.8	10.6	5.6	49.8
Mask steps	7	7	7	4	4	7	6
Number of diffusions	4	4	4	2	1	3	3
1975 Average Price/Gate U.S. cents	8	7	9	4	0.15	0.15	0.1
1980 Projected Price/Gate 1975 U.S. cents	0.5	0.6	2	0.07	0.02	0.02	0.02

Table 4.2 A comparison of the major logic families.

	Bipolar RAM	N-Channel RAM	CMOS RAM
Access Time (nsec)	30-60	150-250	50-60
Cycle Time (nsec)	50-100	350-450	500-600
Power Dissipation/bit (mw)	500	100	20
Density (bits/chip)	1024-4096	1024-8192	256-1024
1975 cost/bit U.S. cents	1.5	0.2	5
1980 projected 1975 U.S. cents	0.5	0.05	0.2

Table 4.3 Semiconductor Memory Technology.

read/write random access memories. Table 4.3 lists the main characteristics of currently available RAM's in bipolar, N-channel, and CMOS technologies. From it we see that choosing a memory technology is somewhat simpler, as the cycle time in the different technologies is complementary, rather than overlapping. These tables merely serve to give the reader an overview. The reader is well advised to check with different manufacturers to obtain the most up to date information before proceeding to design a processor.

We conclude this brief overview of the various technologies with an important remark. Unfortunately it is not straightforward to mix the various logic families. They usually require different power supplies, different clocks, and their output levels are incompatible. While interfaces between them are readily available, it is not obvious when it may be advantageous to mix different logic families in the same system, and designers usually tend to stay with one logic family throughout the system. The only exception is the use of MOS memory in systems where all the logic is TTL. The MOS memory, which is available with power supply and output levels compatible to TTL, is used as the main storage of the system due to its low price and high density. A smaller and faster bipolar memory (known as cache) is used as the working storage. The cache communicates directly with the high speed arithmetic unit, and receives data and transfers back results into the main storage.

We now discuss the main functional units that are commercially available, and can be used as building blocks to implement more complex systems. We will draw all our examples from TTL IC's, as they offer the largest choice of readily available IC's. In many cases the same functional units are available in ECL as well as in CMOS.

Gates

Individual gates, which are available usually 4 per IC, are used mostly in the control part of special processors, in cases in which the intended volume of production does not justify the expense of producing a custom IC containing the gates required for the control. Figure 4.6 depicts the main gates used, their logic symbol, and tables defining their output for all possible combinations of the inputs. Only two input gates are shown; however, gates with up to 8 inputs are readily available.

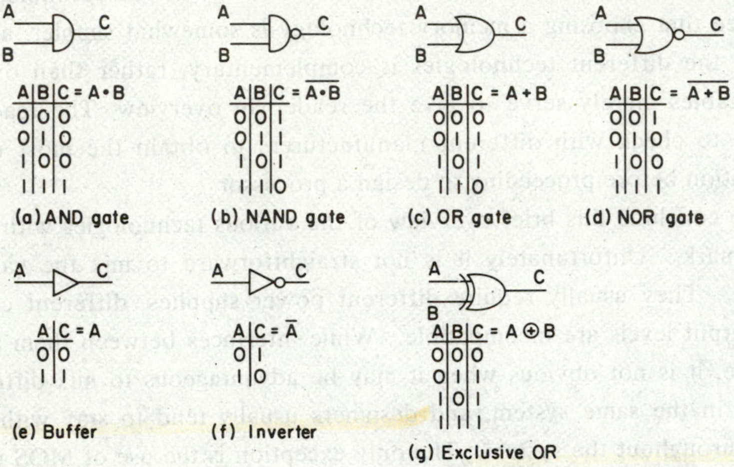

A	B	C = A•B
0	0	0
0	1	0
1	0	0
1	1	1

(a) AND gate

A	B	C = $\overline{A \cdot B}$
0	0	1
0	1	1
1	0	1
1	1	0

(b) NAND gate

A	B	C = A+B
0	0	0
0	1	1
1	0	1
1	1	1

(c) OR gate

A	B	C = $\overline{A+B}$
0	0	1
0	1	0
1	0	0
1	1	1

(d) NOR gate

A	C = A
0	0
1	1

(e) Buffer

A	C = $\overline{A}$
0	1
1	0

(f) Inverter

A	B	C = A ⊕ B
0	0	0
0	1	1
1	0	1
1	1	1

(g) Exclusive OR

Figure 4.6 Some of the more commonly used gates.

Counters

A counter is an element that counts the number of incoming pulses, and has an output from which this number can be read out in parallel. It is most commonly used in the control part of processors. The input to the counter is the system clock, and the counter counts the clock pulses modulo the number of clock pulses that consists the operating period. The counter outputs, indicating the number of clock pulses elapsed, are used as inputs to the control circuitry to determine the appropriate action that has to be taken in each part of the operating period. Figure 4.7 shows a typical 4 bit counter, packaged in a 16 pin IC. This is a synchronous counter, that is, all outputs change coincident with each other upon the change in state of the clock input when counting is enabled. It is possible to preset the counter to any number (between 0 and 15) and counting will start from this value. The table in Figure 4.7 illustrates a typical counting sequence. Using the carry output several such 4 bit counters can be cascaded to count modulo any power of 2. Decimal counters, up/down, and divide-by-n counters are also available.

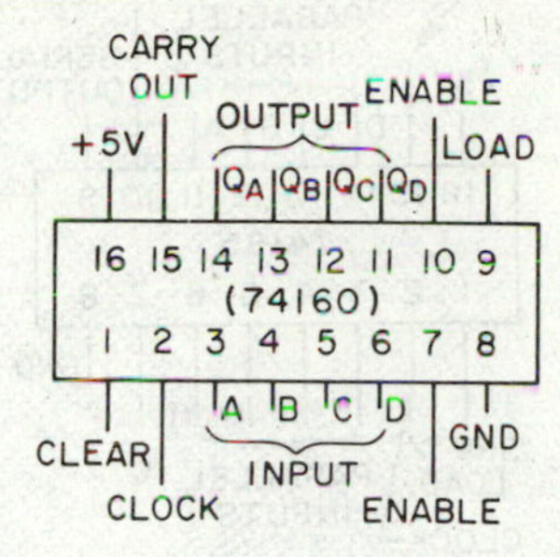

CLOCK	1	2	3	4	5	6	7	8	9	10	11	12	13	14	15	16
CLEAR	0	1	1	1	1	1	1	1	1	1	1	1	1	1	1	1
LOAD	1	0	1	1	1	1	1	1	1	1	1	1	1	1	1	1
ENABLE	0	0	1	1	1	1	1	1	1	1	1	1	1	1	0	0
Q_A	ϕ	1	0	1	0	1	0	1	0	1	0	1	0	1	1	1
Q_B	ϕ	0	1	1	0	0	1	1	0	0	1	1	0	0	0	0
Q_C	ϕ	1	1	1	0	0	0	0	1	1	1	1	0	0	0	0
Q_D	ϕ	0	0	0	1	1	1	1	1	1	1	1	0	0	0	0
COUNT	ϕ	5	6	7	8	9	10	11	12	13	14	15	0	1	1	1
CARRY OUT	0	0	0	0	0	0	0	0	0	0	0	0	1	0	0	0

Typical counting sequence with input preset to: A=1, B=0, C=1, D=0 (5).

Figure 4.7 A 4 bit counter.

Registers

A register is a device used to store and manipulate a binary word. The registers commonly available in TTL are of 4 or 8 bits. The register type mostly used is the shift register. A shift register allows shifting the binary word left or right one position at each clock transition. Depending on the manner in which data can be loaded into and taken out of the shift register there are: parallel in parallel out, serial in serial out, serial in parallel out, and parallel in serial out shift registers. Figure 4.8 shows a typical 8 bit parallel in serial out shift register and its associated controls. Other registers used are buffer registers and register files or stacks, which simply serve as intermediate storage devices to hold operands, instructions or results.

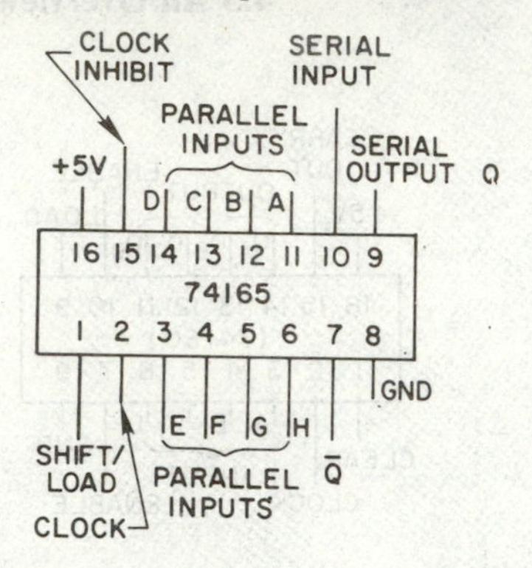

Note:
1. Maximum shifting rate 25MHz.
2. Parallel loading is accomplished by setting the SHIFT/LOAD control to 0.
3. When CLOCK INHIBIT is 1, shifting is inhibited.
4. Upon shifting, output H is the first to appear at the serial output Q.

Figure 4.8 An 8 bit shift register and its associated controls.

Multiplexers

Multiplexers are devices that enable us to route data, instructions, or any other binary word to any one of a number of destinations. They are the exact equivalent of a multiposition mechanical switch. A variety of standard multiplexers are available and can be used to implement any routing scheme. The more common are: 16 line to 1 line, 8 line to 1 line, 4 line to 1 line, 2 line to 1 line and 4 line to 16 line. Figure 4.9 depicts a typical dual 4 line to 1 line multiplexer, with a strobed output. The strobing capability can be used to cascade such multiplexers to implement an N line to 1 line multiplexer.

Arithmetic Logic Units

An arithmetic logic unit (ALU) is a device that accepts two operands and performs any one of 16 arithmetic or logic operations on them (a total of 32 operations). Currently available ALU's in TTL are 4 bit devices, that is, they accept two 4 bit operands and an input carry, and yield a 4 bit result and output carry. Thus such devices can be cascaded to yield ALU's for operands of arbitrary length (if the length is not a multiple of 4 the last ALU will not be fully utilized). A complete list of all possible operations that such an ALU can perform can be found in [7] and will not be given here. Some of the more common operations for which we will use the ALU's are listed in Figure 4.10 together with a typical ALU and its associated controls. Thus we see that we can use ALU's to add or subtract binary numbers, by simply cascading several ALU's and connecting their

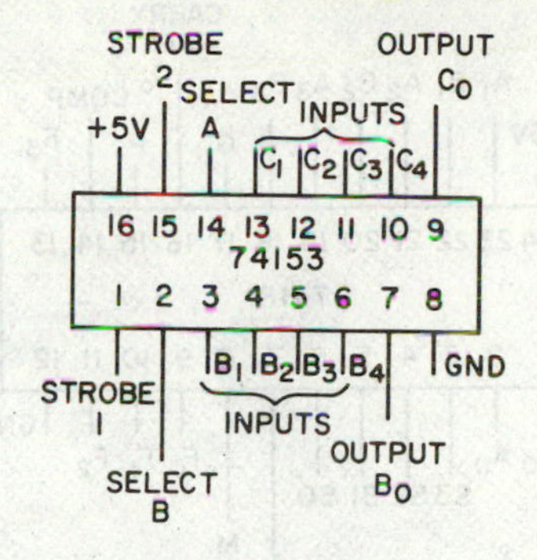

SELECT		OUTPUT	
A	B	C_O	B_O
0	0	C_1	B_1
0	1	C_3	B_3
1	0	C_2	B_2
1	1	C_4	B_4

STROBE 1, 2 = HIGH (1)

Figure 4.9 A dual 4 line to 1 line multiplexer.

carry inputs and outputs. To obtain the result of an addition or subtraction we will have to wait for all carries to propagate. However it is also possible to use look-ahead carry generators together with the ALU's to obtain very fast addition. Table 4.4 gives the typical addition time for two binary numbers for several word-lengths using standard TTL or Schottky TTL ALU's and look-ahead carry generators (for example 74182). Thus we see that using 5 IC's it is possible to add two 16 bit binary numbers in 19 nsec. Several more recent types of ALU's have latching outputs, and others have an additional on chip register to permit accumulation of results.

Multiplier ICs

Multiplier ICs are available from several manufacturers. Usually one IC is a 4 × 4 or 4 × 2 bit multiplier, and any m × n bit multiplier can be made using these ICs to generate partial results in combination with an adder tree (made out of ALUs for example) to add them up in a suitable fashion [7]. We discuss here briefly an IC manufactured by Advanced Micro Devices, Am 2505 which is a 4 × 2 bit 2's complement multiplier. It uses the two bit at a time multiplication algorithm described in Section 4.2 for multiplying directly two 2's complement binary numbers. Furthermore this multiplier actually performs a multiply/add operation of the type $r=x \times y + p$ which is very convenient for computing a sum

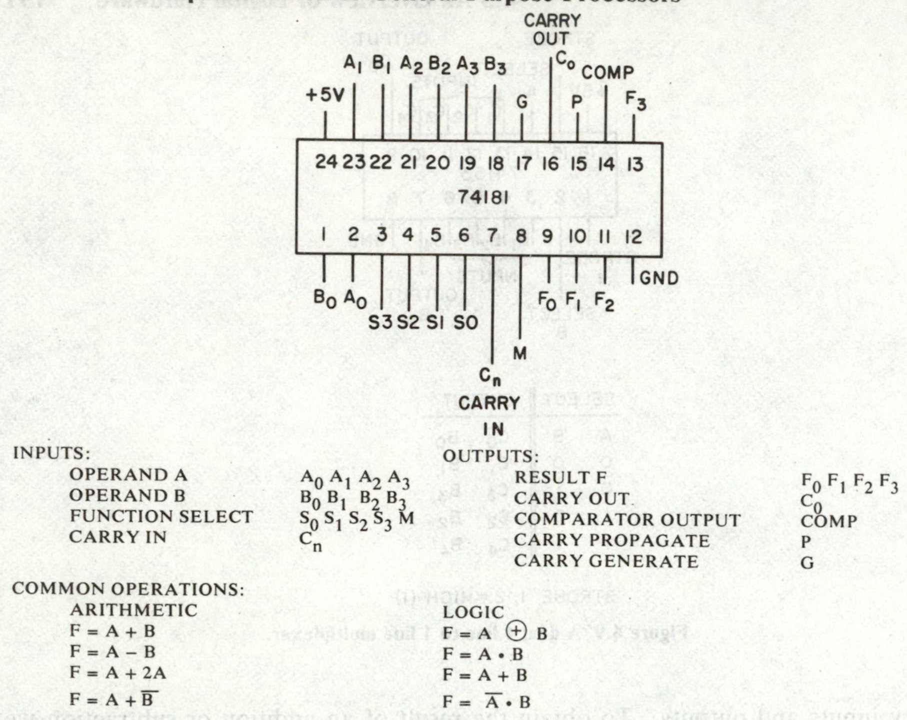

Figure 4.10 A typical arithmetic logic unit (ALU).

of products. These characteristics make this multiplier especially attractive for digital signal processors. The application note [8] for this IC describes several configurations for implementing a complete 2's complement multiplier using Am 2505 ICs and Am 9340 ICs (ALUs). Table 4.5 lists the multiplication time and package count (number of ICs required) to implement a complete multiplier. It should be noted that additional hardware is required to generate the controls necessary to make the multiplier work correctly, and this is not accounted for in Table 4.5.

Bipolar Memory ICs

Bipolar memories are currently available with up to 4096 bits per IC. They are organized as N × B bits, where N is the number of locations addressable in the memory and B is the number of bits stored at each location. The more common organizations are 16×1, 256×1, 1024×1, 16×4, for RAM and 32×8, 256×4, 256×8, 512×4 and 512×8 for ROM. ROM is available in two versions. The first is programmed (the appropriate sequence of 1's and 0's are inserted in the

word length(bits)	TTL	Schottky TTL	ALU's	Carry Look-ahead
1-4	24	11	1	0
5-8	36	18	2	0
9-16	36	19	3-4	1
17-64	60	28	5-16	2-5

Table 4.4 Typical addition time in nsec of two binary numbers and the number of IC's required. (After Texas Instruments TTL Data Book.)

Size n x n Bit	Typical Multiplication Time[nsec]	Number of ICs 2n Bit Full Product	Number of ICs n Bit Truncated Product
8 x 8	135	8	6
12 x 12	205	18	13
16 x 16	275	32	22
16 x 16	180	48	34

Table 4.5 Typical multiplication time and hardware requirements of 2's complement multiplier using Am2505 ICs. (After Advanced Micro Devices Data Book.)

required memory locations) at the factory by generating a suitable mask and using it in the fabrication process. The other version of ROM is field programmable, known as PROM. These can be programmed, using a relatively simple instrument, after the IC has been fabricated. Using PROM is very advantageous in prototype development since it combines the advantages of ROM and still avoids the relatively expensive mask charge that is incurred when ordering a ROM. Figure 4.11 shows a typical RAM IC and its associated controls. The main parameters that characterize a memory performance are access time, writing time, and cycle time. Access time refers to the delay between the time the address is present at the memory input and the time the word contained at this address is present at the memory output. The writing time is the time elapsed between the write enable command is given until usable data is in memory. The cycle time is the minimum time required for the memory to settle after a read/write operation, before another read/write can be initiated. Currently, memories are available that are read while write, and these are especially suitable for scratchpad applications.

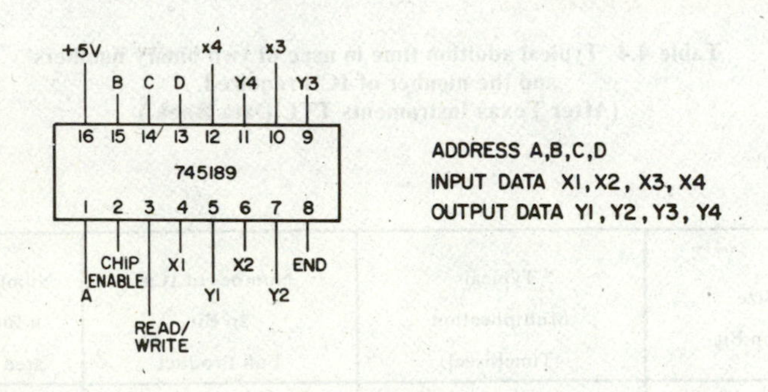

Figure 4.11 A typical 64 bit (16×4) RAM.

A very important property of TTL devices as well as of some CMOS devices is the ability to set the output to a third state in addition to the usual HIGH and LOW. The ICs that have this property are referred to as three state output ICs. The third state is a one in which a high impedance is present at the output, thus not loading other circuits connected to the same wire. This permits connecting many outputs in a wired OR fashion, in bus organized systems. Obviously this property is very convenient when many memory ICs are combined to form a large size memory. The control CHIP ENABLE in Figure 4.11 determines whether the memory outputs are enabled or are in the high impedance state, in which case the output data of a different IC is used.

This section provides only a very brief treatment of this subject. Furthermore, it is highly biased toward providing the bare essentials needed to understand the subsequent chapters. Therefore readers unfamiliar with the topics discussed are urged to consult the references mentioned as well as references [9, 10, 11] for additional details.

4.4 ARCHITECTURAL CONSIDERATIONS FOR DIGITAL SIGNAL PROCESSORS

In this section we discuss some of the more important factors that play a significant role in choosing an architectural structure for digital signal processors. Obviously all the signal processing algorithms we have encountered in the previous chapters can be implemented on a general purpose computer, however, the speed and cost of such *number crunching* applications when implemented on general purpose computers are not particularly attractive. A speed-up of up to a 100 times can be achieved by going to specialized processors, and this also leads to a lower cost per computation.

The motivation to look for specialized structures for digital signal processors lies in the fact that the algorithms such processors are called upon to perform repeatedly are well defined, and have many characteristics that can be exploited to achieve more efficient execution. Two prominent features of signal processing algorithms are:

1. The most common arithmetic operation to be performed is of the type

$$r = \sum_{i=1}^{N} a_i x_i \tag{4-9}$$

   where x_i are data samples, or intermediate results, and a_i are the processing coefficients.

2. The number of input/output (I/O) operations is relatively small compared to the number of arithmetic operations.

These features imply that the main burden falls on the arithmetic unit (AU), and therefore it is essential to design an AU that executes operations of the type of sum of products (Eq. 4-9) efficiently. Another important consideration is the bandwidth of the data transfer (the number of words that can be transferred per second) between memory and the AU. The memory-AU bandwidth and the AU execution rate have to be balanced, otherwise one will be idle waiting for the other to act. We will return to this important point later on in our discussion.

A computer consists of a variety of hardware elements, for example adders, multipliers, memory, control logic, and so on. The architecture of a computer specifies the way in which these elements are interconnected, and defines the communication paths, and the communication protocol for interaction between its various hardware entities. Figure 4.12 depicts the block diagram of a simple computer. From it we see a computer's main elements: the memory that contains the program as well as the data, the control unit, and the arithmetic unit. These three units are communicating through the bidirectional memory bus. The control unit determines which path is active at any given time and transfers information between the units. The program consists of a sequence of instructions residing in memory in contiguous locations. The instruction format is, as seen from Figure 4.12, very simple. The first field contains the operation code (OP) to be performed, and the second field contains the address of the operand. All arithmetic operations are performed between the memory and accumulator, and their result stored in the accumulator. Table 4.6 lists some of the more common OPs such a primitive computer would normally have. Obviously this is a very limited instruction set, and only very simple programs could be written using it. It is only intended to illustrate the basic concepts involved in the operation of a computer. Table 4.7 lists a simple program using these instructions. It computes the absolute value of the difference between two numbers, x and y, which are stored in memory at locations 100 and 101, and puts the result z in location 102.

A more careful examination of Figure 4.12 reveals that there is no provision in this computer for reading in, or writing out, information to the memory from the *outside world*. This is accomplished by introducing one or more registers, called I/O ports, which can be loaded or read from the outside, and adding the commands Store I/O and Load I/O, which transfer data between the memory and I/O port.

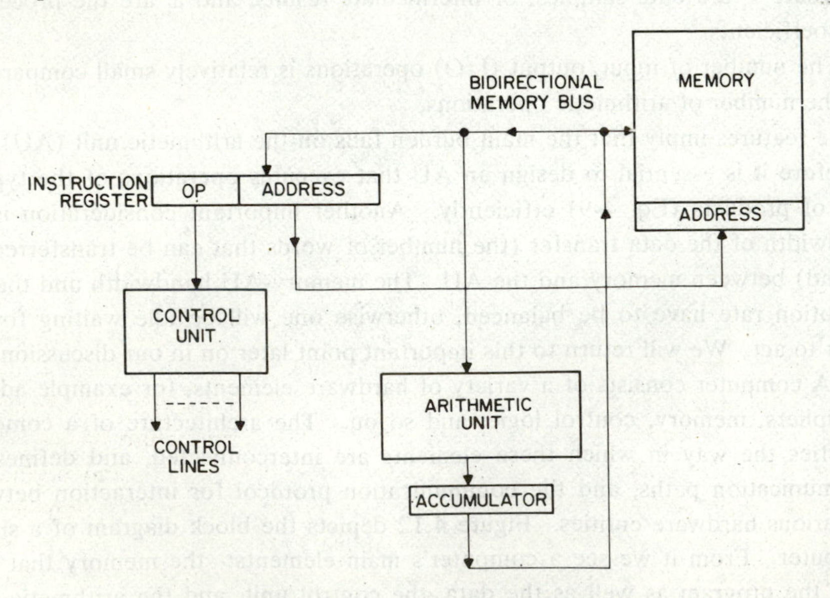

Figure 4.12 The block diagram of a simple computer.

Operation	Mnemonic OP Code	Function
1. Load Accumulator	LAC	ACC ← c(A)
2. Load Immediate Acc.	LIAC	ACC ← A
3. Store Accumulator	STAC	c(A) ← ACC
4. Add Accumulator	ADAC	ACC ← ACC + c(A)
5. Subtract Accumulator	SUAC	ACC ← ACC - c(A)
6. Shift Right Acc.	SRAC	ACC ← ACC × 2^{-1}
7. Jump	JMP	PC ← A
8. Branch	BRPA	PC ← A if ACC>0
9. No Operation	NOP	Idle
10. Halt	HLT	Stop

Note: 1. A denotes the numerical value in the address portion of the instruction.
2. c(A) denotes contents of memory at location A.
3. PC is the value of the program counter which contains the address of the instruction currently executed.

Table 4.6 A Very Simple Instruction Set.

000	LAC	100	ACC ← x
001	SUAC	101	ACC ← ACC - y
002	BRPA	005	Goto 005 if ACC>0
003	LAC	101	ACC ← y
004	SUAC	100	ACC ← ACC - x
005	STAC	102	c(102) ← ACC
006	HLT	000	

Table 4.7 A Simple Program for Computing z= | x−y |.

Let us now consider in some detail the time required to execute an instruction. For simplicity we assume that each basic function, for example memory access, an arithmetic operation, decoding an instruction and setting up the appropriate data path, and so on, require the same time, say 200 nsec. This time is usually referred to as the machine cycle time. The execution of most instructions takes 4 cycles. In the first the instruction is fetched, in the second the OP is decoded, in the third the operand is brought from memory to the arithmetic unit, and in the fourth the arithmetic operation is performed. Thus it will take about 4 msec on our hypothetical machine to do z= | x−y |. This would obviously be irrelevant if we do this operation only occasionally, or if speed is of no importance to our application. On the other hand if the application we consider involves doing this operation most of the time, choosing such a computer for this application would not be a wise choice. Furthermore, if we need to do this operation more than 250,000 times a second, we could not use it and we would have to consider choosing another computer or designing a special computer that does this operation in one instruction.

The example that we discussed above is obviously an oversimplification, and in real life things are never that clear cut. However, it was intended to make a point. The match between the architecture of the computer we choose and the structure of the algorithms that we wish to implement on this computer will determine the efficiency with which the algorithm is executed.

Thus the first and very important lesson to be learned from the above and applied to our problem emerges clearly. When designing a special purpose computer for digital signal processing we have to choose an architecture that best matches digital signal processing algorithms (for example forming the sum of products, Eq. 4.9). This implies choosing an instruction set (OPs) tailored to the types of operations required in programming these algorithms on the machine. It also implies choosing an interconnection scheme for the various hardware entities that will allow efficient execution of the instruction set we chose, in a minimal number of machine cycles. Furthermore, since speed and even more so data throughput are of importance, the architecture should be amenable to the introduction of parallelism in the various execution phases of the instructions, which will greatly enhance the overall performance.

In the next section we present some possible machine organizations for digital signal processors that satisfy the design considerations discussed here.

4.5 ARCHITECTURES FOR DIGITAL SIGNAL PROCESSORS

In this section we present some possible architectures for digital signal processors. These are intended to illustrate the various techniques that can be used in designing digital signal processors to achieve various performance levels. In actual applications various subsets of these techniques may be used, depending on the cost-performance goals the designer is aiming at.

In the previous section we saw a block diagram of a very primitive computer and its instruction set. We start our discussion in this section by introducing a computer architecture that is both more realistic and resembling modern computers, as well as particularly well suited to digital signal processing. It is depicted in Figure 4.13. The differences between it and the one depicted in Figure 4.12 are

Operation	Mnemonic OP Code	Machine Code	Function
Memory Reference Instructions			
1. Load Accumulator	LAC	000001	ACC ← c(EA)
2. Load X Register	LRX	000010	RX ← c(EA)
3. Load L Register	LRL	000011	RL ← c(EA)
4. Load I/O Register	LIO	000100	IO ← c(EA)
5. Load Immediate Acc.	LIAC	000101	ACC ← EA
6. Load Immediate L Reg.	LIRL	000110	RL ← EA
7. Load Immediate Index Reg.1	LIX1	000111	IX1 ← EA
8. Load Immediate Index Reg.2	LIX2	001000	IX2 ← EA
9. Store Accumulator	STAC	001001	c(EA) ← ACC
10. Store L Register	STRL	001010	c(EA) ← RL
11. Store I/O Register	STIO	001011	c(EA) ← IO
12. Store Index Reg.1	STIX1	001100	c(EA) ← IX1
13. Store Index Reg.2	STIX2	001101	c(EA) ← IX2
Arithmetic Instructions			
1. Add Accumulator	ADAC	001110	ACC ← ACC + c(EA)
2. Subtract Accumulator	SUAC	001111	ACC ← ACC - c(EA)
3. Multiply and Accumulate	MAC	010000	ACC ← RX × c(EA)+ACC
4. Multiply	MPY	010001	ACC ← RX × c(EA)
5. Shift Left Acc.	SLAC	010010	ACC ← ACC × 2
6. Shift Right Acc.	SRAC	010011	ACC ← ACC × 2^{-1}
7. Shift Left X Reg.	SLRX	010100	RX ← RX × 2
8. Shift Right X Reg.	SRRX	010101	RX ← RX × 2^{-1}
Branch Instructions			
1. Jump Unconditionally	JMP	010110	NIR ← EA
2. Branch on Zero in L Reg.	BZRL	010111	NIR ← EA if RL≠0
4. Branch on Zero in Acc.	BZAC	011000	NIR ← EA if ACC=0
5. Branch on Positive in Acc.	BRPA	011001	NIR ← EA if ACC≥0
Logical Instructions			
1. AND Accumulator	ANDA	011010	ACC ← ACC ∩ c(EA)
2. OR Accumulator	ORAC	011011	ACC ← ACC ∪ c(EA)
3. Exclusive Or Accumulator	EXAC	011100	ACC ← ACC ⊕ c(EA)
Miscellaneous Instructions			
1. Increment L Register	IRL	011101	RL ← RL + EA
2. Increment Index Reg.1	IIX1	011110	IX1 ← IX1 + EA
3. Increment Index Reg.2	IIX2	011111	IX2 ← IX2 + EA
4. Compare Index Reg.2	CIX2	100000	CP = 1 if IX2 > c(EA)
5. Conditional Subtract IX2	CSX2	100001	IX2 ← IX2 - c(EA) if CP=1
6. No Operation	NOP	100010	Do nothing
7. Halt	HLT	100011	Stop

Note: c(EA) denotes the contents of the data memory at the effective address.

Table 4.8 The Instruction Set of DISP.

immediately noticeable. In this processor there are separate memories for instructions and data, and there is an I/O Register that allows us to communicate between the data memory and the outside world. Figure 4.13 shows in some detail the structure of the control and arithmetic units. We shall refer henceforth to the computer depicted in Figure 4.13 as the DISP computer.

The separation between instruction memory and data memory that characterizes DISP is the main feature that distinguishes it from a general purpose processor. This separation, that requires a certain duplication in hardware (for example there is an adder in the AU and one in the address computation module in the control unit), is rarely found in general purpose computers. The reason is that in addition to requiring more hardware, it may actually offer limited flexibility in dynamically evolving an application program, due to the limited possibility of modifying instructions and program addresses during execution. However, in digital signal processing applications of the type considered in this Chapter, this is no serious limitation, and as we shall see, the separation between instruction memory and data memory will allow a significant gain in throughput. Furthermore, in this case the data word length and instruction word-length are independent, which is desirable in digital signal processing, since data words can be usually much shorter than instruction words.

The DISP instruction format is: a 6 bit field for the OP code, a 2 bit field for designating an index register, and a 12 bit field containing the address of the operand, or of the branch in case of a branch operation. This is depicted in Figure 4.14. Each instruction is assumed to be indexed, that is the effective address (EA) of the operand is obtained by adding the address field in the instruction with the index register indicated by the index field. Setting the index field to 00 has the effect of setting EA equal to the address field in the instruction. As we see from Figure 4.14 the word-length of each instruction is 20 bits. This allows for 64 OP codes, 3 index registers, and addressability of a 2K (2048) word data memory.

The control unit (CU) of DISP, as we see from Figure 4.13, has an instruction register (IR) that holds the instruction to be executed, and a next instruction register (NIR) that contains the address from which the next instruction should be fetched. This address is determined by the program sequencer module. In regular operation the instructions are simply fetched sequentially, and only in the case of branch instructions the address of the next instruction is obtained from the address computation module. Also in the CU are the OP decode module which decodes the machine code of the OP and sets up, through the control lines, the appropriate communication paths and enables the units affected by this operation. The address computation module in the CU computes the effective

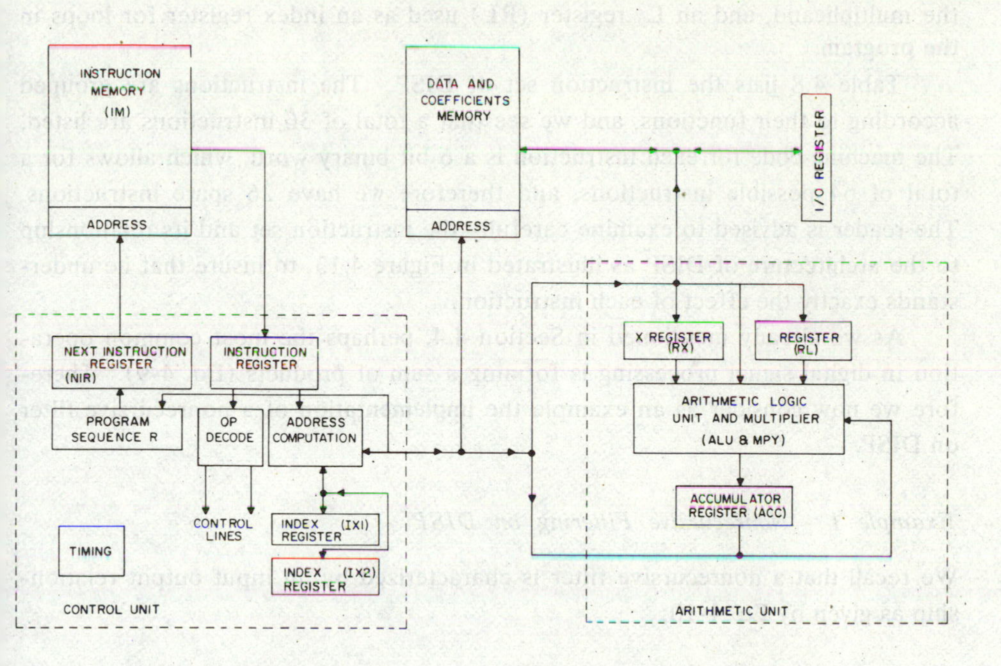

Figure 4.13 The block diagram of DISP.

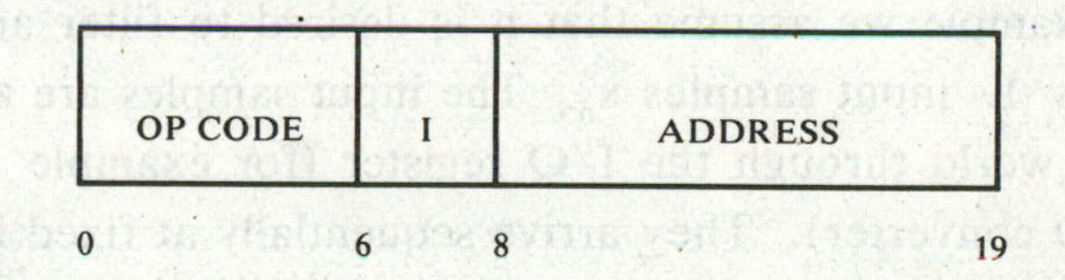

Figure 4.14 DISP Instruction Format.

address (EA) of the operand or branch operation. This is done by adding the contents of the address field in the instruction with the value of one of the two index registers (IX1 and IX2), or with zero, depending on the instruction. As we will see this indirect addressing capability will permit easy manipulation of addresses and will greatly facilitate the programming of our algorithms.

The arithmetic unit (AU), as we see from Figure 4.13, contains an arithmetic module, that consists of an arithmetic logic unit and a multiplier. The AU also has, in addition to the accumulator register (ACC), an X register (RX) to hold the multiplicand, and an L register (RL) used as an index register for loops in the program.

Table 4.8 lists the instruction set of DISP. The instructions are grouped according to their functions, and we see that a total of 36 instructions are listed. The machine code for each instruction is a 6 bit binary word, which allows for a total of 64 possible instructions, and therefore we have 26 spare instructions. The reader is advised to examine carefully the instruction set and its relationship to the architecture of DISP as illustrated in Figure 4.13, to insure that he understands exactly the effect of each instruction.

As we already mentioned in Section 4.4, perhaps the most common operation in digital signal processing is forming a sum of products (Eq. 4-9). Therefore we now consider as an example the implementation of a nonrecursive filter on DISP.

Example 1 - Nonrecursive Filtering on DISP

We recall that a nonrecursive filter is characterized by an input output relationship as given by Eq. 4-10;

$$y_n = \sum_{k=0}^{N} h_k x_{n-k} \tag{4-10}$$

where x_n are the input samples, y_n the output samples, and h_j the filter coefficients. In this example we assume that it is desired to filter an input block of data that contains L input samples x_n. The input samples are assumed to come from the outside world through the I/O register (for example IO may be connected to an A/D converter). They arrive sequentially at fixed intervals, with x_0 leading. In the data memory we will only keep the N most recent input samples (x_n to x_{n-N+1}) that are needed to compute the current output. The output samples are transferred to the *outside world* through the I/O register (for example to a D/A converter) as they are computed. This arrangement is characteristic of real time processors and constitutes a major difference between the way the same task would be handled by a general purpose computer. The later would first read in the whole data block of input points, and then compute L output samples. In our case the computation proceeds concurrently with the arrival of the input data. Furthermore since L may be quite large, say several thousands, and N is usually less than 100, proceeding in the fashion outlined above leads to significant memory savings.

Table 4.9 contains the listing of the DISP program that will perform the nonrecursive filtering as outlined above. The reader is strongly urged to go through the program carefully, executing each command manually to understand fully its operation. While doing this he will realize the importance of the index registers and appreciate the flexibility they provide in manipulating the addresses, so as to make the memory look like a shift register. Table 4.10 shows the changes in memory contents for this example, in the simple case of N−4, as the computation progresses. As we see x_n are maintained in a circular list. To compute the correct address, instructions CIX2 and CSX2 effectively allow us to increment the index register modulo an arbitrary number. This is again an example of the matching required between the architecture and the algorithms.

We see from Table 4.9 that a total of 25 instructions are required for this program. To evaluate the execution time of the program we consider a concrete example, where L=1000, and N=20. Therefore instructions 000 to 004 and 024 are executed once, instructions 005 to 007 and 017 to 023 are executed a 1000 times, and instructions 008 to 016 are executed 20×1000 times each, and hence a total of 190,006 instructions are executed, or an average of about 190 per output. The actual execution time depends on the time required to execute one instruction (if all instructions take the same time).

A total of 190 instructions per output sample seems a large number, considering that all we do is multiply 20 numbers and add them up. A closer examination of the program shows that indeed we only do an average of 40 instructions to do the actual arithmetic, and the other 150 instructions are required to take care of the indexing for looping and memory addressing.

At this point we notice that we could significantly improve the running time of this program if we had some additional instructions in the DISP repertoire. As an example, suppose we add the instruction LRXL whose effect is to load the RX register from memory and decrement the L register by −1. This can obviously be done at the same time, since while we load RX from memory the ALU or the AU can decrement the L register. In this case we could replace instruction 008 by LRXL 1200 and eliminate instruction 010 IRL 0 −1. This will reduce the average number of instructions per output to 170, that is, a more than 10% reduction. We note however that this instruction (LRXL) is somewhat restrictive, in that it does not allow us to specify the size of the increment (it is taken as −1), which makes this program less flexible.

The procedure through which we went on this example, trying to improve the running time of our program by slight perturbations in the DISP architecture, is quite common and is known as tuning the architecture.

```
        * NONRECURSIVE DIGITAL FILTER PROGRAM *
        * THE NUMBER OF TAPS, N, IS IN LOCATION 99 *
        * THE FILTER COEFFICIENTS ARE IN LOCATIONS 200 - 200+N-1 *
        * THE DATA POINTS ARE IN LOCATIONS 100 - 100+N-1 *
        * EACH NEW DATA POINT REPLACES THE N-1 DATA POINT (FIFO) *

000  LIX2   0  100   IX2 ← 100                 Initialize pointers
001  STIX2  0  098   c(98) ← 100               to the top and bottom
002  LAC    0  098   ACC ← c(98)               of the input data list.
003  ADAC   0  099   ACC ← ACC + c(99)
004  STAC   0  098   c(98) ← ACC

005  LRL    0  099   RL ← c(99)                Initialize the loop
006  LIX1   0  000   IX1 ← 0                   and index registers
007  LIAC   0  000   ACC ← 0                   and accumulator.

008  LRX    1  200   RX ← c(200+IX1)           Compute an output
009  MAC    2  000   ACC ← RX × c(IX2)+ACC     and advance pointers
010  IRL    0  -01   RL ← RL - 1               modulo the number
011  JMPL   0  017   Goto 017 if RL=0          of taps.
012  IIX1   0  001   IX1 ← IX1 + 1
013  IIX2   0  001   IX2 ← IX2 + 1
014  CIX2   0  098   CP=1 if IX2>c(98)
015  CSX2   0  099   IX2 ← IX2 - c(99) if CP=1
016  JMP    0  008   Goto 008

017  STAC   0  097   c(97) ← ACC               Store output
018  STIO   2  000   c(IX2) ← IO               sample, load
019  LIO    0  097   IO ← c(97)                new input sample
020  LIAC   0  -01   ACC ← -1                  and restart if
021  ADAC   0  095   ACC ← c(95) - 1           limit is not reached.
022  STAC   0  095   c(95) ← ACC
023  BRPA   0  005   Goto 005 if ACC>0
024  HLT
```

Table 4.9 A Nonrecursive Filter Program on DISP.

We now turn our attention to the timing of the various execution phases of the DISP instruction. This will determine the computational throughput of DISP. As we shall see some of the hardware parallelism in the DISP architecture will allow as to increase its throughput rate.

Let us now consider, in some detail, what exactly occurs during the execution of a DISP instruction. First, the instruction has to be fetched from the IS and stored in the IR. This phase is called instruction fetch (I-FETCH) and the only units involved in its execution are the IS and the IR in the CU. The time required to complete it is determined by the IS access time. The second phase is the one in which the instruction fetched is decoded and the effective address of the operand is computed. This is the OP-DECODE phase, and is performed entirely within the CU. The third phase of the instruction execution sequence is

ADDRESS	DATA MEMORY CONTENTS				
95	5	4	3	2	1
96					
97		y_3	y_4	y_5	y_6
98	104	104	104	104	104
99	4	4	4	4	4
100	x_3	x_3	x_3	x_3	x_7
101	x_2	x_2	x_2	x_6	x_7
102	x_1	x_1	x_5	x_5	x_5
103	x_0	x_4	x_4	x_4	x_4
•					
•					
•					
200	a_0	a_0	a_0	a_0	a_0
201	a_1	a_1	a_1	a_1	a_1
202	a_2	a_2	a_2	a_2	a_2
203	a_3	a_3	a_3	a_3	a_3
Loop	0	1	2	3	4

Table 4.10 Data Memory Contents During the Execution of the Nonrecursive Filter Program, with N = 4.

different for different instruction types. If the instruction is of a type that does not require fetching an operand from memory (for example branch, register increment, etc.) only some very simple control may be set up in this phase. If the instruction involves an operand (for example ADAC, MPY, etc.) an operand is fetched from data memory and loaded into the AU. This is therefore the OP-FETCH phase, and involves the data phase in the execution in performing the actual operation required and putting the result in the target location. Since our instruction set does not include any operations whose target is a memory location (except store) the EXEC phase is also the last phase of the execution cycle, and we are ready to proceed to the next instruction.

The time required to complete each of the four execution phases described above depends on many factors, and will be determined by worst case considerations, that is, by the time required to complete the most complicated task that may come up in the phase. Let us consider a concrete example. The I-FETCH involves only a memory access, and can therefore be completed in 50 to 200 nsec for bipolar memory, depending on the IS size. The OP-DECODE involves decoding a 6 bit word and doing an addition in the address computation as well as setting up a path from the appropriate index register. A quite effective and flexible way of doing the decoding of the OP code is to use a 64 word PROM, in which each word contains the appropriate control lines. This alternative may be more desirable than an equivalent combinatorial circuit since it can be easily adapted to engineering changes or correcting mistakes in the control design. Again using TTL circuitry the OP-DECODE phase can be completed in 60 to

100 nsec. The OP-FETCH also involves only a memory access and can be completed on 100 to 200 nsec. The EXEC phase can in most cases be completed in 100 nsec, the exception being the multiplication. Figure 4.15 depicts the execution cycle in terms of these times, where we used an average reasonable values for each phase. From Figure 4.15 we see that based on our estimates, most instructions will be executed by DISP in about 500 nsec.

```
      I-FETCH    OP-DECODE  OP-FETCH    EXEC
|-----------------|-------|-------------|--------|
                 150     250           400      500
```

Figure 4.15 The instruction execution phases.

A closer examination of the hardware utilization in the various execution phases, reveals that we are doing rather poorly. The AU is active only a fifth of the time (100 nsec out of 500 nsec), and the address computation hardware is also used only a fourth of the time. This is obviously quite wasteful and should be remedied. Fortunately a simple remedy is at hand. It involves pipelining the various execution phases, that is, proceeding to perform several execution phases concurrently. Obviously this can be done since they are carried out by separate and independent hardware entities. As an example we consider a one stage pipelined execution stream, in which I-FETCH and OP-DECODE of the current instruction are overlapped with the OP-FETCH and EXEC of the previous instruction. To illustrate this we consider the timing of the execution of the following program sequence:

```
0    ADAC    1    110
1    ANDA    0    130
2    STAC    0    200
```

The timing is depicted in Figure 4.16. From it we see that now our effective execution time for most instructions is only 250 nsec, that is, two times faster than before, and this is achieved without adding any hardware.

Unfortunately this pipelining of the execution stream is not all roses. A complication arises on Branch instructions, especially if the branch is conditional. In this case we lose on throughput since we cannot start the next instruction, not knowing which it is, that is, if the branch is taken or not. This has the effect of degrading the throughput and makes the improvement of the pipelined execution less than expected. The exact degradation obviously depends on the number of branch instructions in the program. One method often used to diminish the effect of this degradation is to simply assume that the branch will not be taken

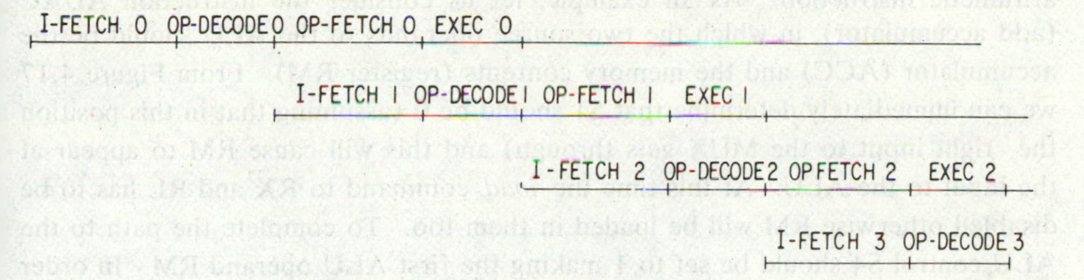

Figure 4.16 Pipelining of the execution phases by overlapping two phases of consecutive instructions.

and proceed with decoding the next sequential instruction. Then if the branch is indeed not taken we incurred no throughput loss.

The reader may by now wonder why this pipelining is not extended to all four phases, which would lead to a 150 nsec instruction throughput rate. This is indeed possible, and may be desirable in some applications. The problem is again on branch instructions, which tend to slow down the execution. Another problem that may occur is determining where a fault has been detected when a machine check or other errors cause an interrupt in the execution. These problems have caused machine designers in the past to restrict the pipelining to one or two stages.

An excellent example of a real time, general purpose digital signal processor is the LDVT [8], which is a processor for real-time speech processing. It is a processor built in ECL technology, and has a two stage pipelined execution stream resulting in an impressive 55 nsec instruction time throughput on most instructions. References [9,10,11] contain descriptions of additional general purpose digital signal processors that are currently operating, and the reader will greatly benefit from consulting these references.

4.6 THE ARITHMETIC UNIT

In this section we discuss in further detail the Arithmetic Unit of DISP using it as an example to illustrate various design techniques and considerations.

Figure 4.17 depicts the AU in more detail, showing explicitly the multiplexers used to set up the appropriate path in the AU to execute any instruction. Control lines S1, S2, S3, S4 that select the multiplexer (MUX) input that will appear at the output, determine the ALU and/or multiplier source operands. Thus we can easily determine the required value of these controls for any given arithmetic instruction. As an example, let us consider the instruction ADAC (add accumulator), in which the two source operands to the ALU should be the accumulator (ACC) and the memory contents (register RM). From Figure 4.17 we can immediately determine that S1 should be 0 (assuming that in this position the right input to the MUX gets through) and this will cause RM to appear at the input to the ALU. At this time the *load* command to RX and RL has to be disabled otherwise RM will be loaded in them too. To complete the path to the ALU control S4 should be set to 1 making the first ALU operand RM. In order to have the ACC as the second source operand to the ALU we should set S2S3 to 01. To actually execute the addition the ALU controls A1 to A4 have to be set to the code that corresponds to addition, say 0010.

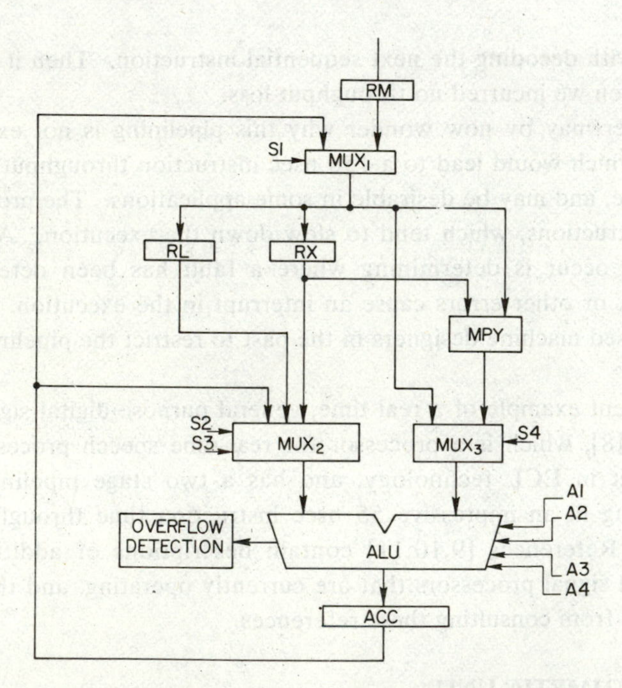

Figure 4.17 The arithmetic unit.

Let us consider as another example the instruction MAC. This should execute ACC ← RX × c(EA) + ACC. Since c(EA) is in RM, we again set S1 to 0 and enable the multiplier. To accumulate the result we set S4 to 0 to make the multiplier output the first ALU operand and set S2S3 to 01 making the ACC the second ALU operand. The ALU code is again set to addition.

Thus we see that for each OP code we can determine in a straightforward manner the required setting for all the controls that govern the operation of the arithmetic unit. A simple way of obtaining all required controls is to use a 64 word ROM in which the required control word to execute an operation resides at the address corresponding to its OP code. This will minimize the control logic and permit engineering changes to be done relatively easily. The control word residing in such a control ROM is called microcode, since it can be thought of as a program code at the lowest level (micro). The reader may benefit from working out this example and constructing a table containing the appropriate control signal for each of the DISP arithmetic instructions. The overflow detection circuit in Figure 4.17 detects an overflow by monitoring the sign bits of the operands and the result as explained in Section 4.2 and Figure 4.2. In the case that an overflow is detected the output of the ALU would be set to + or − the largest value, depending on the direction of the overflow.

4.7 SUMMARY

In this chapter we have presented an introduction to the hardware implementation of digital signal processors. In it we have attempted to highlight and identify the differences between a general purpose computer and a signal processor, and their implications on the machine organization.

The first two sections dealing with binary arithmetic and an overview of digital hardware are included for completeness only and are by no means a sufficiently detailed treatment of the subject. Readers unfamiliar with the subjects should consult the references indicated, and reference [12], which contains a particularly appropriate and timely treatment of the integration of hardware and software design.

The following sections address the various issues arising in the hardware implementation of digital signal processors, mainly through examples intended to illustrate the design philosophy of such processors. We do not go into the details of the hardware design especially in the control and timing but rather choose to discuss the overall design considerations. Chapter 5 contains a more detailed treatment, including actual design examples of special purpose digital signal processors.

EXERCISES

4.1 Represent the numbers ±13 and ±11 in 2's complement notation and compute all possible sums (differences) ±13 ±11 in binary arithmetic.

4.2 Repeat exercise 4.1 with ±17.6875 and ±11.190625.

4.3 Represent the following numbers, first in 2's complement and then in the canonical signed digit code: 11, 31, 15, 125, 181.

4.4 Devise an algorithm for converting serially numbers represented in 2's complement to the canonical signed digit code. (*Hint*: Starting with the least significant bit and proceeding towards the most significant bit, examine two bits at a time and a carry generated in the previous step [2].)

4.5 Multiply the numbers +11 and −13 using 2's complement arithmetic, manually using
(a) a one bit at a time algorithm.
(b) a two bit at a time algorithm.
(c) using array multiplication.
Take +11 to be the multiplier, that is, the one whose bits are examined.

4.6 Show that if two numbers in 2's complement notation having an opposite sign are multiplied directly, and the negative number is the multiplier, the result obtained is incorrect. Use as an example −11 times +13. Show that a simple correction will yield the correct result. (*Hint*: the correction consists of adding an extra 1 bit to the n-th bit of the product, where n is the number of bits used for the terms.)

4.7 Multiply the numbers +11 and −15, first using 2's complement notation for both, and afterwards using the canonical signed digit code for −15. Notice the decreased number of additions required. Repeat this for −117 and +69.

4.8 What does the following DISP program accomplish?

```
LRX   0   60
MPY   0   160
LRX   0   61
MAC   0   161
LRX   0   62
MAC   0   162
STAC  0   260
```

4.9 Rewrite the program in exercise 4.8, using the index register IX1 to increment the addresses of the operands. Which program requires less machine cycles? Which program requires less instructions?

4.10 Write a DISP program to perform the following arithmetic operation.

$$v_k = \sum_{j=0}^{4} x_{k-2j} h_j \qquad k=0, 1, \ldots, 10$$

4.11 Write a DISP program to perform the bit reversal on an array of real data consisting of 512 points stored at locations 100 to 611. Use the algorithm depicted in Figure 3.4.

4.12 Write a DISP program to perform a butterfly computation for the radix 2 decimation-in-time algorithm, Eq. 3-5.

4.13 Write a DISP program to perform an N point FFT where N is a power of 2, $N=2^{n}$ using the radix 2 decimation-in-time algorithm.

REFERENCES

1. G. W. Reitwiesner, *Binary Arithmetic* in *Advances in Computers* vol. 1, edited by F. L. Alt, Academic Press, New York, 1960.
2. A. Peled, *On the Hardware Implementation of Digital Signal Processors*, IEEE Trans. on Acoustics, Speech and Signal Processing, vol. ASSP-24, February 1976, pp. 76-86.
3. I. Flores, *The Logic of Computer Arithmetic*, Prentice Hall Inc., Engelwood Cliffs, N.J., 1963.
4. L. Altman, *The New LSI*, Electronics, July 10, 1975, pp. 81-92.
5. L. Altman, *Technology Update*, Electronics, October 17, 1974, pp. 139-151.
6. E. R. Hnatek, *A User's Handbook of Integrated Circuits*, John Wiley & Sons Inc., New York, 1973.
7. C. S. Wallace, *A Suggestion for a Fast Multiplier*, IEEE Trans. Electron. Comput., vol. EC-13, February 1964, pp. 14-17.
8. P. E. Blankenship, *LDVT: High Performance Minicomputer for Real-Time Speech Processing*, EASCON 75 Record, IEEE Publication 75 CHO 998-5.
9. Y. S. Wu, *Microprogramming Applications to Signal Processing Architecture*, in *Microarchitecture of Computer Systems*, eds. R. W. Hartenstein and R. Zaks, Euromicro, North-Holland Publishing Co., New York, New York, 1975.
10. G. L. Kratz, W. W. Sproul and E. T. Walendziewicz, *A Microprogrammed Approach to Signal Processing*, IEEE Trans. on Comput., vol. C-23, August 1974, pp. 808-817.
11. R. De Mori, S. Rivoira and A. Seria, *A Special Purpose Computer for Digital Signal Processing*, IEEE Trans. on Comput., vol. C-24, December 1975, pp. 1202-1211.
12. T. R. Blakeslee, *Digital Design with Standard MSI and LSI*, John Wiley & Sons, Inc., New York, 1975.

CHAPTER 5

Implementation of Dedicated Hardware Special Purpose Digital Signal Processors

5.1 INTRODUCTION

In many applications, the number of digital signal processing functions that are needed is quite limited and specific. Furthermore, these functions should be performed in real time, and the implementation must be cost effective and competitive with alternate analog implementations.

The term special purpose digital signal processors has been used in the past quite loosely to define a wide range of degrees of dedication and specialization. Therefore it is necessary to define this term more precisely, and thus better focus the scope and range of applications this chapter addresses. A special purpose digital signal processor is defined here as a dedicated hardware entity whose function is to perform a specific, well defined, set of digital signal processing algorithms in real time as a self contained subsystem. Furthermore, it is assumed that any changes in processing coefficients, or algorithms, if made at all, only occurs relatively infrequently as compared to the computational rate. The need for such special purpose processors arises in a variety of fields, such as telecommunications, radar and sonar processing, to mention only a few.

In this chapter we present some hardware implementation alternatives that are suitable for such applications. These mechanizations take advantage of our a priori knowledge of the specific applications to permit efficient and cost effective realizations of the required processing functions. This is achieved, however, at the expense of some loss of flexibility and ease of programmability exhibited by the general purpose processors described in the previous chapter.

We start the discussion in Section 5.2 with a presentation of a mechanization alternative to array multiplication and outline its potential application to digital signal processing. The realization of symmetric nonrecursive digital filters is considered as an example. In Section 5.3 we proceed to discuss in detail the application of these techniques to the implementation of a recursive second order digital filter. The expected performance of a standard TTL MSI implementation of such a filter is evaluated. A modification of the techniques to allow parallel processing is detailed in Section 5.4, and the gain in processing speed and the required increase in hardware are assessed for a TTL realization. In Section 5.5 the implementation of higher order digital filters is discussed, and in Section 5.6 we present some further modifications to the techniques described so far, that are suitable for the implementation of a floating point arithmetic unit of a high speed pipeline FFT processor.

In Appendix 5.1 an actual implementation of a 4-th order digital filter, that was built at Princeton University, is described in detail. The control and timing are explained, and some practical considerations of such an implementation are given. Also included in the Appendix is a PL/1 program for generating the contents of the ROM's or PROM's used in such an implementation.

Appendix 5.2 contains a description of an actual implementation of a 512 point FFT, that was designed and built at Princeton University. This processor uses the basic ideas described in Section 5.6 for implementing the arithmetic unit of a floating point FFT processor. Many practical considerations and implementation details are discussed in detail.

The rapidly changing digital technology, and the continuing emergence of new technologies, make it difficult to assess or recommend any particular one as more suitable for digital signal processing. Therefore, although the examples in this chapter are given in terms of TTL integrated circuits, it is not to be construed as an implied preference. We have chosen TTL only because it is the most established line with the largest repertoire of readily available MSI and LSI functions at the present time. The reader is advised to concentrate on the features of the techniques described which transcend the detailed properties of any one technology and are suitable for implementation in any of today's as well as future technologies.

5.2 MECHANIZATION ALTERNATIVES OF ARRAY MULTIPLICATION

A central operation that a digital signal processor is called on to repeatedly perform is array multiplications of the form

$$y = \sum_{j=1}^{L} a_j x_j \tag{5-1}$$

where a_j is a set of predetermined coefficients, and x_j are data values, or intermediate results. When the data values are scaled so that $|x_j| < 1$ and maintained in the 2's complement code with B bits accuracy in fixed point notation, Eq. 5-1 can be rewritten as

$$y = \sum_{j=1}^{L} a_j \sum_{k=1}^{B-1} x_j^k \, 2^{-k} - x_j^0 \qquad (5\text{-}2)$$

Interchanging the order of summation over indices j and k yields

$$y = \sum_{k=1}^{B-1} 2^{-k} \sum_{j=1}^{L} x^k{}_j a_j - \sum_{j=1}^{L} a_j x_j^0 \qquad (5\text{-}3)$$

We now define a function F with L binary valued arguments as follows;

$$F(x^1, x^2, \ldots, x^L) = \sum_{j=1}^{L} x^j a_j \qquad (5\text{-}4)$$

then we can rewrite Eq. 5-3 in terms of the function F as

$$y = \sum_{k=1}^{B-1} 2^{-k} \, F(x_1{}^k, x_2{}^k, \ldots, x_L{}^k) - F(x_1{}^0, \ldots, x_L{}^0) \qquad (5\text{-}5)$$

Thus given the value of the function $F(\bullet, \bullet, \ldots, \bullet)$ it is possible to compute y by using additions (subtraction for $k = 0$) and shift operations only. Since the arguments of F can take on the values 0 or 1 only, F has a finite number 2^L of possible outcomes, which can either be stored in a memory, or generated by a combinatorial circuit. Obviously the exponential growth of storage requirements with L will limit its value in practical applications, and we shall address this problem in the ensuing discussion. For the moment however, we choose to set aside this concern, and proceed to point out several interesting features implied by Eq. 5-5.

Figure 5.1 shows a schematic block diagram of a possible bit serial mechanization of Eq. 5-5. The data sequence x_j shifts serially into shift registers SR1 to SRL with the least significant bit leading. The bits $x_j{}^k$ are tapped at the output of each shift register and used to address the read only memory (ROM) in which the 2^L values of F are stored. Initially registers R2 and R3 are cleared, and the data bits $x_j{}^k$ start shifting at the clock rate with the least significant bit $x_j{}^{B-1}$ leading. At this time the value in the ROM at address $(x_1{}^{B-1}, x_2{}^{B-1}, \ldots, x_L{}^{B-1})$ is loaded into R1 and added to the value in R2, with the result appearing in R3.

The next clock pulse shifts x_j^{B-2} into the shift registers and generates a new address for the ROM. The value in R3 into R2 with a hardwired 1 bit right shift (skewed parallel connection to provide a multiplication by 1/2). The value in R1, which now is $F(x_1^{B-2},....,x_L^{B-2})$, is added to R2, which now is $2^{-1} F(x_1^{B-1},....,x_L^{B-1})$, to obtain another partial result. The same operation is repeated B times, except for the last time in that we subtract $F(x_1^0,...,x_L^0)$ from the accumulator, and at the end of B clock periods the required result, y, is in R3, and the circuit is ready to compute a new array multiplication.

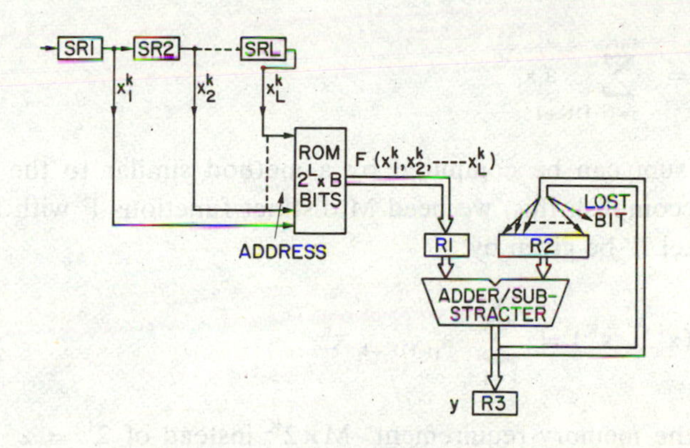

Figure 5.1 A bit serial mechanization of an L term array multiplication.

Figure 5.1 therefore presents a mechanization of array multiplication which at least in principle, that is if memory is assumed to be unlimited and speed is of prime importance, will permit performing L multiplications and L−1 additions in only a fixed number B of addition/memory cycles, B being independent of L.

Unfortunately we have to face reality, sooner or later, and the fact is, that not only is memory capacity limited and costly, but also the memory access time increases with increasing memory size. This will cause the implementation described above to be unrealistic for any L larger than 12, and may even be inefficient for L larger than 5 to 8 depending upon the technology used.

We now proceed to show that this computation can be partitioned in a variety of ways that allow us to trade off an exponential decrease in storage versus either a linear increase in the time required to perform the operation, or a linear increase in the number of adders.

Let L = K M, than we can rewrite Eq. 5-1 as

$$y = \sum_{j=1}^{K} a_j x_j + \sum_{j=K+1}^{2K} a_j x_j + \dots + \sum_{j=(M-1)K+1}^{M \times K} a_j x_j \tag{5-6}$$

or,

$$y = \sum_{i=1}^{M} v_i \tag{5-7}$$

where v_i is the i-th partial sum

$$v_i = \sum_{j=(i-1)K+1}^{i \times K} a_j x_j \tag{5-8}$$

Each partial sum can be computed by a method similar to the one described above. To accomplish this, we need M distinct functions F with K binary arguments each. Let F^i be given by:

$$F^i(x^1,\dots,x^K) = \sum_{k=1}^{K} a_{(i-1)K+k}\, x^k \tag{5-9}$$

thus making the memory requirement $M \times 2^K$ instead of $2^L = 2^{M \times K}$ had we not partitioned the array multiplication as in Eq. 5-6. However an additional $M-1$ additions are needed.

To illustrate the savings in memory introduced, we consider a concrete example. Let $L = 15$, than the direct implementation would require 2^{15} = 32,798 words of storage. If we use the partitioning outlined above, with $M = 3$, $K = 5$, only $3 \times 2^5 = 96$ words of storage are needed; the penalty is two more additions. Such an exchange is indeed worthwhile.

Figures 5.2 and 5.3 illustrate two possible mechanizations of Eq. 5-7. In Figure 5.2 a parallel implementation is shown for the case of $M = 2$, where the results of the individual partial sums are obtained through an adder tree. Here hardware has been added to obtain a high throughput speed, since except for a fixed delay due to the propagation in the adder tree, the result y is obtained every B clock cycles. Figure 5.3 on the other hand shows an implementation where speed of operation has been traded for hardware simplicity, again for $M = 2$. Here partial sums v_i are computed sequentially with the appropriate data bits addressing the corresponding memory locations in which F is located and the sum accumulated for B M clock periods, to provide the result y. In the following sections we will show some examples of possible compromise between those two extremes. These trade a gradual increase in hardware to achieve a computational speed between B to B M clock cycles. In [1] the interested reader may find a discussion on the possible optimum partitioning of L into partial sums for a given technology.

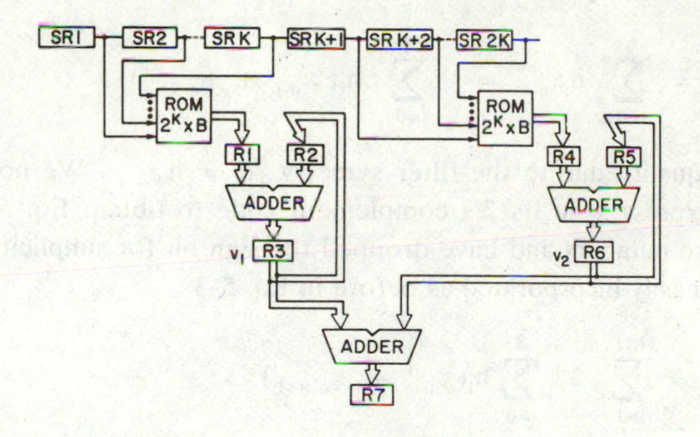

Figure 5.2 The mechanization of a 2×K terms array multiplication by a parallel connection of 2 sections each implementing a K term array multiplication.

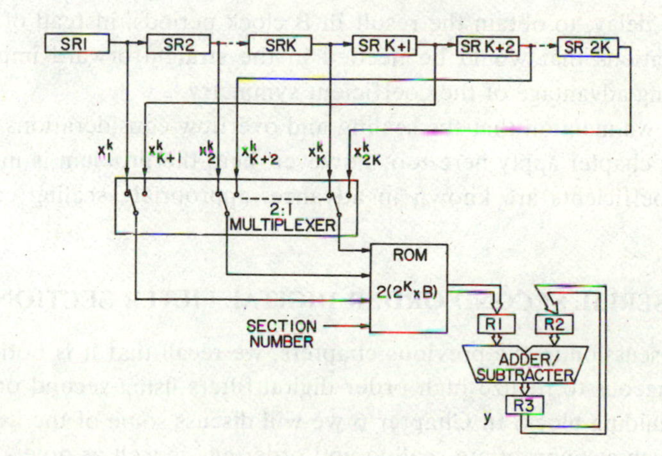

Figure 5.3 The mechanization of a 2×K term array multiplication in a serial fashion.

The above discussion applies directly to the hardware implementation of nonrecursive digital filters, as the alert reader may have noticed by now. A slight simplification is possible for the case of linear phase nonrecursive filters because of the symmetry in the coefficients. To illustrate this we consider a simple example of a 10 tap symmetric nonrecursive filter:

$$z_n = \sum_{i=0}^{9} h_i x_{n-i} = \sum_{i=0}^{4} h_i(x_{n-i} + x_{n-9+i}) \tag{5-10}$$

the second equality due to the filter symetry $h_i = h_{9-i}$. We now proceed as before and express x_i in its 2's complement code to obtain Eq. 5-11. We assumed positive numbers and have dropped the sign bit for simplicity of notation, but it can be easily incorporated as before in Eq. 5-3

$$z_n = \sum_{j=1}^{B-1} 2^{-j} \sum_{i=0}^{4} h_i(x_{n-i}^{\ j} + x_{n-9+i}^{\ j}) \tag{5-11}$$

Now we note that the sum $x_{n-i}^{\ j} + x_{n-9+i}^{\ j}$ can take on only three values 0,1,2, of which the third, that is 2, can be interpreted as a carry to the next bit. This will result in an implementation as depicted in Figure 5.4, where we only need $2^5 = 32$ memory locations and 5 full adders with the carry wrapped around with a one clock period delay, to obtain the result in B clock periods, instead of $2^{10} = 1024$ memory locations that would be needed in the straightforward implementation without taking advantage of the coefficient symmetry.

Finally, we mention that the scaling and overflow considerations discussed in the previous chapter apply here too; however, here the problem is much simpler. Since the coefficients are known in advance, appropriate scaling can be easily determined.

5.3 A BIT SERIAL SECOND ORDER DIGITAL FILTER SECTION

From our discussion in the previous chapters, we recall that it is both convenient and advantageous to realize high order digital filters using second order sections as a basic building block. In Chapter 6 we will discuss some of the aspects of this approach, such as appropriate scaling and ordering, as well as pole - zero pairing for minimum round-off noise. In this section and in subsequent sections we focus on various hardware implementation alternatives of such sections, and present some specific examples of possible TTL implementations.

A second order section digital filter is specified by an input output relationship,

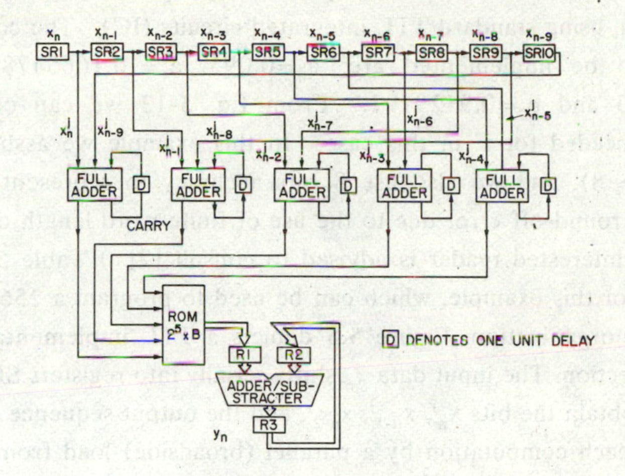

Figure 5.4 A bit serial mechanization of a 10th order symmetric nonrecursive digital filter using only 32×B bits of memory.

$$y_n = a_0 x_n + a_1 x_{n-1} + a_2 x_{n-2} - b_1 y_{n-1} - b_2 y_{n-2} \quad (5\text{-}12)$$

where $\{x_n\}$ is the input sequence, $\{y_n\}$ the output sequence, and $\{a_i\}$, $\{b_i\}$ the coefficients. For sections used in a parallel realization of high order filters $a_2 = 0$. We assume that the input sequence is given in 2's complement code with B bits accuracy and fixed point notation. Furthermore, we assume that appropriate measures are taken that the input sequence is bounded, that is, $|x_n| < 1$. Following the approach of Section 5.2 we define a function F of 5 binary variables as follows:

$$F(x^1,..,x^5) = a_0 x^1 + a_1 x^2 + a_2 x^3 - b_1 x^4 - b_2 x^5 \quad (5\text{-}13)$$

and proceed in a manner analogous to the derivation of Eq. 5-5, that is, we obtain:

$$y_n = \sum_{j=1}^{B-1} 2^{-j} F(x_n{}^j, x_{n-1}{}^j, x_{n-2}{}^j, y_{n-1}{}^j, y_{n-2})^j$$

$$- \quad F(x_n{}^0, x_{n-1}{}^0, x_{n-2}{}^0, y_{n-1}{}^0, y_{n-2})^0 \quad (5\text{-}14)$$

Since the function F has only 32 possible values, we can store them in a ROM of 32×B bits and use a structure similar to the one given in Figure 5.1 to compute y_n [2,3].

We now consider a concrete example, to illustrate the performance of this implementation, using standard TTL integrated circuits (IC). The coefficients of the section to be implemented are: a_0=0.095, a_1=-0.1665478, a_2=0.095 b_1=-1.8080353 and b_2=0.9129197. From Eq. 5-13 we can construct the memory map needed for F in this case. In this example we assume an 8 bit accuracy (B = 8), and we also use 8 bits accuracy to represent F. (For an analysis of the round-off error due to the use of finite word length of this implementation the interested reader is advised to consult [2].) Table 5.1 gives the memory map for this example, which can be used to program a 256 bit memory of a 32×8 bit organization. Figure 5.5 depicts a TTL implementation of this second order section. The input data x_n shifts serially into registers SR1, SR2 that are tapped to obtain the bits x_n^j, x_{n-1}^j, x_{n-2}^j, and the output sequence y_n is loaded at the end of each computation by a parallel (broadside) load from register R6 into SR3 and than shifted out serially into register SR4. The outputs of SR3 and SR4 are tapped to obtain y_{n-1}^j, y_{n-2}^j. The adder/subtracter is 8 bit wide, made of two arithmetic logic units (ALU). Registers R5 and R4 that hold the two words to be added will permit us to overlap the memory and addition cycles, thereby increasing the throughput rate. For simplicity we have chosen all registers as 8 bit bidirectional parallel in, parallel out, capable of a 25 MHz minimum shift rate (35 MHz typical) and a power dissipation of 360 mW. In an actual implementation it would be advisable to use simpler registers for SR1,SR2,SR3, such as Signetics 8277 dual 8 bit serial in serial out. The ROM is SN7488A a 32×8 bit memory with a 30 nsec typical access time (maximum 50 nsec). The ALU's are SN74181 capable of adding two 8 bit numbers in 36 nsec. Thus a 50 nsec clock cycle seems a prudent choice, allowing for worst case conditions. Because of the inherently recursive nature of the computation, that is, y_{n-1} has to be computed and loaded into SR3 before we can start computing y_n , 9 clock cycles, that is, 450 nsec in our case, will be needed to compute each sample. Figure 5.6 gives the timing diagram for the operation of this circuit. It is evident that the control is very simple, and requires only a counter modulo 9 and some gates. For more details on the control the reader is advised to consult Appendix 5.1.

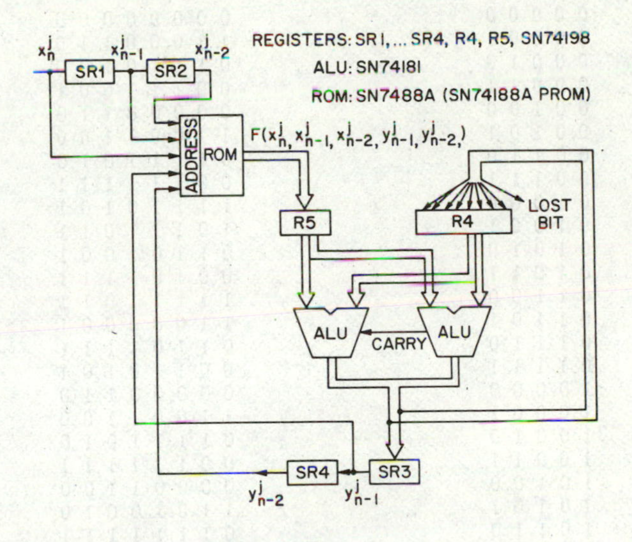

Figure 5.5 A bit serial implementation in TTL technology of a second order section digital filter. (B=8)

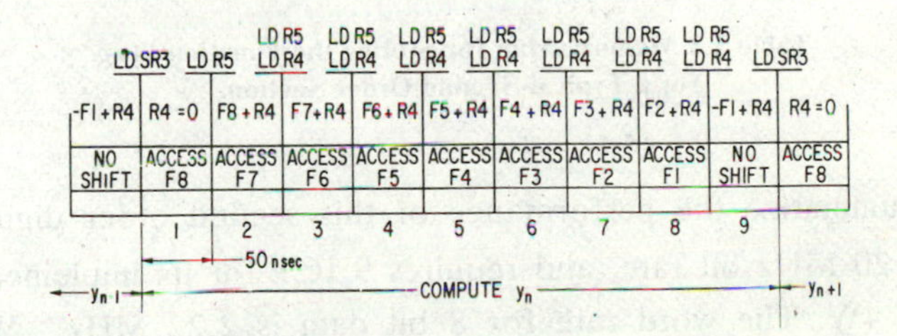

Figure 5.6 Timing diagram of second order section depicted in Figure 5.5.

MEMORY ADDRESS	MEMORY CONTENT
0 0 0 0 0	0 0 0 0 0 0 0 0
0 0 0 0 1	1 1 0 0 0 1 1 0
0 0 0 1 0	0 1 1 1 0 1 0 0
0 0 0 1 1	0 0 1 1 1 0 0 1
0 0 1 0 0	0 0 0 0 0 1 1 0
0 0 1 0 1	1 1 0 0 1 1 0 0
0 0 1 1 0	0 1 1 1 1 0 1 0
0 0 1 1 1	0 0 1 1 1 1 1 1
0 1 0 0 0	1 1 1 1 0 1 0 1
0 1 0 0 1	1 0 1 1 1 0 1 1
0 1 0 1 0	0 1 1 0 1 0 0 1
0 1 0 1 1	0 0 1 0 1 1 1 1
0 1 1 0 0	1 1 1 1 1 0 1 1
0 1 1 0 1	1 1 0 0 0 0 0 1
0 1 1 1 0	0 1 1 0 1 1 1 1
0 1 1 1 1	0 0 1 1 0 1 0 1
1 0 0 0 0	0 0 0 0 0 1 1 0
1 0 0 0 1	1 1 0 0 1 1 0 0
1 0 0 1 0	0 1 1 1 1 0 1 0
1 0 0 1 1	0 0 1 1 1 1 1 1
1 0 1 0 0	0 0 0 0 1 1 0 0
1 0 1 0 1	1 1 0 1 0 0 1 0
1 0 1 1 0	0 1 1 1 1 1 1 1
1 0 1 1 1	0 1 0 0 0 1 0 1
1 1 0 0 0	1 1 1 1 1 0 1 1
1 1 0 0 1	1 1 0 0 0 0 0 1
1 1 0 1 0	0 1 1 0 1 1 1 1
1 1 0 1 1	0 0 1 1 0 1 0 1
1 1 1 0 0	0 0 0 0 0 0 1 0
1 1 1 0 1	1 1 0 0 0 1 1 1
1 1 1 1 0	0 1 1 1 0 1 0 1
1 1 1 1 1	0 0 1 1 1 0 1 1

Table 5.1 Memory Map for Storing the Function F(•) for a Typical Second Order Section.

Let us summarize the performance of this second order digital filter. It operates at a 20 MHz bit rate, and requires 9 IC's for its implementation, consuming about 4W. The word rate for 8 bit data is 2.22 MHz. We note that increasing the required data accuracy to 16 bit will result in a doubling of the hardware and the word rate will be 1.17 MHz. This is so because we can add two 16 bit numbers in the same amount of time by using carry lookahead circuits. To illustrate the efficiency of this implementation we note that if a general purpose multiplier would be used, such as Advanced Micro Devices Am2505, to achieve the same word rate for 16 bit data, 30 IC's would be needed just for the multiplier alone.

5.4 A FULLY PARALLEL HIGH SPEED SECOND ORDER SECTION

In the previous section we have seen a bit serial mechanization of Eq. 5-14 that computes the output samples y_n in B+1 clock periods. But this is only one of many possible mechanizations that trade hardware complexity for speed. The bit serial approach offers the greatest hardware simplicity and therefore the lowest speed. We now present the other extreme, a fully parallel mechanization, requiring the most hardware and offering the highest throughput rate.

In examining Eq. 5-14 we note that if we have available B separate ROM's, each containing the function F(•), then we can generate all summands concurrently and add them up in an adder tree. Figure 5.7 shows such an implementation for the case of B = 8, and a 12 bit accuracy for F. From this figure we see that since we are interested in an output accurate only to 12 bits, we can have decreasing size memory and adders, since their least significant bits are lost anyway. The notation in the figure that shows the first bit entering not only its appropriately shifted location but also all previous positions, is due to the property of dividing negative numbers in 2's complement. We illustrate this by an example: −24 in 2's complement is 101000, −3 which is −24/8, that is, shifted 3 positions, is 111101, and we could only keep in memory 101, and upon input to a 6 bit adder input the first bit 1 to all 3 positions in front of it too.

Let us now estimate the speed of operation and hardware requirements of this mechanization. The inherent recursive nature of the computation in this case, that is y_{n-2} is used in computing y_n, means that we have to wait for the addition result to propagate through the adder tree, before we can proceed to the next computation.

Note that if the computation were not of a recursive nature, that is, a nonrecursive filter, we could include registers in the adders and overlap all addition stages (pipeline structure) to get a $1/T_a$ throughput rate where T_a is the addition time. In the case of TTL this would mean a 20 MHz word rate throughput, and a 50 MHZ rate using ECL.

In the case of a second order digital filter section as depicted in Figure 5.7 we have to wait for $\log_2 B$ addition times and one memory access time, making the throughput $1/(\log_2 B+1)T_a$ (assuming equal times for memory access and addition). In our example using TTL circuits as given in Figure 5.7, the throughput rate would be 8 MHz for 8 bit data words, and the package count 35 IC's. This represents a 3.6 speed up factor versus the bit serial approach, at the expense of a 3.9 factor increase in package count.

The fully parallel implementation of Figure 5.7, for 16 bit data and 16 bit accuracy for F(•), will require 76 IC's to achieve a throughput word rate of 6.6

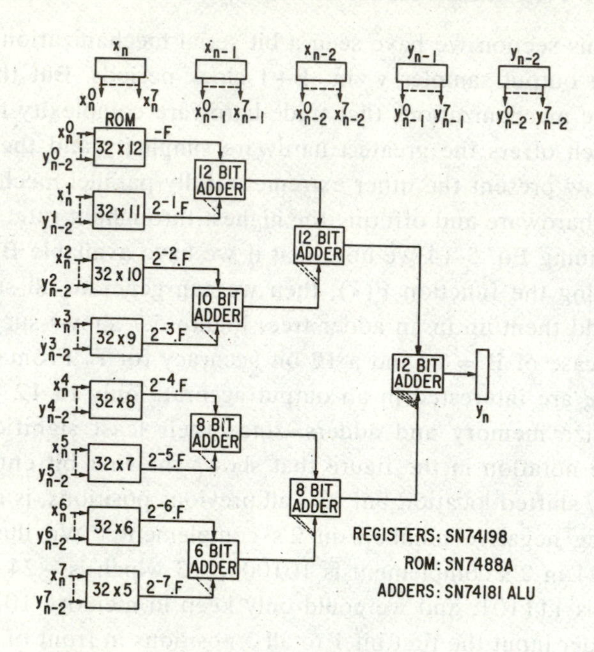

Figure 5.7 A fully parallel implementation of a second order section. (B=8 TTL IC's).

MHz. This represents a speed up factor of 5.8 over the 16 bit serial implementation, at the expense of a 4.2 factor increase in package count. The reader may find it a worthwhile exercise to verify this by drawing the appropriate block diagram.

The fully parallel mechanization presented in this section seems more suitable for low speed high density technologies, such as MOS, CMOS, IIL. In these technologies a fully parallel second order section could be integrated on one custom LSI circuit, taking into account the variety of simplifications that result in nonstandard adders and memories, which cannot be efficiently implemented with standard TTL IC's.

Furthermore, it is easy to see some straightforward modifications of the fully parallel implementation, such as a variety of mixtures of serial parallel implementations. A simple example from which the reader may benefit is to draw a block diagram of an implementation for B = 16 in which 4 bits at a time are used to address 4 parallel memories and adders which are connected in a manner similar to the one shown in Figure 5.7. Here it is possible to obtain a new result every 7

T_a, by suitable timing, (show how) instead of the 17 T_a in a bit serial implementation or a 5 T_a in a fully parallel implementation.

5.5 THE IMPLEMENTATION OF HIGH ORDER DIGITAL FILTERS

As we have seen in Chapter 1, it is advantageous to realize high order digital filters via a cascade or parallel connection of second order sections. In the previous sections we have concentrated on the various implementation alternatives of a second order section. These basic structures that we have derived can now be used to implement high order digital filters.

In implementing higher order digital filters, we have to make the choice between parallel and cascade realizations, not only in the theoretical sense of transfer function decomposition, but also in terms of various degrees of hardware parallelism, that will yield a wide range of throughput rates. This choice will be determined by the required speed of operation, the technology used, and from considerations of round-off error accumulation. The latter are discussed in Chapter 6.

Figures 5.8 and 5.9 depict two of the most obvious possibilities for implementing a sixth order filter. In Figure 5.8 a parallel implementation is shown. First the transfer function is decomposed as a sum of three second order terms. They are represented by the boxes designated as *2-nd order section*, each of which can be internally implemented by any of the methods discussed in the previous sections. Their outputs are denoted by $v_{n,1}$, $v_{n,2}$, $v_{n,3}$, respectively. They must be combined to form the output of the filter y_n, that is,

$$y_n = c_1 v_{n,1} + c_2 v_{n,2} + c_3 v_{n,3} \qquad (5\text{-}15)$$

where c_1, c_2, and c_3 are constants for scaling purposes. This sum is a three term array product, and can be obtained by using the approach described in Section 5.2 . We see that here the term parallel refers to both hardware and transfer function decomposition.

In Figure 5.9 a cascade implementation of a 6-th order filter is shown. Here by cascade we mean a cascade decomposition of the transfer function, and a serial computation from the standpoint of hardware implementation. Again the core of the arithmetic unit is a second order section implemented in bit serial fashion in this example. However, the memory to store F has a 96 words capacity to allow storage of the three distinct sets needed. The memory is addressed through a multiplexer by the appropriate bits, according to which of the three sections is being computed. The throughput of such an implementation would be N times slower than an individual section, for a 2N-th order filter.

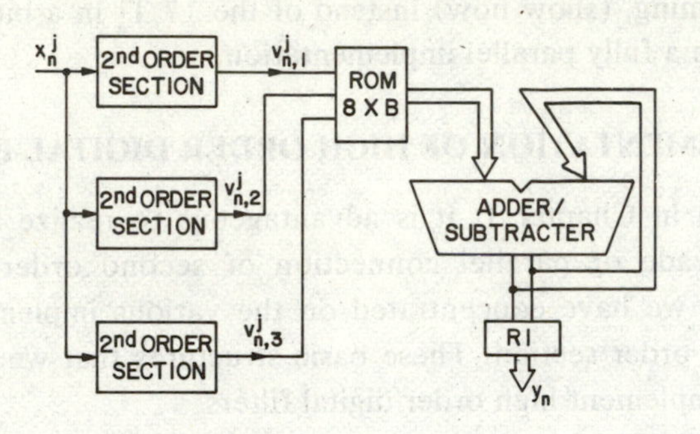

Figure 5.8 A parallel realization of 6-th order digital filter.

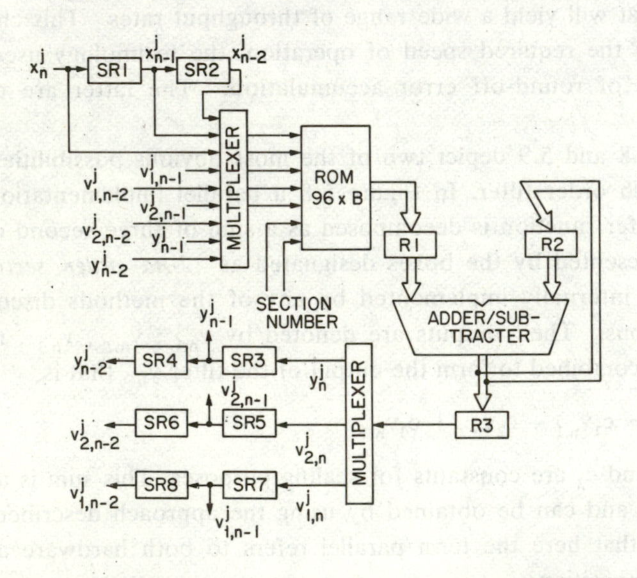

Figure 5.9 A cascade realization of 6-th order digital filter.

It is in the serial implementation that full advantage can be taken of the availability of high density storage IC's. We note that one 512×8 bit ROM will contain the functions F needed for 16 different second order sections. Thus 4 different 8-th order digital filters could be implemented by such a structure, and it would be capable of processing in real time a 137 KHz incoming data stream. It should be noted that more multiplex ports will be needed.

In Appendix 5.1 the actual implementation of a 4-th order digital filter realized in cascade is described, and the reader will find in it many usefull implentation details.

Obviously these 2 represent extreme cases. The reader must recognize at this point the variety of configurations that are available to the designer is quite extensive. With an understanding of the the basic ideas, the reader will be able to make a good choice for his particular application at hand.

5.6 THE IMPLEMENTATION OF THE ARITHMETIC UNIT FOR HIGH SPEED FFT

In Chapter 3 we discussed the various degrees of parallelism that can be introduced to speed up the FFT computation. We have seen that a fully pipelined FFT will have $J = \log_2 N$ stages, where N is the number of points to be transformed, each equipped with an arithmetic unit capable of performing a basic computation. We will refer to this unit as the basic computation unit (BCU). The need for such high speed processors arises mainly in wideband radar aplications [4], where real time processing of data in excess of 10 MHz maybe needed.

In this Section we discuss the possible application of the basic ideas outlined in the previous sections to the hardware implementation of the BCU. The discussion addresses applications in which the number of points in the FFT is not too large, say 1024 or less, fixed in advance, and is not changed during the real time operation of the system.

We also use this section to introduce a modified floating point arithmetic [5] that stores each complex number with a common exponent. This representation of the data permits a wider dynamic range, that is useful for FFT applications. The format that we will use in the examples is 4 bits for the exponent and 8 bits for the mantissa. This has been shown in [5] to provide adequate performance in typical radar applications.

In Chapter 3 we have seen that for a radix 2 decimation-in-time algorithm the BCU will repeatedly be called upon to perform a computation of the form:

$$A' = A + C\,W^k$$

$$C' = A - C\,W^k \tag{5-16}$$

where A and C are two complex numbers representing the input sequence or intermediate results in the FFT algorithm, k is an integer between 0 and N/2, and $W = \exp(-j2\pi/N)$ is the twiddle factor, with $j = \sqrt{-1}$. Thus the BCU accepts an input triple (A,C,k) and produces an output pair (A',C') related to the input by Eq. 5-16.

We assume that the complex data as well as any intermediate results, are represented in a modified floating point format as follows:

$$D = 2^{e_d}(d_R + j\, d_I) \qquad (5\text{-}17)$$

where e_d denotes the value of the common exponent, and subscripts R and I denote respectively the real and imaginary parts. After each operation the result is normalized so that e_d is the lowest possible integer for which it is still true that $|d_R| < 1$ and $|d_I| < 1$. The mantissa itself is represented in 2's complement code, with B bits accuracy.

Let Z be the result of multiplying C by the twiddle factor, that is, $Z = C\, W^k$. Therefore to compute A' and C' we can first compute Z, and then add it to (or subtract it from) A to obtain A' and C'. Eq. 5-18 defines Z in terms of real multiplications and additions;

$$Z = 2^{e_c}\,\{c_R\cos(2\pi k/N) + c_I\sin(2\pi k/N) + j\,[-c_R\sin(2\pi k/N) + c_I\cos(2\pi k/N)\,]\} = 2^{m_c}(z_R + j\, z_I) \qquad (5\text{-}18)$$

We now consider the application of the ideas developed in Section 5.2 on array multiplication to the computation of Z, which in a straightforward implementation requires 4 real multiplications and 2 real additions. We note that the computation of z_R and z_I can be implemented as an array multiplication of 2 terms. However, this case is quite simple and we will not discuss it here. The interested reader can find an evaluation of the expected performance of such an approach in [6]. Instead we choose to present a variation of the approaches discussed so far that involves storing the partial results due to combinations of two bits from each multiplicand.

Let P_k and Q_k be two functions defined as follows:

$$P_k(u,v,w,y) = (2u + v)\cos(2\pi k/N) + (2w + y)\sin(2\pi k/N)$$

$$Q_k(u,v,w,y) = -(2u + v)\sin(2\pi k/N) + (2w + y)\cos(2\pi k/N)$$

$$u,v,w,y = 0 \text{ or } 1\ ;\ k = 0,1,2,\ldots\ldots(N/2 - 1) \qquad (5\text{-}19)$$

Then we can express z_R and z_I in terms of P_k and Q_k as

$$z_R = \sum_{j=1,3,5\ldots} 4^{-(j+1)/2}\, P_k(c_R{}^j, c_R{}^{j+1}, c_I{}^j, c_I{}^{j+1}) \;-\; P_k(0, c_R{}^0, 0, c_I{}^0)$$

$$z_I = \sum_{j=1,3,5\ldots} 4^{-(j+1)/2}\, Q_k(c_R{}^j, c_R{}^{j+1}, c_I{}^j, c_I{}^{j+1}) \;-\; Q_k(0, c_R{}^0, 0, c_I{}^0) \qquad (5\text{-}20)$$

Thus we need here $(B/2)_I$ additions and one subtraction, accumulated with a 2 bit right shift hardwired, to compute z_R or z_I. The storage requirements for P and Q in each of the J stages of the FFT differ, as we recall from the flow graph describing the FFT algorithm discussed in Chapter 3. In the first stage k takes on only one possible value, namely 0, in the second stage k takes on 2 values, namely 0 and N/4, in the third stage k takes on 4 values, and so on. In the last stage, k takes on all possible N/2 values. Since for each k , P_k and Q_k have 16 possible outcomes, the storage requirement S_j for stage j is

$$S_j = 2 \times 2^{j-1} \times (16 \times B) \quad \text{bits} \tag{5-21}$$

where B is the accuracy used to maintain P_k and $_k$ in memory. A useful figure for estimating the storage requirement is the average storage required per stage - S , given by:

$$S_a = (1/J) \sum_{j=1}^{J} S_j = 2\,(N-1)\,(16\,B)/\log_2 N \tag{5-22}$$

The availability of high density storage IC's, such as 512×8 bit ROM's in bipolar technology, or 2048×8 bits in MOS technology, works in our favor in later stages, where one or two IC's will store all functions needed for this stage. However there is some waste in the first few stages where high capacity storage cannot be fully used. Table 5.2 lists for various N and 8 bit accuracy, the average storage requirement per stage, and the total number of IC's required for all stages to store P_k and Q_k. In the table we used the fact that the first two stages require only additions and therefore P_0, Q_0, P_1, and Q_1 are not needed. For the next 4 stages only 2 IC's of storage are needed per stage, one for P_k and one for Q_k. After that the number of IC's for storage grows as a power of 2 for each stage, making it 64 for the 11-th stage. From this fact stems the observation that this implementation would be advantageous only for moderate N, with 1024 probably being the limit for a TTL implementation.

We now consider a concrete example of a 1024 point FFT. The arithmetic precision will be 8 bits for the mantissa, 4 bits for the exponent, and also 8 bits accuracy to represent the functions P and Q . For this case 5 additions (overlapped with memory access) will be needed to obtain z_R and z_I. Figure 5.10 depicts a possible TTL implementation for a module to compute Z. We will refer to this as MODULE Z. The operation of the parts that compute z_R and z_I is similar to the second order section explained in Section 5.3, only here we have to shift two bits at a time. We also choose here to use a basic 50 nsec clock cycle interval as in Section 5.3. To make it possible to shift 2 bits in this interval we

N	Average Storage Per Stage (BITS)	Total Number of IC's
64	336×8	8
128	581×8	12
256	1020×8	20
512	1817×8	36
1024	3274×8	68
2048	5955×8	132

Table 5.2 Storage Requirements for the Functions P and Q for Various N. (8 bit accuracy).

use 4 bit shift registers (SN74S195), capable of shifting at a 100 MHz rate. Here we also can continuously overlap the memory and addition cycle, since the computation is nonrecursive. Thus every 250 nsec a complex number Z is produced and a new complex number C can be accepted.

We note that MODULE Z requires 16 IC's consuming about 7W for the arithmetic, and an additional 7 IC's on the average per stage for storage (N = 1024), making the total 23 IC's consuming about 10W. For comparison we mention that using standard multiplier IC's to achieve approximately the same throughput rate, would require roughly 34 IC's (based on Advanced Micro Devices Am2505, or Texas Instruments SN74285, 4×4 bit multiplier chips). Furthermore, if we are dealing with a 128 point transform instead of 1024, the total number of IC's for MODULE Z would be only 18, making the advantage of such an implementation even more significant.

Figure 5.11 illustrates how MODULE Z can be incorporated in the BCU, to perform the computation required by Eq. 5-16. We now proceed to discuss in some detail the operation of the BCU depicted in Figure 5.11, which incorporates many features that we did not encounter yet, most notably, floating point arithmetic, and serial addition and subtraction.

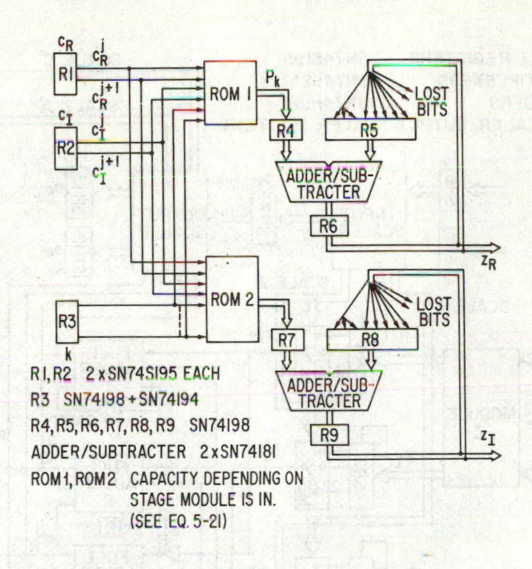

Figure 5.10 A module to implement the complex multiplication required in FFT butterfly, by taking two bits at a time.

First we examine the input scaling operation. The exponents e_a, and e_c, are loaded into the 4 bit registers SR5, and SR6. The input scaler contains a 4 bit ALU (e.g., SN74181), and some logic, to perform the following:

1. The first clock period (50 nsec) the ALU serves as a magnitude comparator betwen e_a and e_c.
2. The next clock period based on the outcome of the comparison the input scaler takes one of the following three actions:
   a. If $e_a = e_c$, then it simply sets $e_z = e_c$, and sets controls SCALE A and SCALE Z equal to zero.
   b. If $e_a > e_c$, then it sets $e_z = e_c$, SCALE A=0, and SCALE Z=$e_a - e_c$.
   c. If $e_a < e_c$, then it sets $e_z = e_a$, SCALE Z=0, and SCALE A=$e_c - e_a$.

Thus controls SCALE A and SCALE Z are 4 bit words, and they are used to control the positions of the 8-line-to-1-line multiplexers (M1,M2) and (M3,M4) respectively. When the value of the control is zero, the least significant bit will appear on the input line to the full adders. When the value is r, the r+1 bit will appear on the input line to the full adders. Now shift registers SR1 to SR4 start shifting serially towards the least significant bit (LSB), with the most significant bit filling the vacancies created. The shifting rate will be 40 MHz (double the basic clock rate) to permit obtaining the 8 bit sum in 4 basic clock periods (200

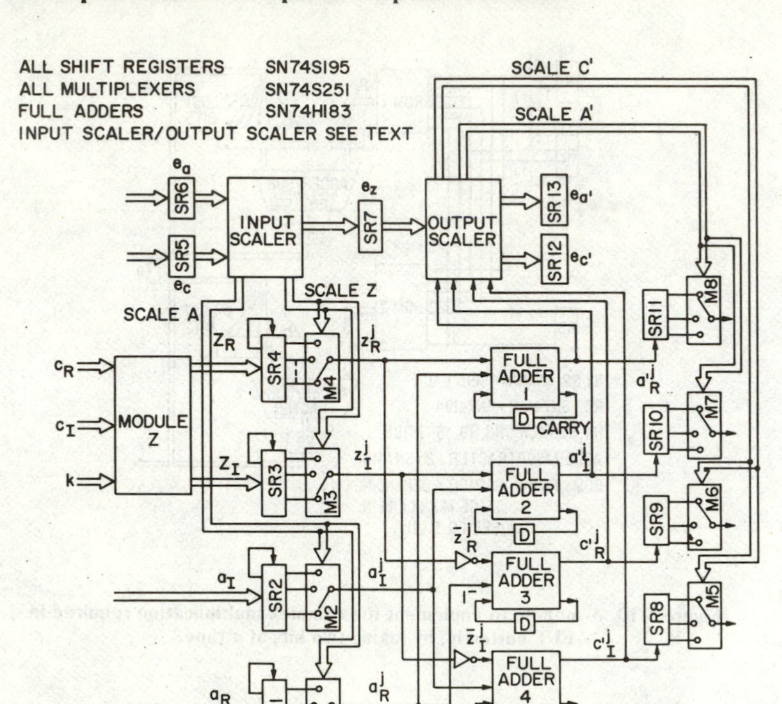

Figure 5.11 A block diagram of the FFT BCU that uses MODULE Z, and has a 4 MHz complex data rate throughput.

nsec). The full adders 1 and 2 generate the sum, that is, the bits of a'_R and a'_I. The full adders 3 and 4 generate the difference, that is, the bits of c'_R, and c'_I. This is accomplished by inverting the incoming bits of z_R and z_I, and taking the first carry to be 1, which has the effect of negating z_R and z_I.

To illustrate how the input scaling and additions/subtractions are done we consider a numerical example. Suppose

$$A = 2^1 [0.5390625 + j(-0.6171875)]$$
$$e_a = 0001 \; ; \; a_R = 01000101 \; ; \; a_I = 10110001$$
$$Z = 2^5 (0.390625 + j0.171875)$$
$$e_c = 0101 \; ; \; z_R = 00110010 \; ; \; z_I = 00010110$$

Initially, at the ALU in the input scaler, $e_c > e_a$, therefore in the next cycle the input scaler sets: $e_z = 0001$, SCALE A = 0100, and SCALE Z = 0000. This will cause the switches in multiplexers M1 and M2 to point to the 5-th bit of a_R and

a_I, and multiplexers M3, M4 to point to the first (LSB) bit of z_R and z_I. This is shown schematically in Figure 5.12. In this figure we also trace the addition and subtraction operation performed. The bits are shown as they appear when clocked out serially and enter the adders, directly or through inverters for subtraction. Also shown are the carries generated by the full adders and the results as they appear serially at the adder outputs, and will be clocked into registers SR8 to SR11. The reader may benefit from checking for himself that the results obtained are correct.

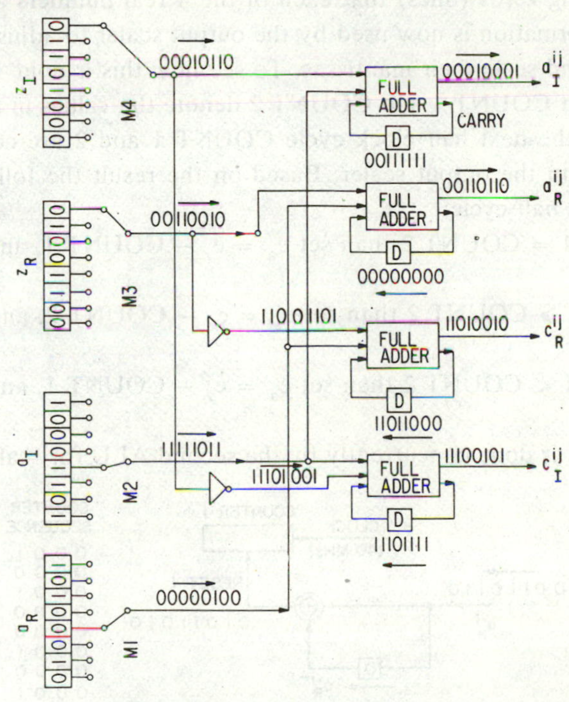

Figure 5.12 Using 8-line-to-1-line multiplexers to achieve scaling, and serial addition/subtraction—a numeric example.

We now return to the explanation of the operation of the BCU depicted in Figure 5.11. At this point the results c'_R, c'_I, a'_R, and a'_I reside in registers SR8 to SR11 respectively. Now the output scaling has to be performed to insure proper normalization of the floating point notation. Let us see how the output scaler accomplishes that. From Figure 5.11 we see that the outputs of the adders are tapped and enter the output scaler. This permits the scaler to make a scaling decision as soon as the addition is completed. This is done in the following manner. Each of the incoming bits (e.g. a'_R) controls the reset of a 4 bit counter that is counting the clock pulses that shift in the addition results into registers SR8 to SR11, so as to detect leading zeroes (or ones in the case of negative numbers). The counter is reset any time that two consecutive bits differ, for example, RESET 1 = $a'^{\,j}_R \oplus a'^{\,j-1}_R$, where $\oplus$ denotes an exclusive or. This is illustrated in Figure 5.13, for the numeric example given above. Thus we see that upon the completion of the addition, each 4 bit counter will contain the number of leading zeros (ones) that each of the 4 real numbers a'_R, a'_I, c'_R, and c'_I has. This information is now used by the output scaler to adjust the exponents for A' and C', and scale their mantissas. To see how this is done we consider the scaling of A'. Let COUNT 1 and COUNT 2 denote the values in the counters for a'_R and a'_I. In the next half clock cycle COUNT 1 and 2 are compared by the ALU contained in the output scaler. Based on the result the following action is taken in the next half cycle:

1. If COUNT 1 = COUNT 2 than set $e_{a'} = e_z$ − COUNT 1, and SCALE A' = COUNT 1.
2. If COUNT 1 > COUNT 2 than set $e_{a'} = e_z$ − COUNT 2, and SCALE A' = COUNT 2.
3. IF COUNT 1 < COUNT 2 than set $e_{a'} = e_z$ − COUNT 1, and SCALE A' = COUNT 1.

Exactly the same is done concurrently by the second ALU for scaling C'.

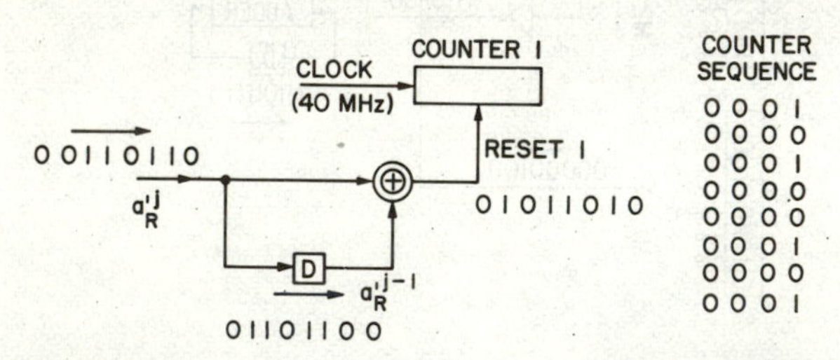

Figure 5.13 Using a 4 bit counter to determine the number of leading zeros (ones).

In the numeric example considered above the value of the counters is:

COUNT 1 = 0001 ; COUNT 2 = 0010 ;

this would lead to setting $e_{a'}$ to

$e_{a'} = e_z$ − COUNT 1 = 0101 − 0001 = 0100

SCALE A' = COUNT 1 = 0001

the effect of which is to connect the output line of multiplexers M7 and M8 to the second bit (counting from the most significant bit in this case). This will make the output A' appear as:

$$e_{a'} = 0100 \; ; \; a'_R = 01101100 \; ; \; a'_I = 00100010$$

$$A' = 2^4\,(0.84375 + j\,0.265625)$$

which is the correct result.

Finally, Figure 5.14 gives the timing diagram that shows the overlap in performing the different operations by the various hardware entities. We see that most units are busy all the time, which indicates a good efficiency. The throughput is seen to be a 4 MHz complex butterfly rate for the TTL implementation discussed. Since a 1024 point FFT will require performing $N/4 = 256$ butterflies in each stage, and since we assumed that each stage has its own BCU, the full transform will be done in 64 μsec. This observation of course applies to the case that high speed memories are used, so that the BCU's can be kept continuously busy. It also does not take into account the setup time and other overhead that may slow this down by as much as 50%.

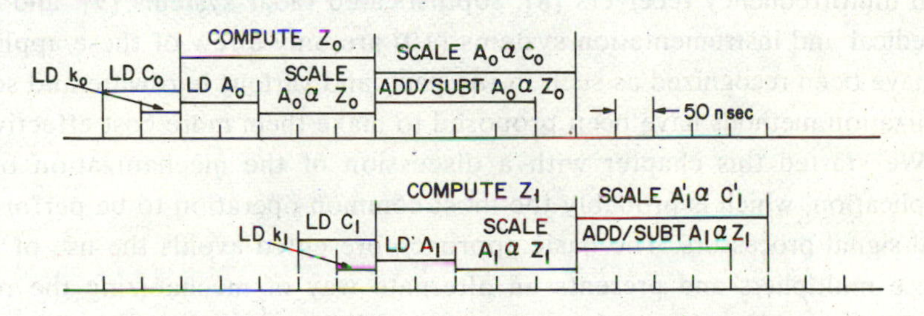

Figure 5.14 Timing diagram of BCU showing the overlap in the operation of the various hardware entities.

We now conclude this section with several remarks. The detailed discussion of various arithmetic and hardware aspects of the BCU was included because we feel that the reader will get a better feeling for the ideas presented. Furthermore, most issues treated here in detail apply equally to the previous sections.

We also emphasize at this point that, although not explicitly stated, the safeguards against overflow explained in Chapter 4 apply equally here and should be undertaken.

A higher throughput rate can be obtained from a BCU similar to the one described by introducing full parallelism to the mechanization of Eq. 5-20. Such an implementation would yield a 25 MHz complex data rate throughput [6].

An FFT processor, even as a dedicated stand alone unit, would be required to perform additional processing that will require multiplication by other than predetermined constants. However, these will usually be at a lower rate, and including one general purpose multiplier in the processor will take care of these operations.

5.7 SUMMARY AND DISCUSSION

In this chapter we have presented several possible hardware implementations of digital signal processors. They are intended for applications where such processors are either stand alone units designed to execute a given, well defined signal processing task. They are specialized peripherals of a larger system included to relieve the general purpose computer of performing fixed repetitive functions that can be performed more efficiently by dedicated hardware. Examples of such applications that fit into this loosely defined class of processors are numerous, with many more becoming apparent with every drop in price and size of digital hardware. Time division multiplex to frequency division multiplex converters [7], digital multifrequency receivers [8], sophisticated radar systems [9], and various biomedical and instrumentation systems [10] are only a few of these applications that have been recognized as such in the past, and various hardware and software optimization methods have been proposed to make them more cost effective.

We started this chapter with a discussion of the mechanization of array multiplication, which is probably the most common operation to be performed by digital signal processors. The basic approach presented avoids the use of general purpose multipliers and presents an alternate way of mechanizing the required multiplications. The proposed approach capitalizes mainly on the availability of high density and high speed memory, and its efficiency lies mainly in the favorable exchange between memory bits and logic gates with respect to cost and integrability. The storage requirement grows rapidly with L, the number of terms in the array multiplication, thus limiting the usefulness of this approach to values of L less than 12. To circumvent this difficulty we show how the problem can be partitioned so as to limit the growth of storage to a linear one at the expense of a linear increase in the number of additions to be performed. Several degrees of parallelism in the hardware implementation were also discussed. The application of these ideas to the implementation of nonrecursive digital filters is shown to be straightforward, and a possible simplification is mentioned for symmetric filters.

In Section 5.3 the implementation of a second order section digital filter in a bit serial fashion was presented in some detail, with a specific evaluation of the expected performance of such a section when implemented in TTL technology. In references [11,12] the reader may find a description of a CMOS implementation of such a second order section used in a special purpose digital control compensator controlling a highly accurate inertial platform. In the next section a high speed fully parallel implementation was presented, and its performance evaluated. Some possible mixtures of serial-parallel realizations were briefly outlined. Section 5.5 included a discussion of the realization of high order digital filters using the basic structures derived in the previous sections.

Finally Section 5.6 treated in detail a possible hardware implementation of the arithmetic unit of high speed FFT processors. The ideas presented before were incorporated here, and applied to a modified floating point arithmetic, to implement the multiplications required.

The techniques presented in this chapter are an attempt to introduce a customization in the design of special purpose digital signal processors that is in some sense equivalent to the nature of the analog elements they are intended to replace. Analog filters for example are a piece of hardware capable of performing a single specific function—shaping the spectrum of the input signal in a predetermined way. An attempt to replace this filter with a general purpose computer (and a fast one is needed), with all the generality it offers that will not be used, may never be cost competitive. On the other hand, designing a special combinatorial circuit that will do the same digitally may also be too expensive, except for very large volume applications, due to the high start-up costs of custom IC's.

The approach presented here is a compromise between the total generality on one hand, and the total customization on the other. It suggests a modular building block approach, that uses standard, well defined hardware entities (e.g., adders, memory, registers, etc.), and its complexity grows with the size of the application.

In conclusion, we observe that obviously there will be a breakover point beyond which the use of a general purpose hardware structure of the type discussed in the previous chapter, will be more efficient. This will be the case when the set of constants we have to multiply by is very large, and the memory requirement will be larger than a comparable hardware multiplier. This point will be determined in each technology by the relative exchange of efficiency in terms of speed-power product between memory bits and logic gates. Based on this exchange the system designer may choose the most efficient implementation for his application.

REFERENCES

1. S. A. White, *On Mechanization of Vector Multiplications* Proc. IEEE, vol. 63, no. 4, April 1975, pp. 730-731.
2. A. Peled and B. Liu, *A New Hardware Realization of Digital Filters* , IEEE Trans. Acoustics, Speech and Signal Processing, vol. ASSP-22, no. 6, December 1974, pp. 456-462.
3. A. Croisier, D.J. Esteban, M.E. Levilion, and V. Rizo, *Digital Filter for PCM Encoded Signals*, U.S. Patent 3777130, December 3, 1973.
4. J.K. Hartt and L. Sheats, *Application of Pipeline FFT Technology in Radar Signal and Data Processing*, IEEE, Electronics and Aerospaces Systems Convention: 71 Record, October 1971, Washington D.C., pp. 216-221.
5. L.W. Martinson and R.L. Smith, *Digital Matched Filtering with Pipelined Floating Point FFT's*, IEEE Trans. Acoustics, Speech and Signal Processing, vol. ASSP-23, April 1975, pp. 222-235.
6. B. Liu and A. Peled, *A New Hardware Realization of High Speed Fast Fourier Transformers*, IEEE Trans. Acoustics, Speech and Signal Processing, vol. ASSP-23, December 1975.
7. S.L. Freeny, R.B. Kieburtz, K.V. Mina and S.K. Tewksbury, *Systems Analysis of TDM-FDM Translator/Digital A-type Channel Bank*, IEEE, Trans. on Communication Technology, vol. COM-19, No. 6, December 1971, pp. 1050-1059.
8. Kunihiko Niwa and Mitsutake Sato, *Multifrequency Receiver for Pushbutton Signalling Using Digital Processing Techniques*, IEEE Int. Conf. on Communication 1974: Conf. Rec. paper 18F, Minneapolis, Minn. June, 1974.
9. R.H. Fletcher and D.W. Burlage, *Initialization Technique for Improved MTI Performance in Phased Array Radars*, Proc. IEEE, vol. 60, No. 12, December 1972, pp. 1551-1552.
10. W. Litwin and F. Begon, *Digital Processing of Phonocardiograms: First Results*, Int. J. Bio-Med. Comput. (GB) vol. 5, No. 1, January 1974, pp. 59-69.
11. B.E. Bona and F.C. Teran, *A Special Purpose Digital Control Compensator*, Eighth Asilomar Conference on Circuits, Systems and Computers, Pacific Grove, Calif., December 1974.
12. S.A. White, *Applications of Digital Signal Processing to Control Systems*, Eighth Asilomar Conference on Circuits, Systems and Computers, Pacific Grove, Calif., December 1974, pp. 278-284.
13. B. Gold and T. Bially, *Parallelism in Fast Fourier Transform Hardware*, IEEE Trans. Audio and Electroacoustics, Vol. AU-21, No. 1, February 1973, pp. 5-16.
14. R. L. Moris and J. R. Miller (eds.), *Designing with TTL Integrated Circuits*, McGraw-Hill, Inc., New York, 1971.
15. A. Barna and D. I. Porat, *Integrated Circuits in Digital Electronics*, J. Wiley & Sons., New York, 1973.
16. K. P. Yiu, *On the Sign Bit Assignment in Vector Multiplication*, Proceedings of IEEE (letter), vol. 64, No. 3, March 1976.

Appendix 5.1

A Fourth Order Digital Filter Implementation

In this appendix we describe an actual implementation of a 4-th order digital filter that was built and tested in the Electrical Engineering Department at Princeton University by Dr. Tran Thong. Although in some cases the choice of components was dictated by budgetary constraints rather than minimum package count and control simplicity, the implementation described will give the reader an insight into many of the details involved in building such a filter.

The fourth order filter is characterized by the following input-output relationship, when implemented in cascade form:

$$v_n = a_{0,1}x_n + a_{1,1}x_{n-1} + a_{2,1}x_{n-2} - b_{1,1}v_{n-1} - b_{2,1}v_{n-2} \qquad (5A\text{-}1)$$

$$y_n = a_{0,2}v_n + a_{1,2}v_{n-1} + a_{2,2}v_{n-2} - b_{0,2}y_{n-1} - b_{2,2}y_{n-2} \qquad (5A\text{-}2)$$

where $\{x_n\}$ is the input sequence, $\{v_n\}$ is the output of the first section, $\{y_n\}$ is the output sequence, and $\{a_{i,j}\}$, $\{b_{i,j}\}$ are the filter coefficients (j the section number).

The filter was built to process analog data and output analog data, so that familiar and readily available measuring instruments could be used to test it, and demonstrate its operation. The word length used was 8 bit, which is quite limited, but an extension to 12 or 16 bits is straightforward. The analog to digital conversion modules that were used are manufactured by Datel Systems Inc.. The modules are; sample and hold (S/H) SHM-3, the A/D converter ADC-D8B that produces 8 bit 2's complement code at a 20 kHz maximum sampling rate and D/A converter DAC-298B. Although the filter itself can operate at a rate of up to 160 kHz, these lower speed A/D's were used for cost considerations.

Figure 5A.1 depicts the filter block diagram. A double line is used for an 8 bit parallel data flow, and a single line for a single serial bit flow. Register RX0 serves as a converter from a parallel to a serial data flow. RX1 and RX2 store x_{n-1} and x_{n-2}, the two last input data words. The timing for computing one output and accepting one input is divided into two main parts. In the first part v_n is computed according to Eq. 5A-1 and in the second y_n is computed according to Eq. 5A-2. In the implementation described it takes 64 clock cycles to compute y_n, and the clock rate can be up to 10 MHz. This is so because each addition takes 4 clock cycles, as no buffer registers are provided for the addends and, furthermore, simple 4 bit adders were used (instead of ALU's), which requires additional exclusive or gates (SN7486) to permit complementing the addend to allow subtraction. The switching implied in the block diagram is performed by the quad two to one multiplexer MUX1 (SN74157).

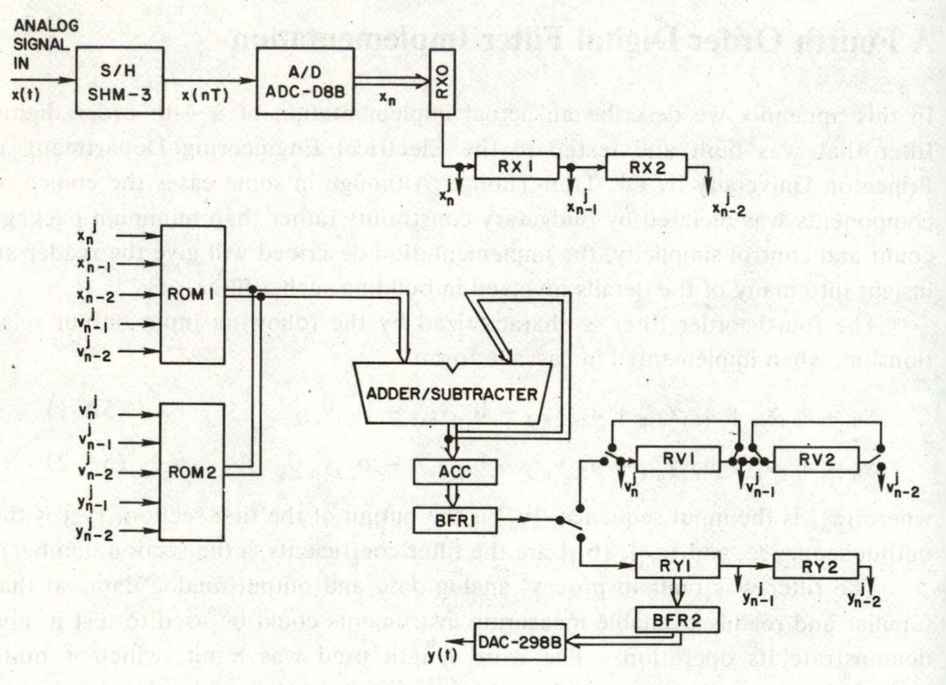

Figure 5A.1 Block diagram of 4-th order filter.

In the first part of the computation period (32 cycles) ROM1 is enabled and ROM2 disabled. ROM1 contains the function F(•,...,•) for the first section Since the bits of v_{n-1} and v_{n-2} which are in registers RV1 and RV2 are needed in both, the computation of v_n and y_n, they are simply recycled into their registers during the computation of v_n, this is indicated by the position of the switches in Figure 5A.1. At the end of 8 addition times (32 cycles), the output v_n accumulated in the 10 bit accumulator register ACC (8202) is truncated and loaded in parallel into buffer register BFR1, which also serves as a parallel to serial converter and during the computation of y_n shifts out v_n into RV1. (During the computation of v_n it shifted out y_{n-1} into RY1.) The next half of the computation period is almost identical, only now ROM2 which contains the function F(•,...,•) of the second section, is enabled, and the position of the MUX1 switches is the opposite of the one shown in the Figure 5A.1.

Figures 5A.2, 5A.3, and 5A.4 give the actual circuit diagrams of the control section, the memory and register section and the arithmetic unit. Figure 5A.5 shows the timing for the main control signals. These control signals are identified, and the boolean expressions that define them in terms of the outputs (A,B,C,D,E,F) of a 6 bit counter (that counts the clock pulses from 0 to 63 and resets) are given in Table 5A.1. Table 5A.2 lists the various events that occur in each part of the computation period. The arithmetic unit requires some further explanation. ACC is a 10 bit parallel in parallel out buffer register whose inputs are the adder outputs (result), and whose outputs are connected back to the adder input with a hardwired 1 bit right shift. In order to avoid shifting in the wrong bit into the sign bit position, in case of an adder overflow, it can be shown that the correct value of the new sign bit is given by performing an exclusive or operation on the sign bits of the two addends with the carry output from the sign position [16], that is, let the sign bits of the addends be a^0 and b^0 and c^0 the carry in the sign bit position from the adder that adds $a+b=d$, than the correct sign bit of the $d\,2^{-1}$ (that is, d shifted right one bit) is:

$$d^0 = a^0 \oplus b^0 \oplus c^0 \qquad (5A\text{-}3)$$

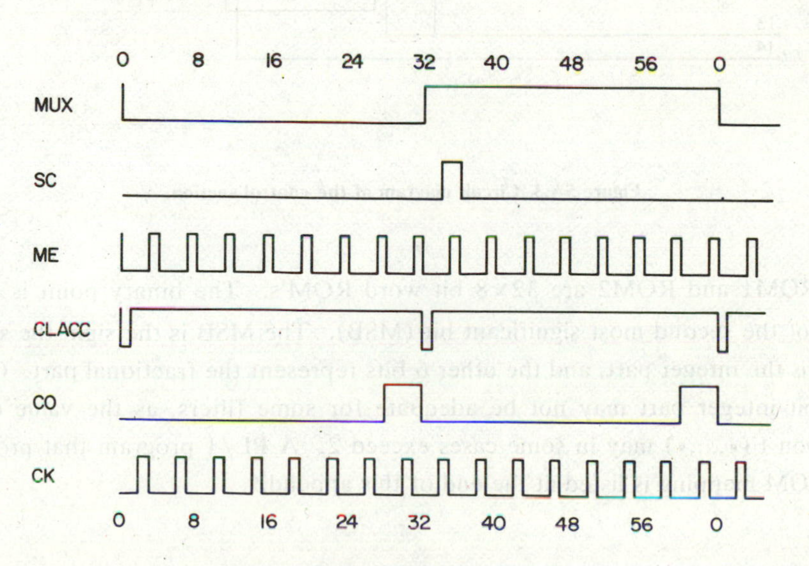

Figure 5A.2 Timing diagram of filter.

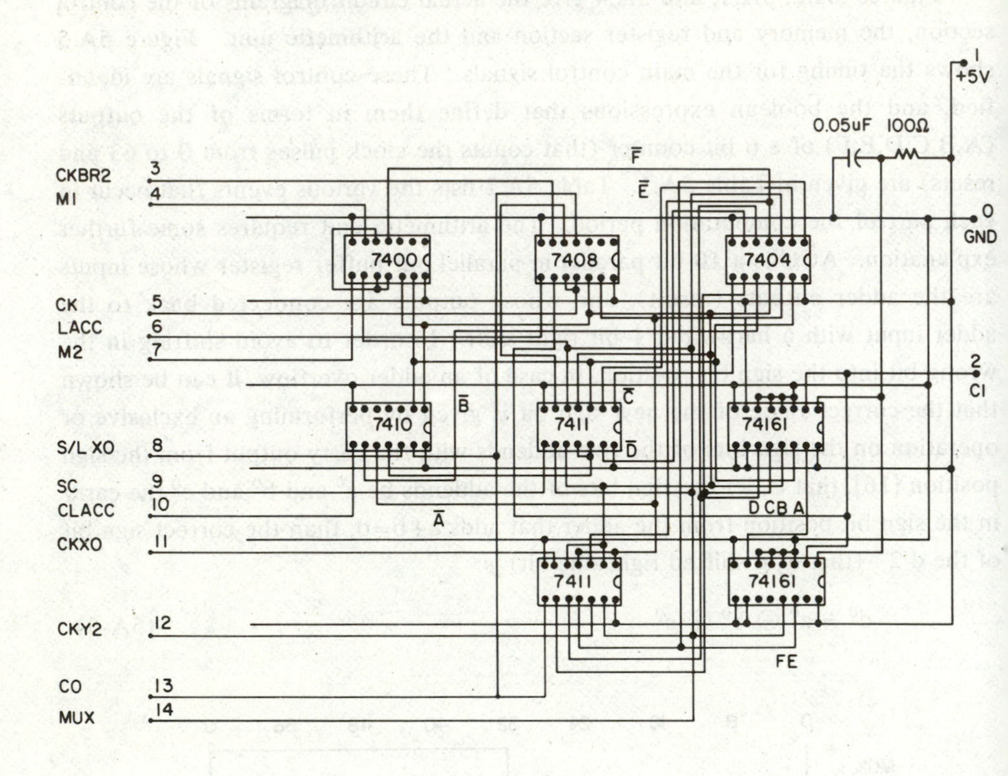

Figure 5A.3 Circuit diagram of the control section.

ROM1 and ROM2 are 32×8 bit word ROM's. The binary point is at the right of the second most significant bit (MSB). The MSB is the sign, the second MSB is the integer part, and the other 6 bits represent the fractional part. Only a one bit integer part may not be adequate for some filters, as the value of the function F(•,...,•) may in some cases exceed 2. A PL/1 program that produces the ROM mapping is listed at the end of this appendix.

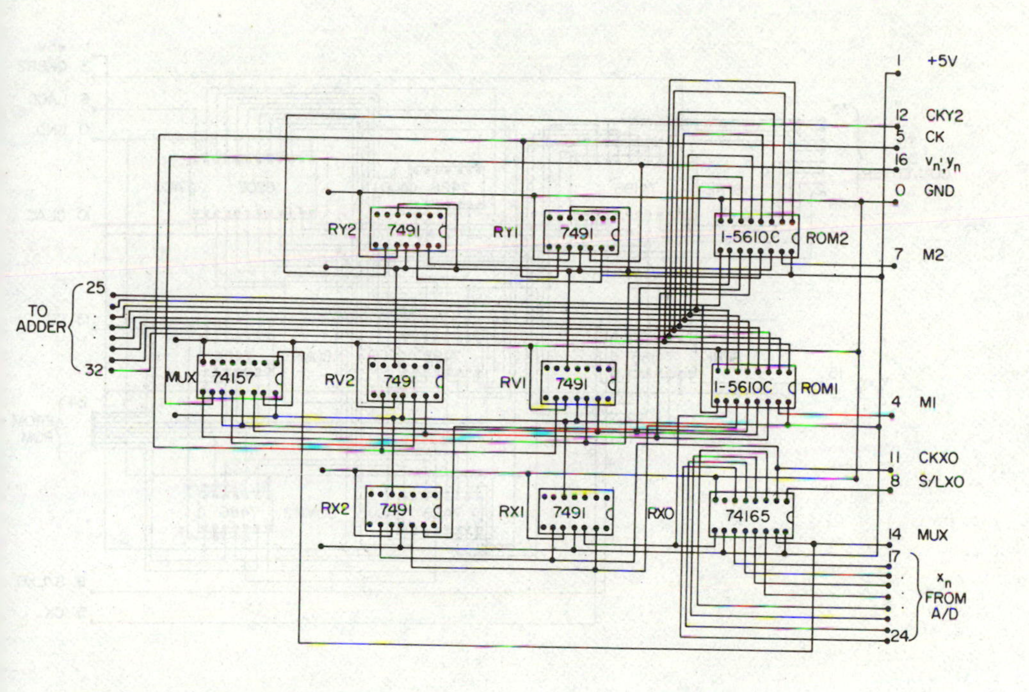

Figure 5A.4 Circuit diagram of the memory and register section.

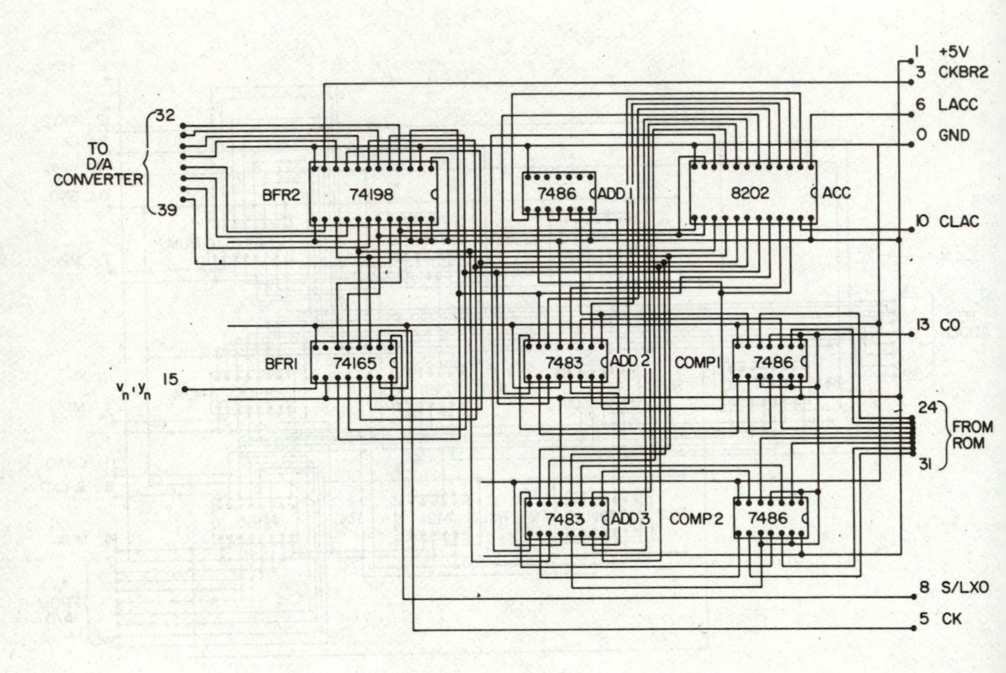

Figure 5A.5 Circuit diagram of the arithmetic unit.

Component	Control signal
A/D Converter:	
Sample & Hold (Datel SHM-3);	$S/H = MUX = F$
A/D (ADC-D8B):	$SC = F \cdot (\overline{E} \cdot \overline{D} \cdot \overline{C}) \cdot B$
Memory:	$ME = A \cdot B$
ROM1 (Intersil IM 5610):	$M1 = F + ME$
ROM2 (Intersil IM 5610):	$M2 = \overline{F} + ME$
Adder:	
Carry-in & complement signal:	$CO = E \cdot D \cdot C$
Round:	$ROUND = E \cdot \overline{D} \cdot C$
Shift Registers:	
RV1,RV2,RY1 (7491):	$Clock = CK = \overline{A} \cdot B$
RX1,RX2 (7491):	$Clock = CKXO = CK \cdot \overline{MUX}$
RX0 (74165):	$Load = S/LXO = \overline{ME \cdot CO}$ $Clock\ inhibit = MUX$ $Clock = CKXO$
RY2 (7491):	$Clock = CKY2 = CK \cdot MUX$
BRF1 (74165):	$Load = S/LXO$ $Clock\ inhibit = GND$ $Clock = CK$
Multiplexer:	
MUX (74157)	$Select = MUX$
Buffer Registers:	
ACC (Signetics 8202)	$Clear = CLACC = \overline{\overline{A} \cdot \overline{B} \cdot (\overline{C} \cdot \overline{D} \cdot \overline{E})}$
	$Load = LACC = CK \cdot CI$
BFR2 (74198):	$Clock = F \cdot CO \cdot ME$

Table 5A.1 A Listing of the Various Control Signals that are Required. (Note that $\overline{A}$ denotes the complement of A.)

Count Computation	*Events*
Period 64	*Cycles*
0-31	Compute v_n
	Sample analog signal
32-63	Compute y_n
34-63	A/D conversion
63	Data is loaded in RX0, BFR2
Computation of v_n/y_n	
Period 32	*Cycles*
28-31	Complement output of ROM's
0	Clear ACC
31	Load BFR1
One Accumulation	
Period 6	*Cycles*
0	Enable memory (ROM1), (ROM2)
2	Load ACC
	Shift (RX0,RX1,RX2),(RV1,RV2,RY1,BFR1),(RY2)
3	Disable memory

Table 5A.2. A Listing of Events that Occur in Each Part of the Computation Period.

A Computer Program for Determining F(•,...,•)

```
ROM:PROCEDURE OPTIONS(MAIN);                                                      010
 /* THIS PROGRAM GIVES THE ROM MAPPING FOR THE COEFFICIENTS OF THE */          020
/* COMBINATORIAL FILTER BUILT IN THE EE DEPT.                      */          030
/* THE ROM USED ARE 256 BIT ROM INTERSIL 5610C.                    */          040
/* THIS PROGRAM WILL ACCEPT ANY NUMBER OF SECOND ORDER SECTIONS.   */          050
/* THE DATA CARD IS PUNCHED EXACTLY AS FOLLOWS:                    */          060
 /* A(0)=... ,A(1)=... ,A(2)=... ,B(1)=... ,B(2)=... ;             */          070
/* THE COEFFICIENTS OF EACH FILTER SECTION IS GIVEN ON ONE OR MORE */          080
/* CARD AND A SEMI COLON MUST BE USED TO MARK THE END OF EACH SET  */          090
/* OF COEFFICIENTS.                                                */          100
/* WRITTEN BY TRAN-THONG, PRINCETON UNIVERSITY , APRIL 1975.      */          110
DCL    (A(0:2),B(2)) BINARY FIXED(15,10),RESULT BINARY FIXED(15,11),           120
   (I_SECTION,ROM_NO,IX0,IX1,IX2,IY1,IY2,I_CHAR1,I_CHAR2,I_BIT)                130
   BINARY FIXED(15),(CHAR_RESULT(12) CHARACTER(1),BIT_RESULT BIT              140
   (16),BIT_TEMP BIT(1) )ALIGNED;                                              150
ON ENDFILE(SYSIN) GO TO TERMINATE;                                             160
ANOTHER_FILTER: DO I_SECTION=1 TO 2;                                           170
ROM_NO=2*(I_SECTION-1)+1;                                                      180
GET FILE(SYSIN) DATA(A,B);                                                     190
PUT FILE(SYSPRINT) EDIT ('SECTION #',I_SECTION,'A(0)-A(2):',A,                 200
   'B(1)&B(2):',B) (PAGE,A,F(2),SKIP,A,3 F(20,12),SKIP,A,                     210
    2 F(20,12)) ('ADDRESS','ROM#',ROM_NO,'ROM#',ROM_NO+1)                     220
   (SKIP(6),X(8),A,X(14),A,F(2),X(19),A,F(2))                                  230
   ('PIN# 14 13 12 11 10','9  7  6  5  4  3  2  1','4  3  2  1')              240
   (SKIP(3),A,X(4),A,X(6),A);                                                  250
LOOP: DO IY2=0 TO 1;                                                           260
DO IY1=0 TO 1;                                                                 270
DO IX2=0 TO 1;                                                                 280
DO IX1=0 TO 1;                                                                 290
DO IX0=0 TO 1;                                                                 300
RESULT=0.1B*(A(0)*IX0+A(1)*IX1+A(2)*IX2-B(1)*IY1-IY2*B(2));                    310
BIT_RESULT=UNSPEC(RESULT);      /* GET BIT PATTERN              */             320
I_BIT=12;        /* THE FOLLOWING DO LOOPS SIMPLIFY THE         */             330
DO I_CHAR=1 TO 8;          /* THE OUTPUT OF THE CONTENT         */             340
BIT_TEMP=SUBSTR(BIT_RESULT,I_BIT,1);                 /* OF THE */              350
CHAR_RESULT(I_CHAR)=BIT_TEMP;    /* RESULT                      */             360
I_BIT=I_BIT-1;                                                                 370
END;                                                                           380
I_BIT=16;                                                                      390
DO I_CHAR2=9 TO 12;                                                            400
BIT_TEMP=SUBSTR(BIT_RESULT,I_BIT,1);                                           410
CHAR_RESULT(I_CHAR2)=BIT_TEMP;                                                 420
I_BIT=I_BIT-1;                                                                 430
END;                                                                           440
PUT FILE(SYSPRINT) EDIT (IY2,IY1,IX2,IX1,IX0,CHAR_RESULT)                      450
(SKIP    ,X(4),5 F(3),X(4),8 A(3),X(4),4 A(3));                                460
IF (RESULT>=1)|(RESULT<-1) THEN PUT FILE(SYSPRINT) EDIT                        470
('OVERFLOW IN ABOVE RESULT ') (SKIP,A);                                        480
END LOOP;                                                                      490
END;                                                                           500
PUT EDIT('MSB OF FUNCTION IN ROMS IS ON PIN 1 IN ROM@ 1,',
      'LSB ON PIN 4 IN ROM 2')(SKIP(4),A,A);
 PUT EDIT('ADDRESS IS FORMED AS FOLLOWS:')(SKIP(4),A);
 PUT EDIT('BITS FROM Y(N-2) GOTO PIN 14')(SKIP,A);
 PUT EDIT('Y(N-1)','13')(SKIP,X(10),A,X(10),A);
 PUT EDIT('X(N-2)','12')(SKIP,X(10),A,X(10),A);
 PUT EDIT('X(N-1)','11')(SKIP,X(10),A,X(10),A);
 PUT EDIT('X(N)  ','10')(SKIP,X(10),A,X(10),A);
GO TO ANOTHER_FILTER;                                                          510
TERMINATE: END ROM;                                                            520
```

Appendix 5.2

The Princeton FFT Processor

A Design Case Study

This appendix describes a low-cost FFT processor designed at Princeton University, which employs techniques described in this Chapter. The processor was designed and built by Mr. Alan Hochberg, a senior in the department of Electrical Engineering. The processor performs radix-2 decimation-in-frequency basic computations on data in a random-access memory (RAM), which is accessible to a computer. Its main characteristics are summarized in Table 5A.3. The organization of the processor is taken from Gold and Bially [13]. The processor adds and multiplies concurrently. In other words, it reads two data points, A_1 and C_1, from the RAM. Then, it computes $A_1 + C_1$ and $A_1 - C_1$, while reading A_2 and C_2 from the RAM. Next, it multiplies $A_1 - C_1$ by some power of W, at the same time computing $A_2 + C_2$ and $A_2 - C_2$. This is referred to as *pipelined processing*, and the intermediate registers are referred to as a *pipeline*, even though this processor does not contain a pipeline of BCU's operating concurrently as described in Chapter 3.

The RAM contains 512 complex data points. Each point comprises a 4-bit binary exponent, an 8-bit, 2's-component real mantissa, and an 8-bit 2's-complement imaginary mantissa. The memory is arranged so that two words share an 8-bit address which specifies an A and a C as input to a basic computation. This is the representation described in Section 5.6, Eq. 5-17. The block diagram of the processor is depicted in Figure 5A.6.

Chapter 3 shows how an FFT algorithm can be represented by a *flow graph*. The flow diagram (Figure 5A.7) shows a 16-point FFT. The number pairs and associated W's represent inputs to a basic computation. Coming from each dot, the results of the two basic computations are returned to the RAM in a transposed order. As Figure 5A.8 shows, if the results of a pair of computations come out of the pipeline in the order:

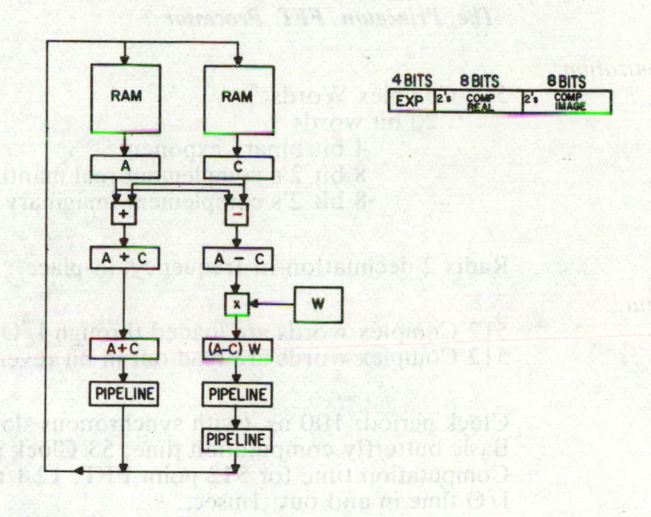

Figure 5A.6 Block diagram of the Princeton FFT processor.

A B

C D

Then, except on the last pass of the FFT, the results will be stored as:

A C

B D

so, while C and D share a double register in the pipeline, B and D will share a double word in the RAM. $J=\log_2 N$ passes over the entire memory are needed to compute the FFT.

(a) *Implementation of the Arithmetic Unit*

The processor employs a circuit similar to that of Figure 5.10 to multiply (A−C) by W. The circuit described in Section 5.6 is for computation using two bits at a time; the Princeton FFT processor computes using one bit at a time, to reduce memory requirements. To describe this scheme, let P_k and Q_k of Eq. 5.19 be redefined as:

The Princeton FFT Processor

Data Organization:	
	512 Complex Words 20 bit words 4 bit binary exponent 8 bit 2's complement real mantissa 8 bit 2's complement imaginary mantissa
Algorithm:	
	Radix 2 decimation-in-frequency, in-place.
Input/Output:	
	512 Complex words are loaded through I/O port. 512 Complex words are read out in bit reversed order.
Timing:	
	Clock period; 100 ns. (with synchronous slowdown). Basic butterfly computation time; 53 Clock periods. Computation time for 512 point FFT; 12.1 msec. I/O time in and out; 1msec.
Technology:	
	MSI and SSI Standard TTL and MOS static memory. 170 Integrated Circuit Packages
Physical:	
	Power Consumption: 5 VDC 7 A. Packaged on 3 Wirewrap boards, 9×16×1.5 inch.

Table 5A.3. The Main Characteristics of the Princeton FFT Processor.

$$P_K(u,w) = u\cos(2\pi K/N) + w\sin(2\pi K/N)$$

$$Q_K(u,w) = -u\sin(2\pi K/N) + w\cos(2\pi K/N) \tag{5A-4}$$

Letting

$$H = A - C = \left[\sum_{j=1}^{B} 2^{-j}h_R{}^j - h_R{}^0\right] + \left[\sum_{j=1}^{B} h_I{}^j\, 2^{-j} - h_I{}^0\right] \tag{5A-5}$$

yields expressions for z_R and z_I as follows:

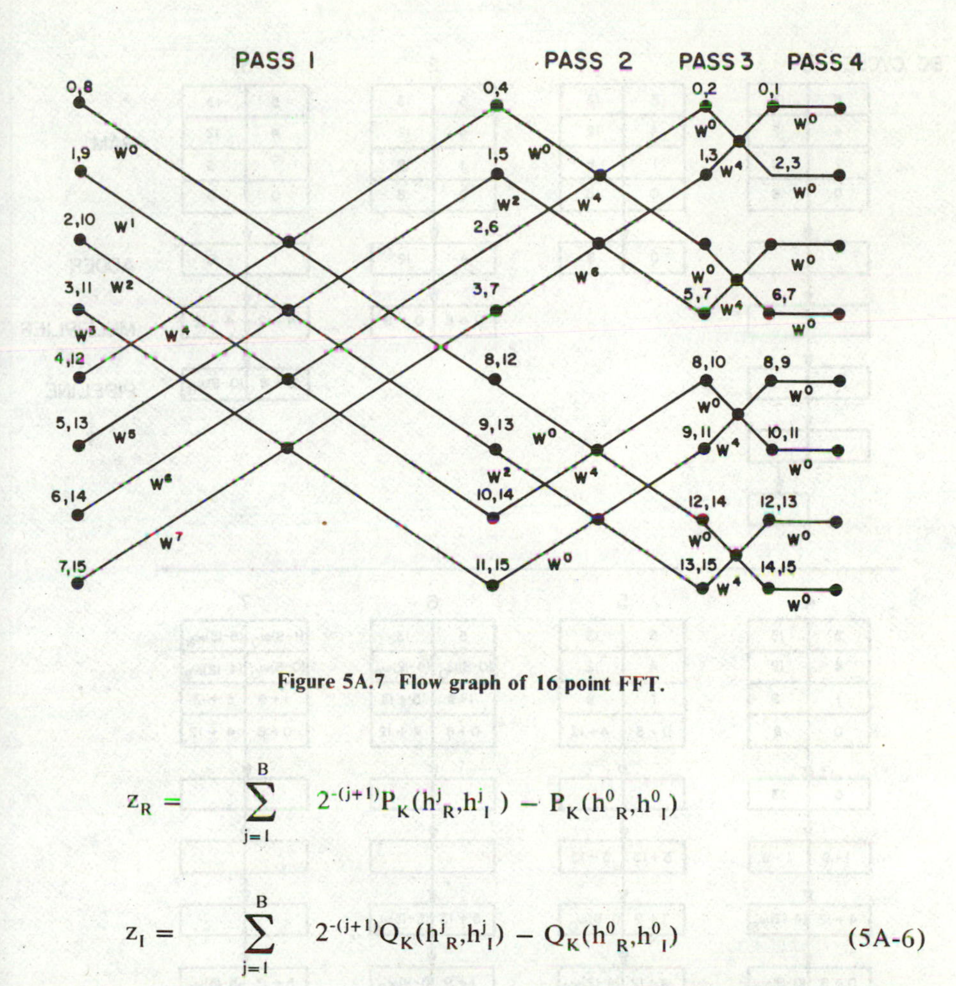

Figure 5A.7 Flow graph of 16 point FFT.

$$z_R = \sum_{j=1}^{B} 2^{-(j+1)} P_K(h^j_R, h^j_I) - P_K(h^0_R, h^0_I)$$

$$z_I = \sum_{j=1}^{B} 2^{-(j+1)} Q_K(h^j_R, h^j_I) - Q_K(h^0_R, h^0_I) \quad (5A\text{-}6)$$

This scheme thus requires (B-1) additions and one subtraction to compute z_R and z_I, with a hardwired one-bit right shift accumulation. The circuit is similar to Figure 5.10, with $h_R{}^j$ and $h_I{}^j$ replacing $c_R{}^j$ and $c_I{}^j$ respectively, and the connections to $c_R{}^{j+1}$ and $c_I{}^{j+1}$ not included.

The processor contains only one BCU, so only one set of memories is required. This memory contains P_k and Q_k for all values of K=0,1,2,...,225. As the flow diagram shows, only the first pass over the RAM actually uses all 256 values of P_k and Q_k.

Figure 5.10 shows ROM's addressed by the bit string $(k^0, k^1, ... k^J, h_R{}^j, h_I{}^j)$. To

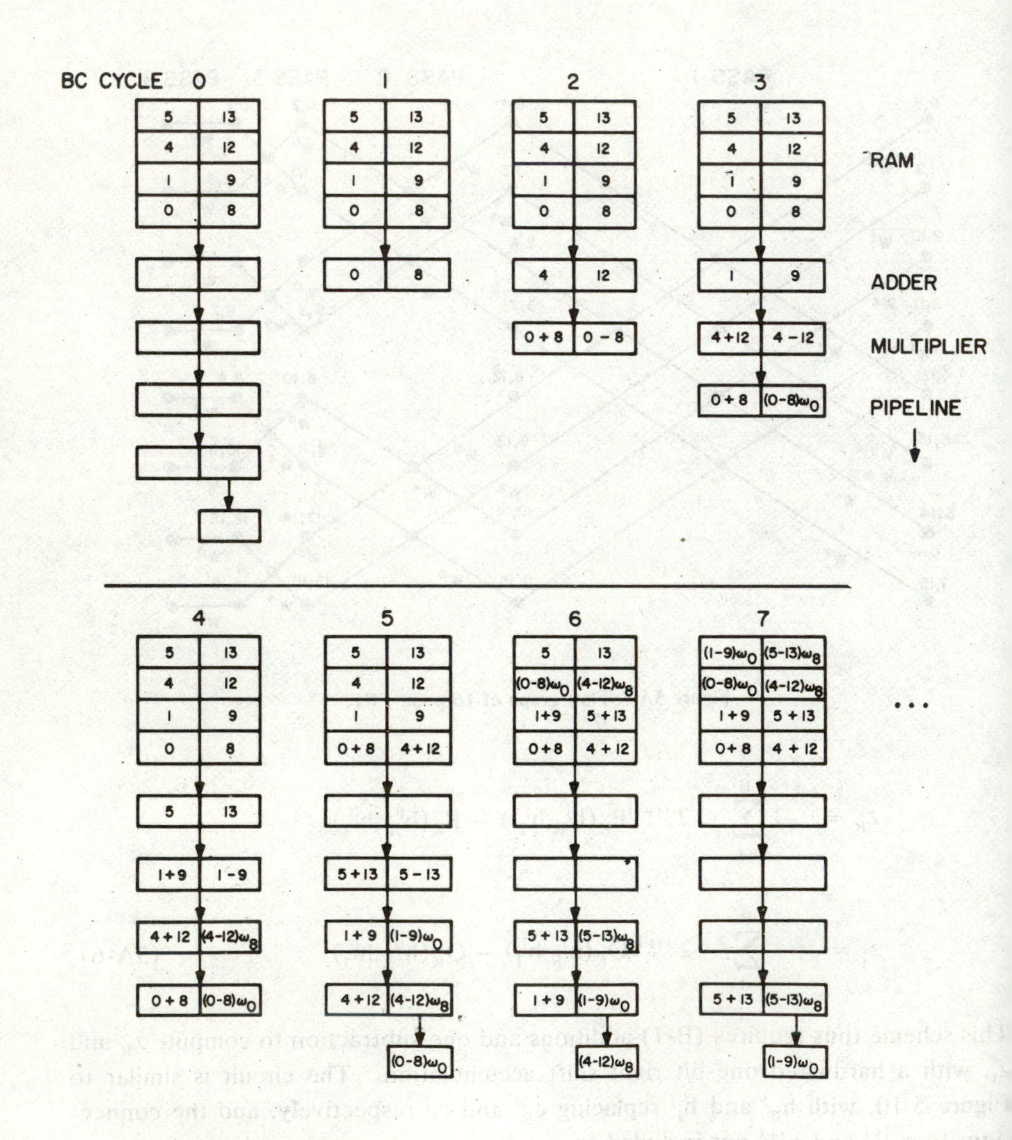

Figure 5A.8 Flow of data through the pipeline.

perform a basic computation, then, the ROM is accessed B times. A faster circuit has been designed, in which all the coefficients are not stored in the same ROM. The modified scheme is shown in Figure 5A.9. Here, at the beginning of each basic computation, $P_k(u,w)$ and $Q_k(u,w)$ are stored, for all four possible combinations of u and w, in a TTL register called a cache. The outputs of the caches are wired together onto a bus. The bits u and w are connected to the *OUTPUT ENABLE* pins of the cache IC's, so that only the cache whose *OUTPUT ENABLE* code corresponds to the values of u and w will assert the value it contains, $P_k(u,w)$ or $Q_k(u,w)$, onto the bus.

One advantage of this modified scheme is that a basic computation now requires only one reading of the ROM, plus B accesses to the cache. Since the cache will typically be more than ten times faster than the ROM, this scheme can greatly increase the speed of computation. Table 5A.4 illustrates this increase. A second advantage derives from the fact that:

$$P_k(0,0) = Q_k(0,0) = 0 \quad \text{for all } k \tag{5A-7}$$

This allows substitution of a grounded wire for a more expensive 256×8 ROM! This advantage is also reflected in Table 5A.4.

One problem that arises in designing the cache is that, if the cache outputs are ordinary TTL, they cannot simply be wired together into a bus. If an output in a '1' state is wired to an output in a '0' state, the IC in the '1' state can be damaged. This problem is solved by type 74173 *tri-state* TTL registers. In addition to data inputs and outputs, this IC has two *ENABLE* control inputs. When both *ENABLE* inputs are low, the register outputs are normal TTL '0's and '1's. Otherwise, the register is effectively *disconnected* the outputs merely present a high-impedance load to the bus.

Tri-state TTL is also used in the *2×2 matrix transposing* circuitry of this processor, shown at the bottom of Figure 5A.9. The outputs of the pipeline are gated through type 74368 tri-state buffers onto busses to either word of the double-word RAM. Input from the computer to load the RAM can also be asserted onto this bus.

(b) *Design Details*

2's-complement arithmetic for the FFT is not hard to implement. The type 74181 Arithmetic-Logic Unit (ALU) IC provides addition and subtraction circuitry for four bits and a carry bit. To rescale a number after an arithmetic operation, the exponent is placed in a type 74161 up-counter, while the mantissas are loaded into type 74199 shift registers, as shown in Figure 5A.10. While the

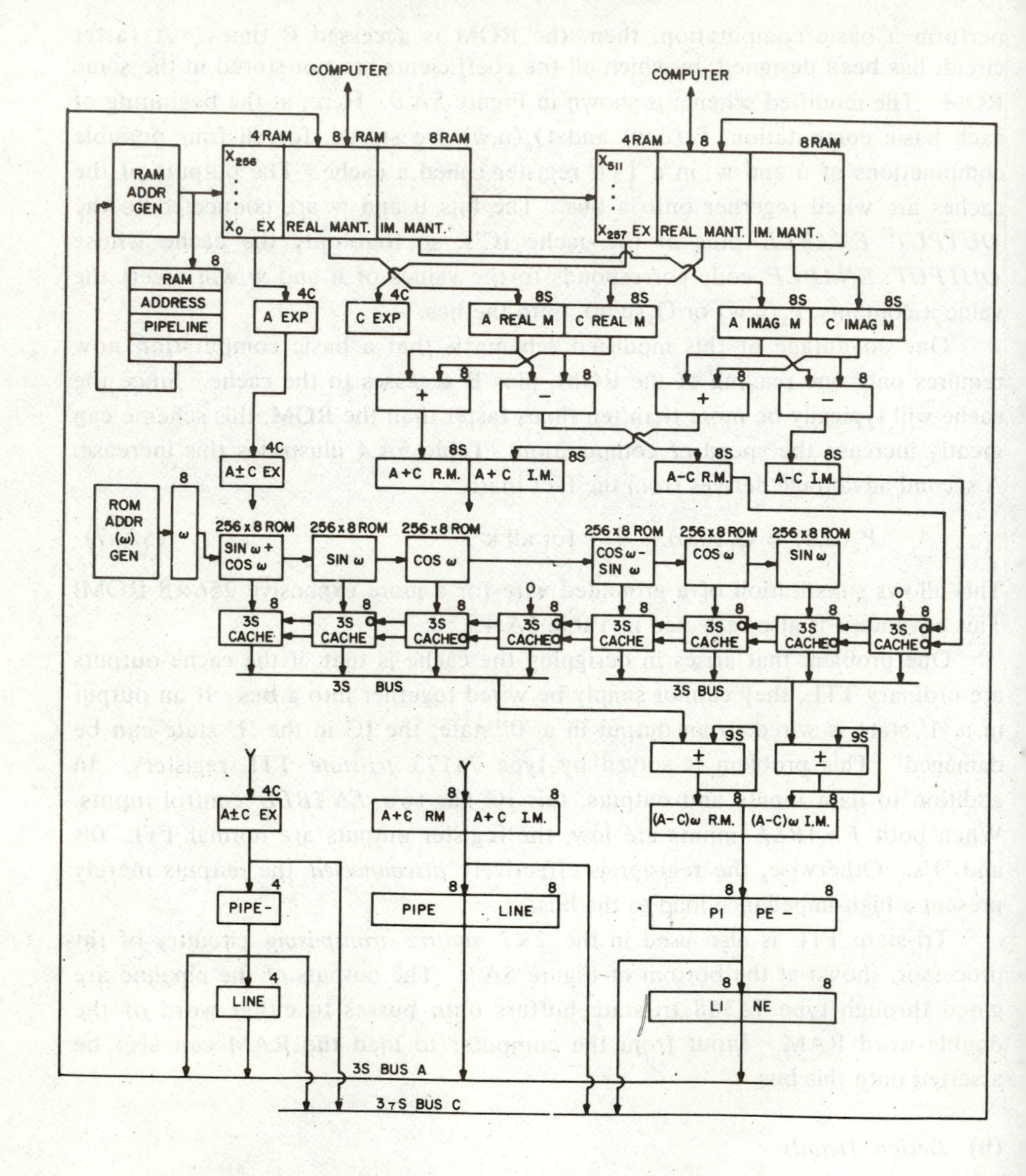

Figure 5A.9 Diagram of processor and arithmetic unit organization.

	1 Bit No Cache	2 Bit No Cache	1 Bit Cache	2 Bit Cache
#Mem. Access	8	4	1	1
#Cache Access	0	0	8	4
Time μs	8.8	4.4	2.04	1.52
Mem. Size	16×1024	16×4096	$6 \times 8 \times 256$	$30 \times 8 \times 256$
Cache Size	0	0	64	256
Cost (U.S.$)	240	960	204	996

Table 5A.4. A Case for Using a Cache in the FFT.
(Assumptions: 1. 1 μs MOS memory, 30 nsec cache, 100 nsec compute.
2. Memory cost 30$/256 words, Cache cost 3$/8 bits.)

74181 provides a carry bit, it provides no indication of arithmetic overflow. Overflow of the operation C=A+B is detected by the logic equation:

$$[\text{SIGN}(A) = \text{SIGN}(B) \cap (\text{SIGN}(A) \oplus \text{SIGN}(C)] \rightarrow \text{OVERFLOW} \quad (5A\text{-}8)$$

Similarly, the equation:

$$[\text{SIGN}(A) = \overline{\text{SIGN}(B)} \cap (\text{SIGN}(A) \oplus \text{SIGN}(C)] \rightarrow \text{OVERFLOW} \quad (5A\text{-}9)$$

can be applied to detect an overflow from 2's-complement subtraction.

Designing an FFT processor invariably involves designing circuitry to generate a large number of control pulses of varying widths and frequencies. A brute-force approach to this sometimes calls for a chain of one-shot multivibrators, counters, and gates. One-shot multivibrators, however, are always plagued with drift, which may lead to race conditions. Also, use of one-shots precludes the ability to *single-step* the processor at slow speeds for maintenance and debugging.

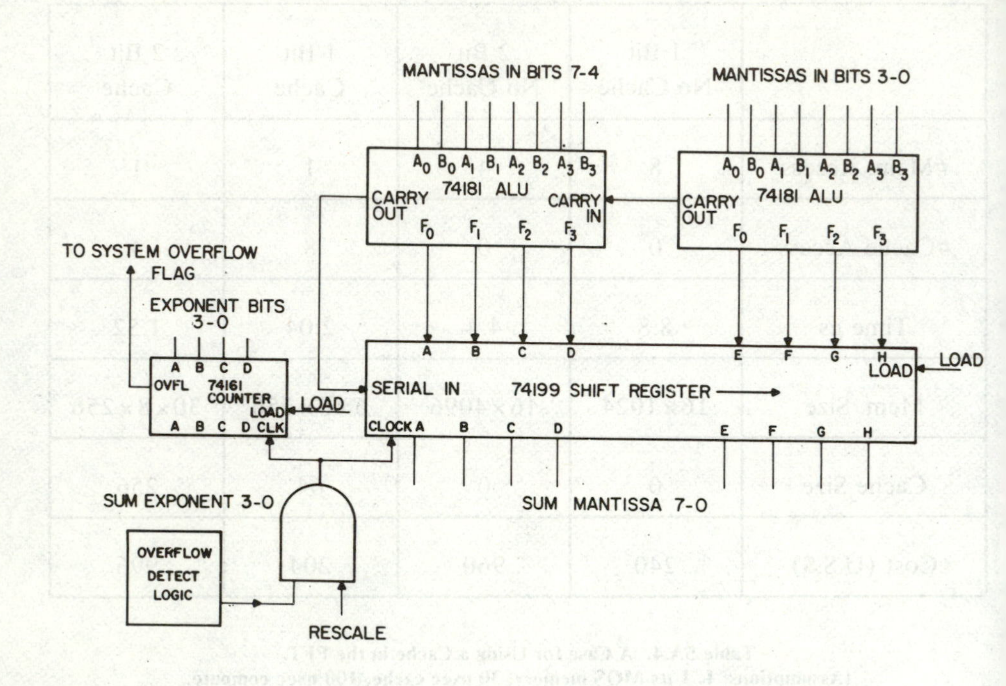

Figure 5A.10 Accumulation and overflow detection.

A better-organized approach to pulse generation is shown in Figure 5A.11. The shift register contains a single '1', which constantly rotates. Sequential pulses can be *tapped* from the shift register outputs. For wider pulses, the single pulse is used to set, and later clear, a set/reset flip-flop. To control its basic computation cycle, the Princeton FFT processor employs a 53-bit shift register, of which each bit is pulsed on for 100 nsec of the 53μs. basic cycle time. Approximately twenty control signals are derived from this register, so the entire FFT is under the control of one time base. The time base frequency can vary freely from DC to 10 MHz without affecting the computations.

(c) *Addressing the RAM and ROM*

In addition to control circuitry, the Princeton FFT processor includes a circuit to generate the proper sequence of RAM and ROM addresses (k's) to perform the FFT. Expressions for these numbers are messy, but the circuitry is surprisingly simple. Once a RAM address is generated, it is stored and travels along with the

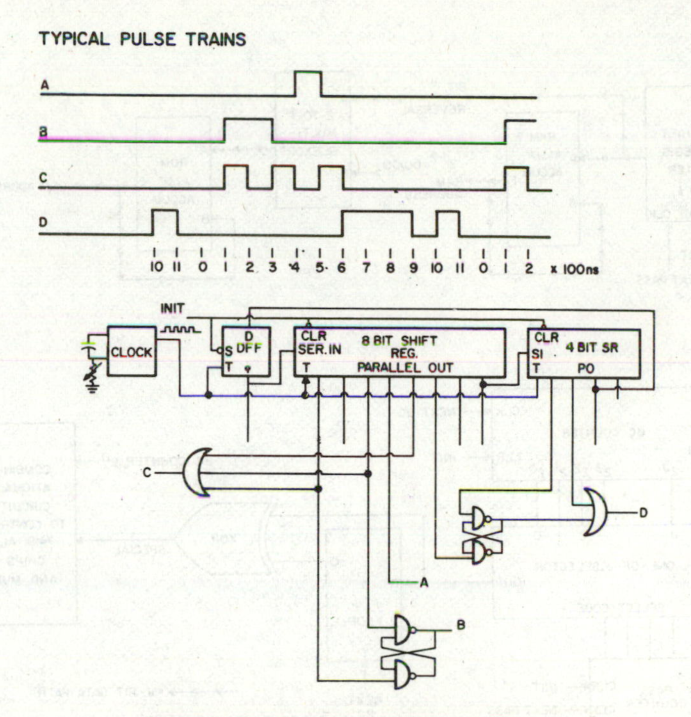

Figure 5A.11 Generation of timing pulses.

data through the pipeline, so that data can be written back into the locations from which it was read, without generating the address again.

The addressing of the RAM and ROM is relatively straightforward, and can be easily derived from the explanation of the FFT algorithm in Chapter 3. For an N point FFT, the processor makes $J = \log_2 N$ passes over the entire RAM. The addresses are generated by the circuit shown in Figure 5A.12.

An incoming pulse clocks a J-bit counter for each new address. The low-order bit of this counter provides the 1,0,1,0,... signal to control the alternating difference values. The (J+1)th bit becomes a '1' at the start of a new pass, and is used to clear the RAM and ROM accumulators.

Except for some special cases, the RAM address increment is always $+2^{J-p-1}$ or $-(2^{J-p-1}-1)$. These values can be derived from a shift register that is initially loaded with '0011...11', and is shifted right one place after each pass. The ROM address increment is either $+2^{J-2}$, that may be hardwired, or a number which turns out to be a bit reversed version of the RAM address shift register mentioned above. The same register can thus be used in both places. The only

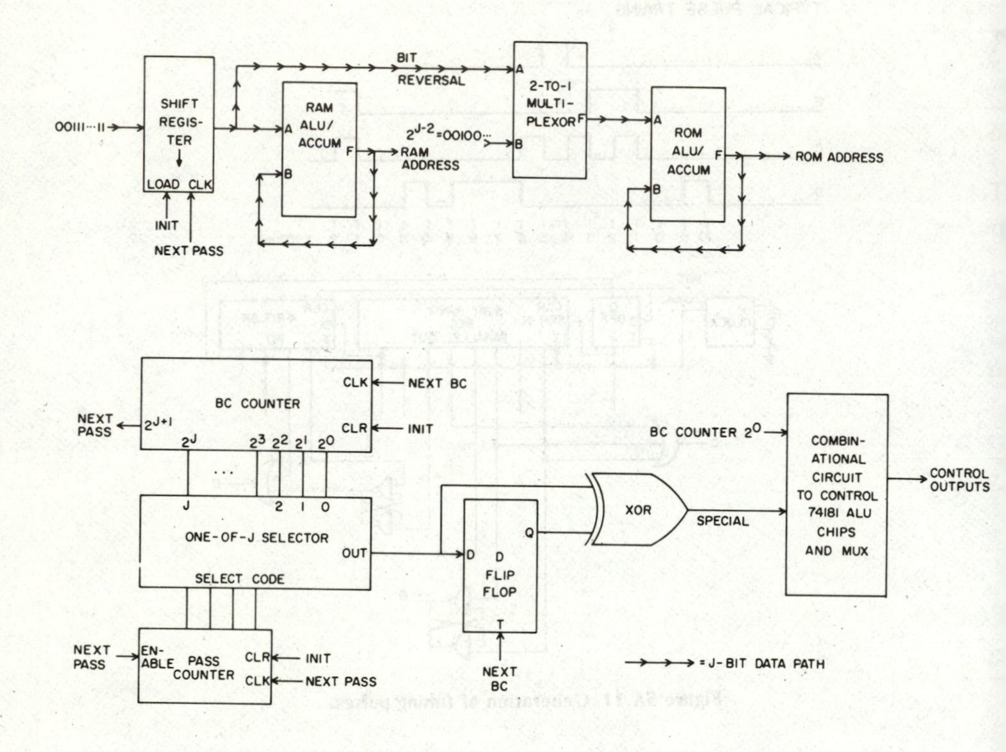

Figure 5A.12 Generation of addresses.

remaining problem is to generate a pulse (called *SPECIAL*) to take care of these special cases. These operations always involve incrementing the RAM address and clearing the ROM address. A mask or selector samples the proper bit, and the XOR gate around the D-type flip-flop provides the pulse *SPECIAL*.

Design of this circuit is greatly facilitated by the availability of the type 74181 ALU IC, which can produce the outputs:

$$F \leftarrow A + B + 1 \tag{5A-10}$$
$$F \leftarrow A + B$$
$$F \leftarrow A - B$$
$$F \leftarrow A + 1$$
$$F \leftarrow 0$$

required for the accumulators. Standard logic techniques provide control of the RAM address ALU as follows:

$$\mathit{SPECIAL} \rightarrow (\text{RAM ALU CTRL: } F \leftarrow A+1) \qquad (5A\text{-}11)$$

$$\overline{\mathit{SPECIAL}} \cap (\text{BC COUNTER } 2^0) \rightarrow (\text{RAM ALU CTRL: } F \leftarrow A+B+1)$$

$$\overline{\mathit{SPECIAL}} \cap (\overline{\text{BC COUNTER } 2^0}) \rightarrow (\text{RAM ALU CTRL: } F \leftarrow A-B)$$

The ROM address ALU and the multiplexer that chooses between its two possible inputs are controlled according to the equations:

$$\mathit{SPECIAL} \rightarrow (\text{ROM ALU CTRL: } F \leftarrow 0) \qquad (5A\text{-}12)$$

$$\overline{\mathit{SPECIAL}} \cap (\text{BC COUNTER } 2) \rightarrow (\text{ROM ALU CTRL: } F \leftarrow A+B,\ \text{MUX CTRL: } F \leftarrow A)$$

$$\overline{\mathit{SPECIAL}} \cap (\overline{\text{BC COUNTER } 2}) \rightarrow (\text{ROM ALU CTRL: } F \leftarrow A-B,\ \text{MUX CTRL: } F \leftarrow B)$$

Several other integrated circuits types have facilitated the design of the Princeton FFT processor these are listed in Table 5A.5, with special uses indicated.

IC Type	Function	Usage
7483	4 bit adder	Adder-Subtractor
74100	8 bit buffer	Pipelines
74157	Quad 2:1 Mux	Address generator
74161	4 bit counter	Rescalable exponent
74173	Quad 3-S buffer	Computation
74181	4 bit ALU	Computation, addr. gen.
74199	8 bit shift reg.	Rescalable mantissa
74368	Hex 3-S gate	2×2 matrix transposition
Intel 1702	PROM	Twiddle factors
GI RA/3/4256B	256x4 RAM	Data memory

Table 5A.5. Integrated Circuits Used.

(d) *Interfacing*

The FFT processor interfaces to a computer with standard done/busy signals. The computer asserts a signal called *DATA AVAILABLE* to send data to the FFT processor. The processor samples this signal, strobes the data in from an I/O bus, and asserts *DATA ACCEPTED*. This tells the computer to drop *DATA AVAILABLE* and put the next sample on the I/O bus. A similar sequence occurs to send the transformed data back to the computer. The I/O interface signals are shown in Figure 5A.13.

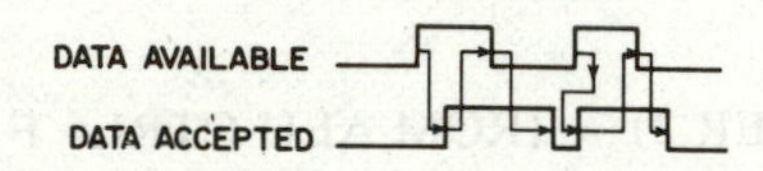

Figure 5A.13 Computer interfacing protocol.

The processor may be modified to not include its own random access memory. Instead, it would perform the FFT on data stored within the main memory of a general purpose computer, using direct-memory-access and cycle-stealing techniques. This would eliminate the overhead involved in loading and reading the RAM.

(e) *Summary*

This appendix has presented some of the design techniques used in the Princeton FFT processor. An efficient memory organization was explained, and tri-state TTL logic was shown to simplify its implementation. General methods of overflow rescaling and control pulse generation were discussed. Finally, the circuitry used to generate the *butterfly* pattern of computation was discussed. The information in this appendix will be helpful to anyone designing hardware for the fast Fourier transform. Additional information on the design with TTL circuits may be found in [14,15].

CHAPTER 6

Some Additional Implementation Considerations

6.1 INTRODUCTION

We have presented in the previous chapters the basic theory of digital signal processing as well as the implementation of digital filters and FFT processors. It has been noted that because of the finite length words used, certain phenomena that occur can seriously affect the operation of these systems. These undesirable effects must be kept under control. Some of the problems have been discussed in the previous chapters, Sections 1.8 and 2.5 for example. In the present chapter, we shall discuss a number of additional topics that are important from the practical point of view.

In Section 6.2 we analyze the round-off error caused by the use of finite length words. We deal only with the case of fixed point arithmetic with rounding. Using some elementary results from random signal theory, we derive expressions for the round-off error in the case of direct, parallel and cascade realization of digital filters. Section 6.3 takes up the problem of properly scaling the signal levels at various intermediate points so as to prevent overflow, and choosing a good pole zero pairing so as to minimize the round-off error in cascade realizations. Appendix 6.1 contains a PL/1 computer program that automatically generates such assignments. Section 6.4 is devoted to an introductory treatment of the problems of limit cycle and overflow oscillations. Finally, Section 6.5 deals with the accuracy and scaling problems arising in the computation of the FFT using fixed point arithmetic.

6.2 ROUND-OFF ERROR IN DIGITAL FILTERS

In Section 1.6, we have identified three sources of errors due to the finite word length; namely,

1. Quantization of input.
2. Quantization of filter coefficients.
3. Round-off in arithmetic operations.

Of these three errors, the round-off error is usually the most serious one. Although no detailed discussion has been devoted so far to the input quantization problem, nor to the arithmetic round-off problem, we have already seen in Section 2.9 that the quantization of filter coefficients causes the frequency response to deteriorate from that obtained in the original design. Coefficient quantization may also result in instability of the filter, as we shall see later.

For some of the discussion to be presented later in this section, we will need a result from random signal theory. Consider a linear system with transfer function H(z) and a stationary random input signal $\{u_n\}$. Let the output be denoted by $\{v_n\}$ which can be shown also to be stationary. Suppose that the input $\{u_n\}$ is white, has zero mean and mean squared value σ_u^2, that is,

$$\langle u_n \rangle = 0 \tag{6-1}$$

$$\langle u_n^2 \rangle = \sigma_u^2 \tag{6-2}$$

where the symbol $\langle \bullet \rangle$ is used to denote statistical average. Then the stationary output $\{v_n\}$ has zero mean,

$$\langle v_n \rangle = 0 \tag{6-3}$$

and a mean squared value σ_v^2 is given by[1]

$$\sigma_v^2 = \langle v_n^2 \rangle = \sigma_u^2 \frac{1}{2\pi j} \oint H(z)H(1/z)\frac{dz}{z}$$

$$= \sigma_u^2 \frac{1}{2\pi} \int_0^{2\pi} |H(e^{j\lambda})|^2 d\lambda \tag{6-4}$$

In the general case where $\{u_n\}$ is not white, but stationary with zero mean and power spectral density $\Phi_{uu}(z)$. Then we have

---

[1] Unless otherwise specified, the contour integration with respect to z is along the unit circle $|z| = 1$.

$$\sigma_u^2 = \langle u_n^2 \rangle = \frac{1}{2\pi j} \oint \Phi_{uu}(z) \frac{dz}{z} \tag{6-5}$$

The output $\{v_n\}$ is also stationary with zero mean, its power spectral density is

$$\Phi_{vv}(z) = H(z)\, H(1/z)\, \Phi_{uu}(z) \tag{6-6}$$

and the mean squared value is

$$\begin{aligned} \sigma_v^2 = \langle v_n^2 \rangle &= \frac{1}{2\pi j} \oint \Phi_{vv}(z) \frac{dz}{z} \\ &= \frac{1}{2\pi j} \oint H(z)\, H(1/z)\, \Phi_{uu}(z) \frac{dz}{z} \\ &= \frac{1}{2\pi} \int_0^{2\pi} |H(e^{j\lambda})|^2\, \Phi_{uu}(e^{j\lambda})\, d\lambda \end{aligned} \tag{6-7}$$

Equation 6-7 reduces to 6-4 when $\{u_n\}$ is white. The derivation of these results can be found in many text books dealing with random signals [1].

(a) *Input Quantization*

Quantization of the input signal takes place in the analog-to-digital converter, Figure 6.1, which is the first component in the system of Figure 1.9. Here the input analog signal is sampled periodically and each sample is coded into a digital word of, say, B bits. Typically B is about 12 bits, although it may be as low as 8 bits or as high as 20 bits. Before a sample x_n can be coded, it must be properly modified so that it takes on only one of the 2^B possible values. This is accomplished by means of a quantizer that is a nonlinear transducer with an input-output transfer characteristic as shown in Figure 6.2. The output of the quantizer x_n^Q takes on the value $k\Delta$ if the input x_n is between $(k-1/2)\Delta$ and $(k+1/2)\Delta$. Let the full dynamic range be from -1 to 1. Then the step size Δ is 2^{-B+1} It is clear that

$$|x_n - x_n^Q| < \Delta/2 \tag{6-8}$$

we may also write

$$x_n^Q = x_n + e_n \tag{6-9}$$

where e_n is the error due to quantization. From Eq. 6-8,

$$-\Delta/2 < e_n < \Delta/2 \tag{6-10}$$

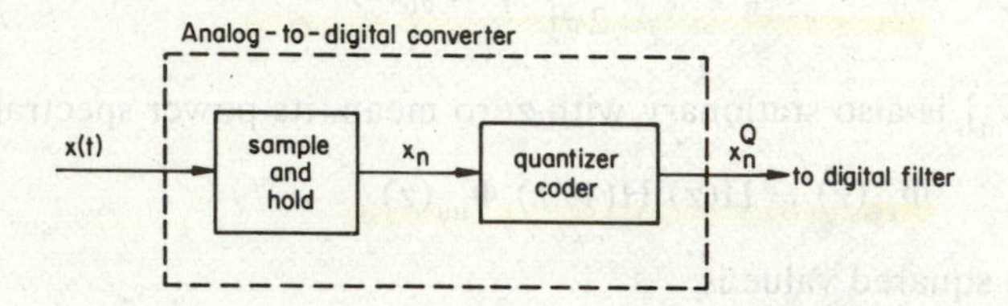

Figure 6.1 Block diagram of analog-to-digital converter.

That is, the error due to quantization is bounded by half the step size. It has been shown [2], that for most input signals the sequence $\{e_n\}$ can be regarded as a *white noise* that is statistically independent of the original signal sequence $\{x_n\}$. For this reason, we may call $\{e_n\}$ quantizing noise. It can also be shown that e_n has zero mean and its mean squared value is $\Delta^2/12$.

$$\langle e_n \rangle = 0 \qquad (6\text{-}11)$$

$$\langle e_n^2 \rangle = \Delta^2/12 = 2^{-2B}/3 \qquad (6\text{-}12)$$

The last result follows from the fact that $\Delta = 2^{-B+1}$.

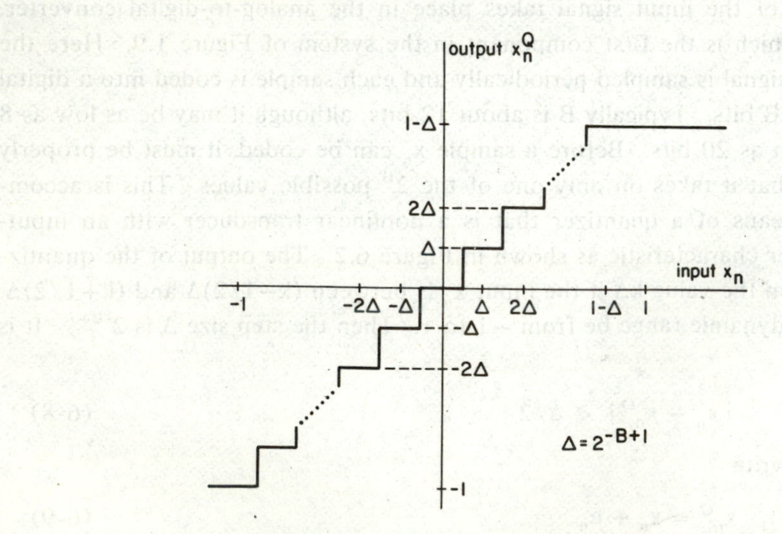

Figure 6.2 Transfer function of the quantizer.

From this discussion, we see that the actual input to the digital filter is $\{x_n^Q\}$ rather than $\{x_n\}$. According to Eq. 6-9, $\{x_n^Q\}$ has two additive components, the original $\{x_n\}$ and a quantizing noise $\{e_n\}$ (see Figure 6.3). Superposition holds for the digital filter, it being a linear system. So the output also consists of two additive components. The one due to $\{x_n\}$ as input is the desired or ideal output $\{y_n\}$, and the other component due to $\{e_n\}$ as input is the error. Let us denote the output error by $\{\epsilon_n\}$. Since $\{e_n\}$ is white noise, its mean squared value as given by Eq. 6-12 is $2^{-2B}/3$. According to Eqs. 6-3 and 6-4, the error ϵ_n has zero mean and its mean squared value is

$$<\epsilon_n^2> = \left(\frac{2^{-2B}}{3}\right)\frac{1}{2\pi}\int_0^{2\pi} |H(e^{j\lambda})|^2 \, d\lambda \qquad (6\text{-}13)$$

where H(z) is the transfer function of the filter. There are a number of short cuts to evaluate this integral [3] that will not be discussed here.

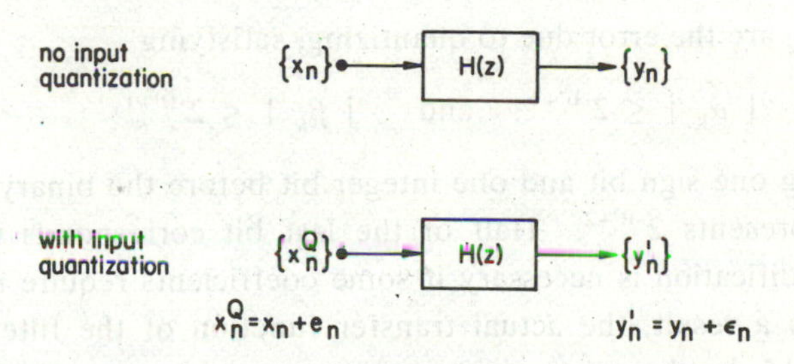

Figure 6.3 The effect of input quantization.

Equation 6-13 gives the statistical mean squared error. It is also possible to show that the absolute error $|\epsilon_n|$ is bounded by

$$|\epsilon_n| \leq 2^{-B} \sum_{n=0}^{\infty} |h_n| \qquad (6\text{-}14)$$

where $\{h_n\}$ is the impulse response of the filter. However, this bound is overly pessimistic to be of much practical use.

(b) *Coefficient Quantization*

This problem has already received some attention in Chapter 2 where we have seen that when a filter with transfer function

$$H(z) = \frac{\sum_{k=0}^{M} a_k z^{-k}}{1 + \sum_{k=1}^{L} b_k z^{-k}} \tag{6-15}$$

is actually implemented, the coefficients a_k and b_k have to be quantized. Suppose B' bits are used to represent the coefficients, and let the actual coefficients be denoted by $[a_k]_{B'}$ and $[b_k]_{B'}$. Typically B' is 6 or higher. From our previous discussion, we see that

$$\begin{aligned} [a_k]_{B'} &= a_k + \alpha_k \\ [b_k]_{B'} &= b_k + \beta_k \end{aligned} \tag{6-16}$$

where α_k and β_k are the error due to quantizing, satisfying

$$|\,\alpha_k\,| \le 2^{-B'+1} \quad \text{and} \quad |\,\beta_k\,| \le 2^{-B'+1} \tag{6-17}$$

This is assuming one sign bit and one integer bit before the binary point, so that the last bit represents $2^{-B'+2}$. Half of the last bit corresponds to $2^{-B'+1}$. An appropriate modification is necessary if some coefficients require more than one integer bit. As a result, the actual transfer function of the filter is no longer given by Eq. 6-15, but becomes,

$$[H(z)]_{B'} = \frac{\sum_{k=0}^{M} [a_k]_{B'}\, z^{-k}}{1 + \sum_{k=1}^{L} [b_k]_{B'}\, z^{-k}} \tag{6-18}$$

The frequency response of the actual filter can now be calculated by using Eq. 6-18. The reader may recall from the examples in Chapter 2, Section 2.9, that the actual filter characteristics depend on the number of bits used in the implementation, and the effect of coefficient quantization can be quite serious in some cases.

One can also calculate the movements of the poles and zeros of the transfer function due to coefficient quantization. Suppose the poles in H(z) of Eq. 6-15 are at p_i, $i=1,2,...,L$. Then it can be shown [4] that the poles of $[H(z)]_B$, of Eq. 6-18 are moved to $p_i + \Delta p_i$, where

$$\Delta p_i \approx \sum_{k=1}^{L} \frac{p_i^{k+1}}{\prod_{\substack{m=1 \\ m \neq i}}^{L} (1 - p_i/p_m)} \beta_k \tag{6-19}$$

Similar results hold for the movement of zeros. From these movements, the changes in the filter frequency response can be studied.

When the filter has high Q poles, that is, poles that are close to the unit circle in the z plane, instability may occur as a result of the pole movement. This problem can be quite serious when the sampling rate is relatively high, even for moderately low-order filters simulated on the computer in the direct form [5]. It has been shown [4] that for an L-th order low-pass filter operating at a sampling rate of 1/T with distinct poles at $e^{-z_k T}$, $k=1,2,...,L$, stability is guaranteed if the number of bits used, B' satisfies

$$B' > \left[-\log_2\left(\frac{5\sqrt{L}}{2^{L+2}} \prod_{k=1}^{L} z_k T\right)\right]_I + 1 \tag{6-20}$$

where the bracket $[\ ]_I$ here denotes the integer part of the quantity inside.

Although results of this type give an absolute guarantee for the filter to be stable, the actual performance of the filter would have deteriorated so much as to be unacceptable before stability becomes a problem. For this reason, we have not presented the details in these discussions. The interested reader should work out some of the problems related to this topic at the end of this chapter, or consult the references for the details.

In general, the effect of coefficient inaccuracy is more pronounced for a high order filter when it is realized in the direct form than in the parallel or cascade form. As a rule, therefore, the parallel or cascade form should be used for high order filters whenever possible. The saving in the number of coefficient bits can be quite substantial.

(c) *Round-off Error in Digital Filters* [6]

Consider a digital filter with input-output relationship

$$y_n = \sum_{k=0}^{M} a_k x_{n-k} - \sum_{k=1}^{L} b_k y_{n-k} \tag{6-21}$$

its transfer function H(z) is given by Eq. 6-15. To focus our attention on the round-off error, let us first assume that the input samples x_n are between ± 1, quantized and coded to B bit words including the sign bit, and that each filter coefficient is represented by a B' bit word. Let us also assume that the filter calculates the output sample y_n following Eq. 6-21 strictly. That is the filter has a multiplier to calculate the individual products $a_k x_{n-k}$ and $b_k y_{n-k}$ and has an accumulator to add (subtract) these products to form y_n.

The quantity $a_k x_{n-k}$, being the product of a B bit word and a B' bit word, has in general B+B' bits. In order to fit it into a data register that has room only for B bits, the product must be shortened. There are two ways to accomplish this, *chopping* and *rounding* [6,7]. In chopping, only the most significant B bits are kept, whereas in rounding, a 1 is first added to the B-th bit if the (B+1)-th bit is 1 and the result is then chopped to B bits. Needless to say, rounding is more accurate, but more hardware or time is needed. The analysis of round-off error when chopping is used is somewhat involved [8]. We will treat here only the rounding case.

The process of rounding a (B+B') bit product to B bit is very similar to quantization discussed in the beginning of this section. It introduces an error that is bounded in absolute value by 2^{-B}, according to Eq. 6-8. The error can also be regarded as a random noise with zero mean and mean squared value $2^{-2B}/3$, according to Eqs. 6-11 and 6-12. In the calculation of y_n by Eq. 6-21, a total of M+L+1 products are formed, and each introduces such a quantization error. It should be noted that if a particular coefficient, a_k or b_k, is either 0 or 1, then no rounding error is introduced in calculating that particular term. It is easy to see that if the additions (subtractions) in Eq. 6-21 do not cause the result to overflow the B bit data register, then no error is introduced in the addition. This can be accomplished by properly scaling the input $\{x_n\}$, as described in the next section. Under these conditions, the only error introduced in computing y_n is the (M+L+1) quantizing errors due to multiplication. Let us denote the actual computed output by y'_n. From the above discussion, we have

$$y'_n = \sum_{k=0}^{M} a_k x_{n-k} - \sum_{k=1}^{L} b_k y'_{n-k} + \delta_n \tag{6-22}$$

where δ_n is the sum of (M+L+1) quantizing errors, each with mean zero and mean squared value of $2^{-2B}/3$. Since the errors are independent and white, $\{\delta_n\}$ is also a white noise with

$$<\delta_n> = 0 \tag{6-23}$$

$$<\delta_n^2> = (M+L+1)2^{-2B}/3 \tag{6-24}$$

Let us define the difference between the actual output y'_n and the ideal output y_n as the output error ϵ_n, that is,

$$\epsilon_n = y'_n - y_n \tag{6-25}$$

By subtracting Eq. 6-21 from Eq. 6-22 and by using Eq. 6-25, we have

$$\epsilon_n + \sum_{k=1}^{L} b_k \epsilon_{n-k} = \delta_n \tag{6-26}$$

This equation describes a linear system with input $\{\delta_n\}$ and output $\{\epsilon_n\}$, and the transfer function is

$$\frac{1}{1 + \sum_{k=1}^{L} b_k z^{-k}}$$

as shown in Figure 6.4. Note that the transfer function is simply $1/D(z)$ where $D(z)$ is the denominator of the transfer function of the original filter $H(z)$. Since the input to this system in a white noise, Eqs. 6-3 and 6-4 apply, we obtain

$$<\epsilon_n> = 0 \tag{6-27}$$

$$<\epsilon_n^2> = (M+L+1)\left(\frac{2^{-2B}}{3}\right) \frac{1}{2\pi j} \oint \frac{dz}{D(z)D(1/z)\,z} \tag{6-28}$$

where

$$D(z) = 1 + \sum_{k=1}^{L} b_k z^{-k} \tag{6-29}$$

Thus we see that because of arithmetic round-off, the output of the digital filter has an additive error which has zero mean and a mean squared value given by Eq. 6-28.

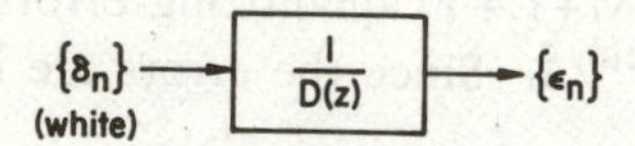

Figure 6.4 Round-off error accumulation model.

From Eqs. 6-21, 6-25 and 6-26, we can arrive at the block diagram of Figure 6.5 for the digital filter including the effect of round-off. This block diagram will facilitate our discussion in later sections.

Although Eq. 6-21 describes a recursive filter, our result also applies to nonrecursive filters. In this case, $b_1=b_2=...=b_L=0$, so $D(z)=1$. Equations 6-23 to 6-29 still hold but with $L=0$. The integral becomes simply $2\pi j$. Thus the round-off error at the filter output has mean zero and mean squared value

$$<\epsilon_n^2> = (M + 1)\, 2^{-2B}/3 \tag{6-30}$$

The reader may recall from Chapter 1 that there are three common realization forms of a high order digital filter: the direct form, the parallel form, and the cascade form. We have treated the direct form realization, the other two realizations are discussed in the next two subsections.

(d) *Round-off Error in Parallel Realization*

As discussed in Section 1.4, the transfer function $H(z)$ is written as a partial fraction expansion for parallel realization of the filter. Thus we write

$$H(z) = \sum_{i=1}^{K} H_i(z) \tag{6-31}$$

where each $H_i(z)$ is a second or first order transfer function

$$H_i(z) = \frac{N_i(z)}{D_i(z)} = \frac{a_{0i} + a_{1i}z^{-1}}{1 + b_{1i}z^{-1} + b_{2i}z^{-2}} \tag{6-32}$$

Although this $H_i(z)$ has the appearance of a second order transfer function, this form includes the possibility of a single pole by setting $b_{2i}=a_{1i}=0$. It also includes the possibility of a constant setting $a_{0i}=b_{2i}=b_{i1}=0$. The parallel form of realization is shown in Figure 1.6 where the transfer function $H_i(z)$ given by Eq. 6-32 and the K intermediate outputs $\{y_{in}\}$, $i=1,2,...,K$, are calculated first and then summed to form the final output $\{y_n\}$.

From the discussion in the previous subsection, we may represent the i-th subfilter $H_i(z)$ of Figure 1.6 by the block diagram of Figure 6.5 to include the round-off error generated in the i-th subfilter with y_n replaced by y_{in}, ϵ_n replaced by ϵ_{in}, and y'_n replaced by y'_{in}. The entire filter then has the block diagram shown in Figure 6.6 The total round-off noise at the filter output ϵ_n is the sum of the individual ϵ_{in}'s.

$$\epsilon_n = \sum_{i=1}^{K} \epsilon_{in} \tag{6-33}$$

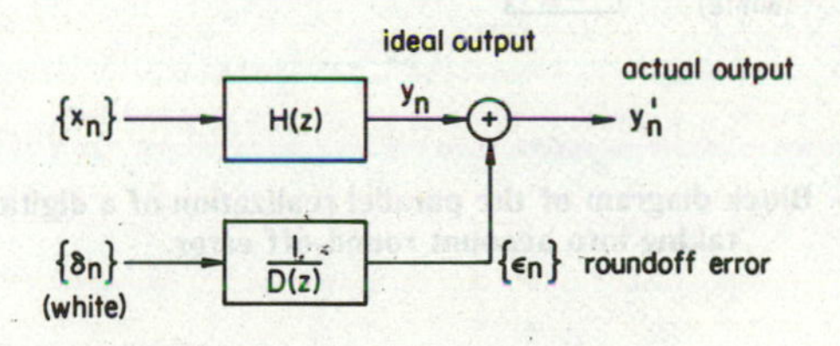

Figure 6.5 Block diagram of digital filter that includes the effect of round-off error.

The mean and the mean squared value of ϵ_{in} are given by Eqs. 6-27 and 6-28 with M=1 and L=2

$$<\epsilon_{in}> = 0 \tag{6-34}$$

$$<\epsilon^2_{in}> = 4\left(\frac{2^{-2B}}{3}\right)\frac{1}{2\pi j}\oint \frac{dz}{D_i(z)D_i(1/z)\,z} \tag{6-35}$$

Since the ϵ_{in}'s are statistically independent, the mean squared value of ϵ_n is simply the sum of all the $<\epsilon_{in}{}^2>$. So

$$<\epsilon_n> = 0 \tag{6-36}$$

$$<\epsilon_n{}^2> = \sum_{i=1}^{K} <\epsilon_{in}{}^2> \tag{6-37}$$

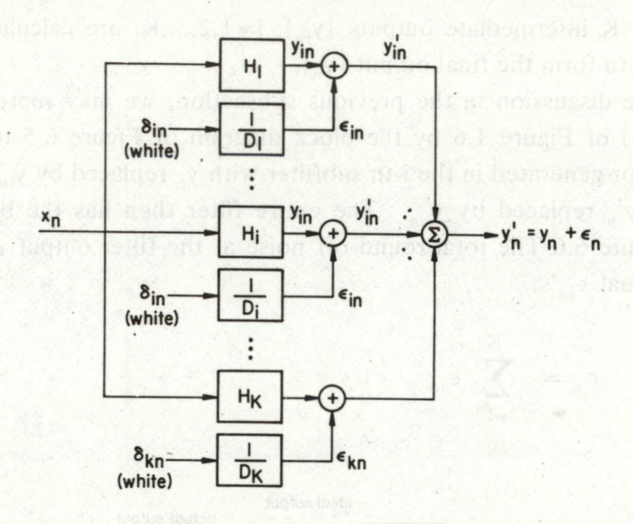

Figure 6.6 Block diagram of the parallel realization of a digital filter taking into account round-off error.

(e) *Round-off Error in Cascade Realization*

To realize the digital filter in the cascade form, the transfer function H(z) is written as

$$H(z) = c \prod_{i=1}^{K} H_i'(z) \tag{6-38}$$

where

$$H_i'(z) = \frac{N_i'(z)}{D_i'(z)} = \frac{1 + a_{1i}z^{-1} + a_{2i}z^{-2}}{1 + b_{1i}z^{-1} + b_{2i}z^{-2}} \tag{6-39}$$

and c is a constant that for the time being will be taken as 1 for simplicity. Note that the numerator of $H_i'(z)$ is different from that of $H_i(z)$ of Eq. 6-32. Although this form of $H_i'(z)$ has the appearance of having two zeros and two poles, it includes the possibility of a single pole by setting $b_{2i}=0$. The number of zeros may also be 1 if $a_{2i}=0$ or zero if $a_{2i}=a_{1i}=0$. The cascade form of realization is shown previously in Figure 1.7.

Again, we can replace the i-th section $H_i'(z)$ in Figure 1.7 by a block diagram of the form of Figure 6.5 to include the effect of round-off. The entire filter may then be represented by the block diagram of Figure 6.7. Each $\{\delta_{in}\}$, $i=1,2,...,K$ will produce an error component at the output. The transfer function from $\{\delta_{in}\}$ to the filter output is

$$\frac{1}{D_i'(z)} H_{i+1}'(z) \ldots H_K'(z) \qquad i = 1,2,\ldots,K-1$$

$$\frac{1}{D_K'(z)} \qquad i = K \tag{6-40}$$

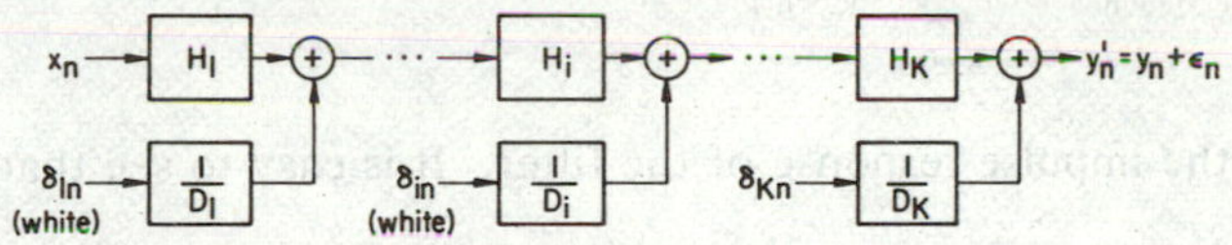

Figure 6.7 Block diagram of the cascade realization of a digital filter taking into account round-off error.

So the i-th error component has a mean squared value of

$$\sigma_i^2 = \begin{cases} 4\left(\dfrac{2^{-2B}}{6\pi j}\right) \displaystyle\oint \frac{dz}{D_i(z)D_i(1/z)z} \prod_{m=i+1}^{K} H_m(z)H_m(1/z) & i = 1,2,\ldots,K-1 \\ 4\left(\dfrac{2^{-2B}}{6\pi j}\right) \displaystyle\oint \frac{dz}{D_K(z)D_K(1/z)z} & i = K \end{cases} \tag{6-41}$$

These errors components are statistically independent of one another. So the mean squared value of the total error is the sum

$$<\epsilon_n^2> = \sum_{i=1}^{K} \sigma_i^2 \tag{6-42}$$

6.3 SCALING AND POLE ZERO PAIRING OF CASCADE FILTERS

In the analysis of round-off error presented in the preceding section, we see that errors are generated at the multipliers and that no round-off occurs at the adder provided the sum does not overflow. Serious consequences such as oscillations of large amplitude may occur if overflow does happen [9]. It is therefore important to prevent overflow by properly scaling the signal level.

(a) *Second Order Filter*

Shown in Figure 6.8*a* is a second order filter with transfer function
$H(z) = N(z)/D(z) = (a_o + a_1 z^{-1} + a_2 z^{-2})/(1 + b_1 z^{-1} + b_2 z^{-2})$.
To prevent adder overflow, we must scale the input $\{x_n\}$ so that the output $|y_n| < 1$ for all n, or at least most of the times so that the distortion introduced is small. There are a number of approaches that can be used to accomplish this [10,11]. Let us begin with the most conservative one. We know that

$$y_n = \sum_{k=0}^{\infty} h_k x_{n-k} \tag{6-43}$$

where $\{h_k\}$ is the impulse response of the filter. It is easy to see that

$$\begin{aligned} |y_n| &\leq \sum_{k=0}^{\infty} |h_k| \, |x_{n-k}| \\ &\leq (\max |x_k|) \sum_{k=0}^{\infty} |h_k| \end{aligned} \tag{6-44}$$

Therefore, by making $\max |x_k| = 1 / \Sigma |h_k|$, overflow of y_n can be prevented. If $|x_k| < 1$, then a_o, a_1, and a_2 must be multiplied by $s = 1 / \Sigma |h_k|$. The actual filter implemented takes the form of Figure 6.8*b* where we have also shown the round-off error $\{\delta_n\}$ due to the multipliers. Experiments have shown that this scaling policy is too pessimistic and does not allow good use of the full dynamic range of the filter.

A second approach would be to assume that the input is a sinusoidal signal and that the output sinusoidal's amplitude be required to be less than 1. This is equivalent to requiring $|H(e^{j\lambda})| \leq 1$ for $-\pi \leq \lambda \leq \pi$. This can be accomplished by multiplying the coefficients a_0, a_1, and a_2 by a scaling constant s so that

$$\max_{-\pi \leq \lambda \leq \pi} \left| s \frac{a_0 + a_1 e^{-j\lambda} + a_2 e^{-2j\lambda}}{1 + b_1 e^{-j\lambda} + b_2 e^{-2j\lambda}} \right| < 1 \tag{6-45}$$

This scaling policy is often too optimistic, that is to say, it may lead to overflow too frequently. So if this approach is used, it would be advisable to change the right side of Eq. 6-45 to a constant c less than 1. The choice of c would depend upon the type of signal to be processed by the filter. One way of determining a

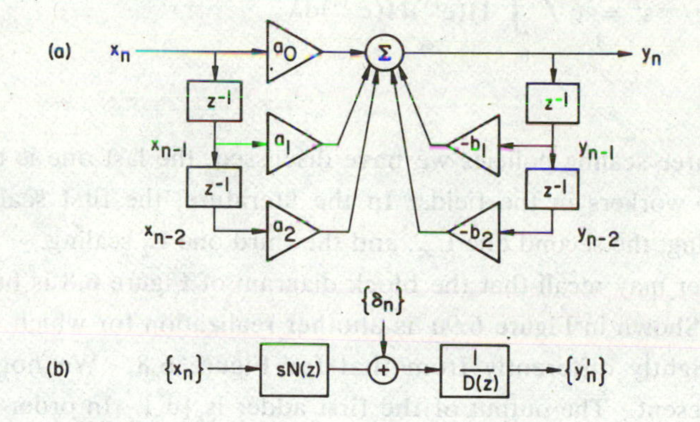

Figure 6.8 (a) Second order digital filter. (b) Model taking into account round-off error and scaling.

proper value of c is by carrying out a simulation using the appropriate input signal and study the effect of various choices of c.

A third approach begins with the relationship Y(z) = H(z)X(z). The *energy* of the output signal is Σy_k^2 which is $\int Y(e^{j\lambda})Y(e^{-j\lambda})d\lambda/2\pi$ by the Parseval relationship. By using the Schwartz inequality, we have

$$\sum_k y_k^2 = \int_0^{2\pi} H(e^{j\lambda})H(e^{-j\lambda})X(e^{j\lambda})X(e^{-j\lambda})\, d\lambda/2\pi$$

$$\leq \left(\int H(e^{j\lambda})H(e^{-j\lambda})d\lambda\right)\left(\int X(e^{j\lambda})X(e^{-j\lambda})d\lambda/2\pi\right)$$

$$= \left(\sum_k x_k^2\right) \int_0^{2\pi} H(e^{j\lambda})H(e^{-j\lambda})d\lambda \qquad (6\text{-}46)$$

One may therefore use a scaling constant s satisfying

$$\int_0^{2\pi} s^2 H(e^{j\lambda})H(e^{-j\lambda})d\lambda = 1 \qquad (6\text{-}47)$$

to keep the output signal from becoming too large. Again, in many applications, this policy would be too optimistic, and it is advisable to change the right side to a constant c less than 1. The value of s is given, therefore, by

$$s^2 = c \Big/ \int_0^{2\pi} H(e^{j\lambda})H(e^{-j\lambda})d\lambda \qquad (6\text{-}48)$$

for some $c<1$.

Of the three scaling policies we have discussed, the last one is the preferred one by many workers in the field. In the literature, the first scaling policy is called L_1 scaling, the second one L_∞, and the third one L_2 scaling.

The reader may recall that the block diagram of Figure 6.8 is but one of the realizations. Shown in Figure 6.9*a* is another realization for which the scaling is introduced slightly differently from that for Figure 6.8. We notice that two adders are present. The output of the first adder is $\{u_n\}$. In order to prevent it from overflow, a scaling multiplier s is inserted just before the filter, the transfer function from $\{x_n\}$ to $\{u_n\}$ is

$$H'(z) = \frac{U(z)}{X(z)} = \frac{s}{1 + b_1z^{-1} + b_2z^{-2}} \qquad (6\text{-}49)$$

so the three scaling policies just discussed can be applied here to determine the value s. The scaling of the second summer is not important, because the output $\{y_n\}$ is not used as input to any multiplier. The situation is different when one filter in question is one section in a cascade connection, as will be discussed later. With scaling, the filter can be represented by the block diagram of Figure 6.9*b* which also includes the round-off errors. $\{\delta_n'\}$ represents the error due to multiplication by $(-b_1)$ and by $(-b_2)$, whereas $\{\delta_n''\}$ represents the error due to multiplications by a_0, a_1, and a_2.

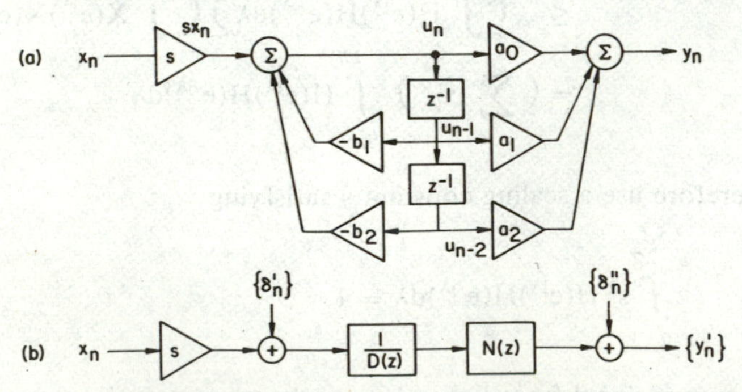

Figure 6.9 (a) An alternate realization of second order filter including prescaling. (b) Model taking into account round-off error and scaling.

(b) *Cascade Realization of High Order Filters* [12]

To realize a filter in the cascade form, its transfer function H(z) is factored as

$$H(z) = a_0 \frac{\prod_{k=1}^{K} (a_{k0} + a_{k1}z^{-1} + a_{k2}z^{-2})}{\prod_{k=1}^{K} (1 + b_{k1}z^{-1} + b_{k2}z^{-2})} \tag{6-50}$$

Let the k-th denominator factor be paired with the n-th numerator factor to form the i-th section in the cascade realization. Eq. 6-50 may be written as

$$H(z) = s_0 \prod_{i=1}^{K} s_i H_i(z) \tag{6-51}$$

where

$$H_i(z) = \frac{N_i(z)}{D_i(z)} = \frac{1 + a_{i1}z^{-1} + a_{i2}z^{-2}}{1 + b_{i1}z^{-1} + b_{i2}z^{-2}}$$

where s_i is the scaling constant associated with the i-th section so as to avoid overflow. Obviously $s_0 s_1 ... s_K = a_0$. The realization is shown in Figure 6.10 where we have assumed that each second order section is realized by Figure 6.9.

In Figure 6.10, s_0 is introduced to prevent the overflow of the output of $1/D_1(z)$, s_1 is introduced to prevent the overflow at the output of $1/D_2(z)$, and so on. If we use the third scaling policy discussed in the previous section, it is easy to see that

$$s_0^2 = \frac{1}{\frac{1}{2\pi j} \oint \frac{dz/z}{D_1(z)D_1(1/z)}}$$

$$s_1^2 = \frac{1}{\frac{s_0^2}{2\pi j} \oint \frac{N_1(z)N_1(1/z)\, dz/z}{D_1(z)D_1(1/z)D_2(z)D_2(z)D_2(1/z)}}$$

$$\vdots$$

$$s_i^2 = \frac{1}{\frac{s_0^2 \, ... \, s_{i-1}^2}{2\pi j} \oint \frac{N_1(z)N_1(1/z)...N_i(z)N_i(1/z)dz/z}{D_1(z)D_1(1/z) \, . \, . \, . \, D_{i+1}(z)D_{i+1}(1/z)}} \tag{6-52}$$

$$i = 1,2, . . .,K$$

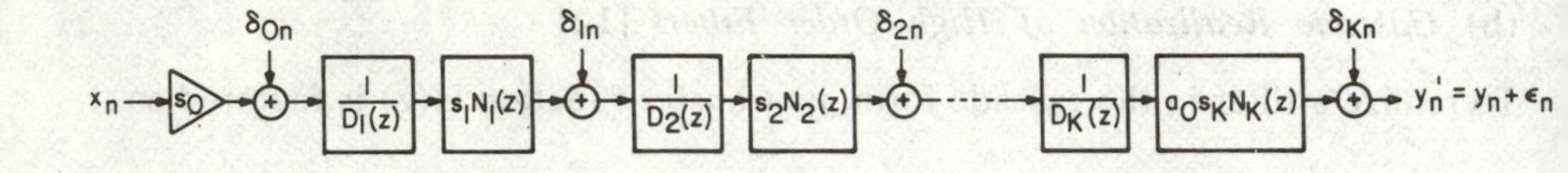

Figure 6.10 Scaling of a cascade realization to prevent overflow.

It is seen that $\{\delta_{0n}\}$ is due to the multiplication by s_0, and the b_{k1} and b_{k2} for $D_1(z)$, $\{\delta_{in}\}$ is due to the multiplication in $N_1(z)$ and $D_2(z)$, a total of 5 multipliers, etc. From these discussions, we may write down an expression for the mean squared output noise E^2.

$$E^2 = \frac{a_0^2 2^{-2B}}{12} \cdot \Big\{ A_K + \sum_{i=1}^{K} A_{i-1} \frac{1}{2\pi j} \oint \prod_{k=1}^{i} s_{k-1}^2 \frac{N_{k-1}(z)N_{k-1}(1/z)dz/z}{D_k(z)D_k(1/z)}$$

$$\times \frac{1}{2\pi j} \oint \prod_{k=i}^{K} s_k^2 \frac{N_k(z)N_k(1/z)dz}{D_k(z)D_k(1/z)\,z} \Big\} \tag{6-53}$$

where $N_k(z) = D_k(z) = 1$ if $k \leq 0$. A_n is the number of noninteger multipliers in the second order section and is 5 for most n, except $A_0 = 3$ and $A_K = 2$.

Eq. 6-53 shows clearly that the output round-off error depends on the ordering of the numerator and denominator factors in the cascade structure. All in all, there are $(N!)^2$ such possible orderings. To evaluate Eq. 6-53 for all the $(N!)^2$ possibilities in order to pick the best design is too time consuming even for moderate N. Dynamic programming approach [13] can be used to find the optimum solution with a reduced number of error evaluations. But the number is still too high in many cases. For example, 1695 evaluations are needed when $N=5$. We present here a *heuristic* approach to find a near optimum solution which requires a greatly reduced number of error evaluations. The approach calls for choosing a random ordering and performing a local optimization. This is repeated a fixed number of times and the best of these local optima is taken as the near optimum solution. The approach consists of the following four steps.

Step 1: *Generate a Random Assignment*: A random ordering for the zeros $\{N_k(z)\}$ and a random ordering for the poles $\{D_k(z)\}$ are generated.

Step 2: *Perform a Local Optimization*: By keeping the zero ordering $\{N_k\}$ fixed, interchange all possible pairs of pole sections $\{D_k\}$ in the random assignment generated in *Step 1*. Find the permutation that gives the least output error E^2. This is the local optimum from this particular random start, and this step requires $(N^2-N)/2+1$ evaluations of the expression in Eq. 6-53.

Step 3: Repeat *Steps 1* and *2* a prescribed number of times, say M.

Step 4: The best *local optimum* assignment found among the M random starts is taken as the *near optimal* assignment.

Thus, the total number of evaluations of Eq. 6-53 is approximately $MN^2/2$.

Usually, from M=5 to 10 random would be adequate to produce a good near optimum solution. A PL/1 program which implements this optimization is included in Appendix 6.1. Further details on the use of this approach and its extension to the scaling of nonrcursive filters can be found in [12]. Following are a few examples of its use.

This procedure is applied to an eighth order filter (K=4), and the result is listed in Table 6.1.

Each line represents a run of 5 random starts (M=5). For this 35 error evaluations are needed, as compared to 125 evaluations if dynamic programming is used. The running time on an IBM 360/91 computer is 1.6 seconds. The noise gain is defined as $E^2/(2^{-2B}/3)$. For this example, the optimum solution has been found by dynamic programming to have a noise gain of 32.03 dB. From Table 6.1, it is seen that the *near optimum* solution is never more than 0.5 dB worse than the true optimum. Since 6 dB corresponds to one bit, a difference of 0.5 dB would translate to a small fraction of a bit.

Table 6.2 shows the result obtained when this approach is applied to a 22nd order bandstop filter (K=11). Each line represents the near optimum solution from 5 random starts and takes 23 seconds on an IBM 360/91 computer. For this case, a dynamic programming approach would require 352715 error evaluations. We see that, even with 5 random starts, the spread among the noise gains is 7 dB, corresponding to a little over one bit. With 10 random starts, the spread is narrowed to 2 dB or less than half a bit and the computing time is 46 seconds. So for such high order filter, it is advisable to use 10 random starts.

6.4 LIMIT CYCLES AND OVERFLOW OSCILLATIONS

When the effect of finite word length was treated in Section 6.2, the arithmetic round-off was modelled as an independent additive white noise component, and the digital filter was considered essentially as a linear system. However, for special types of inputs, the round-off *noise* is highly dependent upon the signal, making it necessary to adopt a different approach to analyze the effects. To see this, let us consider a simple example of a first order filter.

```
********************************************************************************
                        RECURSIVE FILTER OF ORDER    8
A0 =    0.01023751
A( 1,1) =   1.00000000   A( 1,2) =   -0.35800000   A( 1,3) =   1.00000000
B( 1,1) =   1.00000000   B( 1,2) =   -0.68404000   B( 1,3) =   0.85862000
A( 2,1) =   1.00000000   A( 2,2) =   -0.18961000   A( 2,3) =   1.00000000
B( 2,1) =   1.00000000   B( 2,2) =   -0.90191000   B( 2,3) =   0.66204000
A( 3,1) =   1.00000000   A( 3,2) =    0.35800000   A( 3,3) =   1.00000000
B( 3,1) =   1.00000000   B( 3,2) =   -1.17844000   B( 3,3) =   0.42357000
A( 4,1) =   1.00000000   A( 4,2) =    1.61851000   A( 4,3) =   1.00000000
B( 4,1) =   1.00000000   B( 4,2) =   -0.59971000   B( 4,3) =   0.96329000

         BEST ASSIGNMENT                                     NOISE GAIN        DB
         FOUND IN 5 RANDOM STARTS
   1         3  1  2  4
             4  2  3  1                                        1688.2       32.27
---------------------------------------------------------------------------------
   2         3  1  4  2
             4  2  1  3                                        1727.2       32.37
---------------------------------------------------------------------------------
   3         2  4  3  1
             1  4  2  3                                        1654.5       32.19
---------------------------------------------------------------------------------
   4         1  3  4  2
             1  3  2  4                                        1736.5       32.40
---------------------------------------------------------------------------------
   5         2  4  3  1
             1  2  3  4                                        1682.9       32.26
---------------------------------------------------------------------------------
   6         2  4  3  1
             2  1  4  3                                        1660.8       32.20
---------------------------------------------------------------------------------
   7         2  4  3  1
             1  3  2  4                                        1687.1       32.27
---------------------------------------------------------------------------------
   8         2  4  3  1
             1  4  2  3                                        1654.5       32.19
---------------------------------------------------------------------------------
   9         1  3  4  2
             1  3  2  4                                        1736.5       32.40
---------------------------------------------------------------------------------
  10         1  3  4  2
             1  4  2  3                                        1781.4       32.51
---------------------------------------------------------------------------------
  11         3  4  2  1
             4  1  3  2                                        1747.5       32.42
---------------------------------------------------------------------------------
  12         2  4  3  1
             2  1  4  3                                        1660.8       32.20
---------------------------------------------------------------------------------
  13         2  4  3  1
             1  2  4  3                                        1599.3       32.04
---------------------------------------------------------------------------------
  14         3  1  2  4
             4  3  1  2                                        1673.7       32.24
---------------------------------------------------------------------------------
  15         3  1  2  4
             4  3  2  1                                        1650.5       32.18
---------------------------------------------------------------------------------
```

Table 6.1 Typical Output of Heuristic Optimization Program for Pole Zero Pairing Listed in Appendix 6.1.

$$y_n = x_n + 0.5y_{n-1} \tag{6-54}$$

with $x_n=0$ and $y_0=0.5$. It is easy to see that the output $\{y_0,y_1,y_2,y_3,...\}$ is $\{0.5,0.25,0.125,0.0625,0.03125,0.015625,...\}$, so eventually y_n approaches zero. However, suppose the data registers have 3 bits plus a sign bit. Then we see $y_0=0.100$, $y_1=0.010$, $y_2=0.001$, $y_3=0.0001$ before rounding. After rounding,

```
*******************************************************************************
                    RECURSIVE FILTER OF ORDER    22
A0 =     0.03633350
A( 1,1) =    0.99890799       A( 1,2) =    0.41839524      A( 1,3) =    0.99890799
B( 1,1) =    1.00000000       B( 1,2) =    0.42851755      B( 1,3) =    0.99890799
A( 2,1) =    1.14043461       A( 2,2) =    0.13804272      A( 2,3) =    1.14048461
B( 2,1) =    1.00000000       B( 2,2) =    0.11064725      B( 2,3) =    0.99888373
A( 3,1) =    0.99572787       A( 3,2) =    0.41294743      A( 3,3) =    0.99572787
B( 3,1) =    1.00000000       B( 3,2) =    0.43175125      B( 3,3) =    0.99572787
A( 4,1) =    1.27046937       A( 4,2) =    0.15924493      A( 4,3) =    1.27046937
B( 4,1) =    1.00000000       B( 4,2) =    0.10637363      B( 4,3) =    0.99563068
A( 5,1) =    0.98829190       A( 5,2) =    0.39891142      A( 5,3) =    0.98829190
B( 5,1) =    1.00000000       B( 5,2) =    0.44149478      B( 5,3) =    0.98829190
A( 6,1) =    1.72339736       A( 6,2) =    0.23586369      A( 6,3) =    1.72339730
B( 6,1) =    1.00000000       B( 6,2) =    0.09407400      B( 6,3) =    0.98800730
A( 7,1) =    0.96614183       A( 7,2) =    0.36664231      A( 7,3) =    0.96614183
B( 7,1) =    1.00000000       B( 7,2) =    0.46635063      B( 7,3) =    0.96614183
A( 8,1) =    1.93520596       A( 8,2) =    0.31317083      A( 8,3) =    1.93520510
B( 8,1) =    1.00000000       B( 8,2) =    0.06162631      B( 8,3) =    0.96518140
A( 9,1) =    0.87422935       A( 9,2) =    0.29283818      A( 9,3) =    0.87422935
B( 9,1) =    1.00000000       B( 9,2) =    0.51860380      B( 9,3) =    0.87422935
A(10,1) =    1.99293732       A(10,2) =    0.41339035      A(10,3) =    1.99293732
B(10,1) =    1.00000000       B(10,2) =   -0.02045647      B(10,3) =    0.86946006
A(11,1) =    1.95055011       A(11,2) =    0.52953266      A(11,3) =    1.95055011
B(11,1) =    1.00000000       B(11,2) =    0.20745922      B(11,3) =    0.52836505

                 BEST ASSIGNMENT                                     NOISE GAIN       DB
                 FOUND IN 5 RANDOM STARTS
  1       10  3 11  6  4  8  5  2  1  7  9
           7  8  4  5  6  9  2  3  1 11 10                           35028.9        45.44
-----------------------------------------------------------------------------------------
  2       11  6  7  2  4  8  3  5  1 10  9
           6  1  2  8  4  9  7  3 10  5 11                           15515.5        41.91
-----------------------------------------------------------------------------------------
  3        6  3  9  7  4 11  2  1  8  5 10
           9  5 10  2  1  6  4  3 11  8  7                           24779.2        43.94
-----------------------------------------------------------------------------------------
  4        8 11  2  7  4  6  3  9  5  1 10
           4  8  2  6  9  5  1  3 11  7 10                           26125.5        44.17
-----------------------------------------------------------------------------------------
  5        9  3  6 10 11  1  8  4  5  2  7
           1 10  8  2  3  9  7  4  5 11  6                           16375.5        42.14
-----------------------------------------------------------------------------------------
  6        9  4  1 10  5  7 11  8  6  3  2
          11  3  5  4  1  8  9  2 10  7  6                           15451.6        41.89
-----------------------------------------------------------------------------------------
  7        8  6  9  7  3  2 11  5  1  4 10
           6  2  1  7  3 11  4  9  5  8 10                           21725.9        43.37
-----------------------------------------------------------------------------------------
  8        9  1 11  8  4  6  7  3  2  5 10
           3  9 10  2  4  7  5  6  1  8 11                            6603.4        38.20
-----------------------------------------------------------------------------------------
  9        9  7 11 10  1  6  5  4  3  2  8
           5 10 11  3  7  2  9  1  4  8  6                           14589.0        41.64
-----------------------------------------------------------------------------------------
 10        9  4  1  7  8 10  2  6  5 11  3
           1  6  9  7  8  4  2  5 10  3 11                           21190.1        43.26
-----------------------------------------------------------------------------------------
 11        7 11  9  5  2  8  3  4  1 10  6
          10  1  3  7  8  4  5  6  9 11  2                           29189.9        44.65
-----------------------------------------------------------------------------------------
 12        7  6 10  9  2  3 11  4  8  5  1
           5 10  2 11  4  3  9  8  6  1  7                           13592.0        41.33
-----------------------------------------------------------------------------------------
 13       11  2  9  1  8  5  3 10  4  6  7
           6  9 11  5  1  7  2  8 10  3  4                           13111.0        41.18
-----------------------------------------------------------------------------------------
 14        8  4  5  2  9  3 11  7  1  6 10
           2  8  4  5  3  7  9  1 10  6 11                           15254.7        41.83
-----------------------------------------------------------------------------------------
 15        6 11  5  2  9  7  8  1  4 10  3
           9  8  6  5  4 11  3  2  1  7 10                           17120.1        42.34
-----------------------------------------------------------------------------------------
```

Table 6.2 Pole Zero Assignment for a 22-nd Order Filter Using the Computer Program Listed in Appendix 6.1.

$y_3=0.001$ which is 1/8 or 0.125, same as y_2. The next output sample y_4 is 0.0001 before rounding but becomes 0.001 after rounding, again same as y_3. In fact $y_n=1/8$ for all $n \geq 2$. That is, y_n does not settle to 0 no matter how large n is. Let us look at the round-off error $\{\epsilon_n\}$ in this case. It is clear that $\epsilon_1=0$, $\epsilon_2=0$, $\epsilon_3=-1/8$, $\epsilon_4=-1/8$,... . Obviously such a sequence cannot be regarded as white noise, the heavy dependence on the signal $\{y_n\}$ is evident.

Let us change the example slightly by letting

$$y_n = x_n - 0.5y_{n-1} \tag{6-55}$$

but with the same input and initial condition. It is easy to verify that for 3 bits the output $\{y_0, y_1, y_2,\}$ becomes in this case $\{0.5, -0.25, 0.125, -0.125, 0.125, \ldots\}$, an oscillating sequence. So is the sequence of round-off errors $\{\epsilon_n\}$. The reader can easily verify that, if the data registers have 4 bits plus a sign bit, then the output would oscillate between +0.0625 and −0.0625 and that, if n bits plus a sign bit are used, the output would oscillate between $(0.5)^n$ and $-(0.5)^n$. Thus the oscillation can be reduced by using more bits.

This type of nonzero output in the absence of input is called zero input limit cycle oscillations and is due to the nonlinear nature of the arithmetic quantization. It is therefore important to use enough bits in the data registers so that these oscillations, when they occur, are kept small enough. We will derive a bound on the amplitude of these oscillations for second order filters. The results can be extended to higher order filters [14].

Consider the second order filter

$$y_n = a_0x_n + a_1x_{n-1} + a_2x_{n-2} - b_1y_{n-1} - b_2y_{n-2} \tag{6-56}$$

For zero input, the actual output sequence is given by

$$y'_n = -b_1y'_{n-1} - b_2y'_{n-2} + \delta'_n + \delta''_n \tag{6-57}$$

where δ'_n and δ''_n are the errors due to the rounding of the products $b_1y'_{n-1}$ and $b_2y'_{n-2}$ respectively. So $|\delta'_n| < 2^{-B}$ and $|\delta''_n| < 2^{-B}$. Eq. 6-57 describes a linear system with output $\{y'_n\}$ and input $\{\delta_n\}$ where

$$\delta_n = \delta'_n + \delta''_n \tag{6-58}$$

It is clear that $|\delta_n| < 2^{-B+1}$.

We may write the output as a convolution

$$y_n' = \sum_{k=-\infty}^{n} h_k \, \delta_{n-k} \tag{6-59}$$

where $\{h_k\}$ is the impulse response of the system whose transfer function is $1/(1 + b_1z^{-1} + b_2z^{-2})$. For oscillations of period p,

$$\begin{aligned} y'_{n+p} &= y'_n \\ \delta_{n+p} &= \delta_n \end{aligned} \tag{6-60}$$

Equation 6-59 can be written as a double sum by setting $k=m+ip$ with $0 \leq m \leq p-1$,

$$y_n' = \sum_{i=0}^{\infty} \sum_{m=0}^{p-1} h_{m+ip} \, \delta_{n+ip-m}$$

$$= \sum_{m=0}^{p-1} \delta_{n-m} \sum_{i=0}^{\infty} h_{m+ip} \tag{6-61}$$

where we have used $\delta_{n+p} = \delta_n$. By using $|\delta_n| < 2^{-B+1}$, we see that

$$|y_n'|_{max} \leq 2^{-B+1} \sum_{m=0}^{p-1} \left| \sum_{i=0}^{\infty} h_{m+ip} \right| \tag{6-62}$$

This is a bound on the amplitude of oscillation of period p. By evaluating explicitly $\{h_n\}$, it can be shown [14] that for p=1 and p=2, inequality 6-62 reduces to

$$|y_n'|_{max} \leq \frac{1}{1 + b_1 + b_2} \qquad p = 1 \tag{6-63}$$

$$|y_n'|_{max} \leq \frac{1}{1 - |b_1| + |b_2|} \qquad p = 2 \tag{6-64}$$

For general p, inequality 6-62 becomes

$$|y_n'|_{max} \leq \frac{2^{-B}}{c} \sum_{m=0}^{p-1} \left| \frac{(-b_1/2+c)^{m+1}}{1 - (-b_1/2+c)^p} - \frac{(-b_1/2 - c)^{m+1}}{1 - (-b_1/2 - c)^p} \right| \tag{6-65}$$

where $c=(b_1^2/4 - b_2)^{-1/2}$, if the roots of $z^2 + b_1z + b_2 = 0$ are distinct. For repeated roots Eq. 6-65 becomes

$$|y_n'|_{max} \leq \sum_{m=0}^{p-1} |(-b_1/2)^m [\frac{m}{1-(-b_1/2)^p} + \frac{1+(p-1)(-b_1/2)^p}{[1-(-b_1/2)^p]^2}]| \quad (6\text{-}66)$$

These bounds depend on the period p. A bound that is independent of p can be derived by noting that

$$\sum_{m=0}^{p-1} | \sum_{i=0}^{\infty} h_{m+ip} | \leq \sum_{k=0}^{\infty} |h_k| \quad (6\text{-}67)$$

Therefore

$$|y_n'|_{max} \leq 2^{-B+1} \sum_{k=0}^{\infty} |h_k| \quad (6\text{-}68)$$

When the roots of $z^2+b_1z+b_2=0$ are real, the right side of this inequality can be summed straightforwardly. The result is

$$\sum_{k=0}^{\infty} |h_k| = \begin{cases} \dfrac{1}{1-|b_1|+b_2} & \text{distinct real roots} \\ \dfrac{1}{(1-b_1/2)^2} & \text{double roots } (b_1^2=4b_2) \end{cases} \quad (6\text{-}69)$$

In case of complex roots, it can be easily verified that

$$h_k = r^k \sin(k+1)\theta/\sin\theta \quad (6\text{-}70)$$

where $r = \sqrt{b_2}$ and $\theta=\cos^{-1}(-b_1/2\sqrt{b_2})$. Two bounds of $\Sigma\ |h_k|$ can be obtained as follows:

$$\sum_{k=0}^{\infty} |h_k| \le \sum_{k=0}^{\infty} \frac{r^k}{|\sin\theta|} = \frac{1}{(1-r)\sin\theta}$$

$$= \frac{1+\sqrt{b_2}}{(1-b_2)\sqrt{1-b_1^2/4b_2}} \qquad (6\text{-}71)$$

or

$$\sum_{k=0}^{\infty} |h_k| \le \sum_{k=0}^{\infty} (k+1)r^k = \frac{1}{(1-\sqrt{b_2})^2} \qquad (6\text{-}72)$$

Combining Eqs. 6-68, 6-69, 6-71 and 6-72, we obtain

$$|y_n'|_{max} \le \begin{cases} \dfrac{1}{1-|b_1|+|b_2|} & b_2<0 \text{ or } b_2>0 \text{ and } 2\sqrt{b_2} \le |b_1| \\ \dfrac{1}{(1-\sqrt{b_2})^2} & b_2>0 \text{ and } 2b_2(2/\sqrt{b_2}-1)^{1/2} \le |b_1| \le 2\sqrt{b_2} \\ \dfrac{\sqrt{b_2}}{(1-b_2)\sqrt{1-b_1/4b_2}} & b_2>0 \text{ and } |b_1| \le 2b_2(2/\sqrt{b_2}-1)^{1/2} \end{cases} \qquad (6\text{-}73)$$

Another bound on the limit cycle that has been derived concerns the rms value. It has been shown that [15]

$$\left(\frac{1}{p}\sum_{m=0}^{p-1} y'^{\,2}_m\right)^{1/2} \le \begin{cases} \dfrac{1}{1-|b_1|+b_2} & b_2 \le 0 \text{ or } b_2>0 \text{ and } |b_1| \ge 4b_2/(1+b_2) \\ \dfrac{1}{(1-b_2)\sqrt{1-b_1^2/4b_2}} & b_2>0 \text{ and } |b_1| < 4b_2/(1+b_2) \end{cases} \qquad (6\text{-}74)$$

where p is the period of oscillation. This is an important result because limit cycle oscillations for longer period often have a sinusoidal shape. Figure 6.11 shows the absolute bound according to Eq. 6-73, the rms bound according to Eq. 6-74 and the actual amplitude of oscillation for $b_1=1.0$ and $b_2=0.9$.

In addition to the type of oscillations just discussed, digital filters implemented with 2's complement or 1's complement arithmetic have been observed to produce large oscillations the cause of which can be traced to the overflow of adders in these filters rather than the arithmetic round-off.

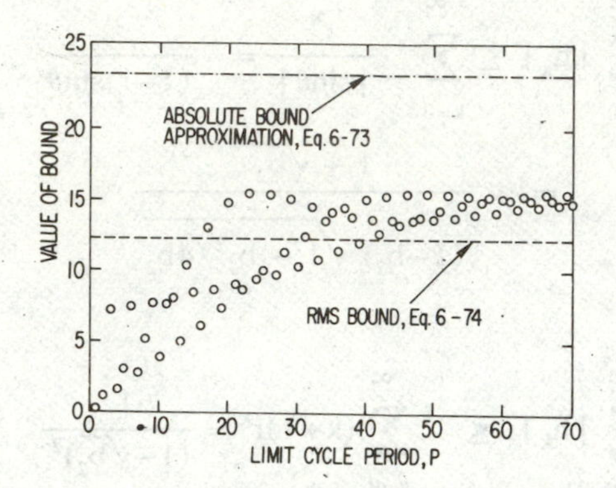

Figure 6.11 Bounds for limit cycle oscillations.

We recall that in 2's complement arithmetic, a sign bit is attached to the left of the binary point. A 1 signifies that the number is negative. Suppose two positive fractional numbers add up to larger than 1. Then a carry bit would result to the left of the binary point, and the *sign bit* would become 1 and the sum would be interpreted as a negative number.

Consider a second order filter with zero input,

$$y_n = -b_1 y_{n-1} - b_2 y_{n-2} \tag{6-75}$$

For such a filter to be stable, the roots of $z^2+b_1z+b_2=0$ must lie inside the unit circle. It can be shown that for this to be true, b_1 and b_2 must take on values inside the stability triangle shown in Figure 6.12. Let us ignore the round-off error for the time being, but assume 2's complement arithmetic is used. Suppose both terms on the right side of Eq. 6-75 are positive fractions but their sum is larger than 1. The addition would cause a carry of 1 to the sign bit position and the sum would be interpreted as a negative number. Thus a 2's complement adder behaves like a nonlinear transformation on the true sum as indicated by the transfer characteristic f(•) shown in Figure 6.13. As can be seen, a positive sum between 1 and 2 would be interpreted as a negative fraction. Thus the true output of the filter in question is

$$y_n = f\,[-b_1 y_{y-1} - b_2 y_{n-2}] \tag{6-76}$$

instead of Eq. 6-75.

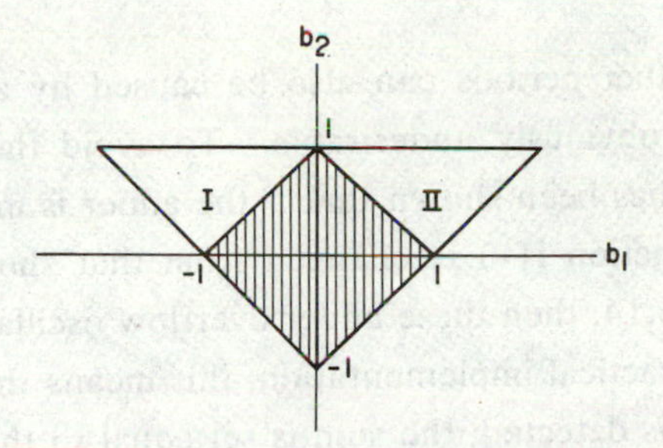

Figure 6.12 The stability triangle.

It is clear that if $|-b_1y_{n-1}-b_2y_{n-2}|<1$, then y_n will be correctly interpreted. Suppose $|y_{n-1}|<1$ and $|y_{n-2}|<1$, then a sufficient condition is

$$|b_1| + |b_2| < 1 \tag{6-77}$$

which is shown in Figure 6.12 as the square inside the stability triangle with vertical hatching.

We can demonstrate that if Eq. 6-77 is not satisfied, it is easy to find periodic solutions of $\{y_k\}$ satisfying Eq. 6-76. Suppose b_1 and b_2 satisfy

$$b_1 + b_2 > 1 \tag{6-78}$$

which correspond to the small triangle marked II in Figure 6.12. Then the number $A=2/(1+b_1+b_2)$ is positive but less than 1. It can be seen that $y_n=y_{n-1}=y_{n-2}=A$ is a solution of Eq. 6-76. Since in this case $-b_1y_{n-1} - b_2y_{n-2} = -(b_1+b_2)\ A = 2[1-1/(1+b_1+b_2)] = A-2$, so $f[-b_1y_{n-1}-b_2y_{n-2}] = A - y_n$. Thus we see that if b_1 and b_2 take on value in region II of Figure 6.12, the filter can support a constant output of value $A=2/(1+b_1+b_2)$ without any input. Depending on the values of b_1 and b_2, A can be quite close to 1.

Now suppose

$$b_2 - b_1 > 1 \tag{6-79}$$

which corresponds to the small triangle I in Figure 6.12. It can be shown that the filter can support an oscillation of the form $y_n=(-1)^nA$ with $A=2/(1-b_1+b_2)$. The amplitude of oscillation A can also be quite close to 1 with certain values of b_1 and b_2. Therefore, we have seen that unless the condition of Eq. 6-77 is satisfied, the use of 2's complement adder may cause oscillations arbitrarily close to the maximum values the output even in the absence of input.

Oscillations of other periods can also be caused by adder overflow. Such large oscillations are obviously undesirable. To avoid them, it is necessary to modify the adder. It has been shown that if the adder is modified to *saturate* so that the nonlinear function f(•) is changed from that shown in Figure 6.13 to that shown in Figure 6.14, then these adder overflow oscillations can be eliminated [9]. In terms of practical implementation, this means that overflow has to be monitored and if one is detected, the sum is set equal to the maximum allowable value. The proof of this result is somewhat involved and will not be presented here.

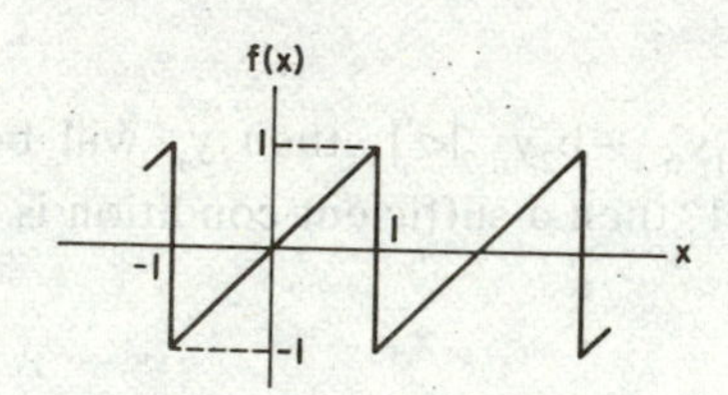

Figure 6.13 Transfer characteristic of 2's complement adder that does not saturate.

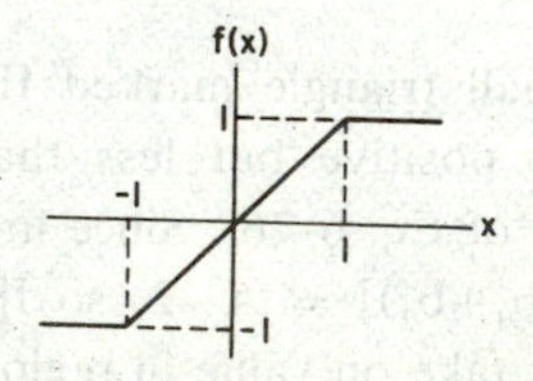

Figure 6.14 Transfer characteristic of 2's complement adder that does saturate.

6.5 SCALING AND ROUND-OFF ERROR IN FIXED POINT FFT COMPUTATION [7,16,17]

To compute the DFT of a data sequence by using one of the FFT algorithms and fixed point arithmetic, either on a general purpose computer or with a special purpose processor, scaling is usually necessary to prevent overflow. This can be seen from the Parseval relationship, Eq. (1-85),

$$N \sum_{n=0}^{N-1} |x_n|^2 = \sum_{p=0}^{N-1} |A_p|^2 \tag{6-80}$$

It implies that the numbers in the transformed array $\{A_p\}_{p=0,N-1}$ are in general larger than those in the original data array $\{x_n\}_{n=0,N-1}$. We recall that the FFT calculates M arrays of complex numbers, $N=2^M$. In each iteration, only the numbers from the previous array are used. Consider the basic butterfly computation of the i-th stage in a radix 2, decimation-in-frequency FFT

$$\begin{aligned} A_k^{i+1} &= A_k^{\,i} + A_r^{\,i} \\ A_r^{i+1} &= (A_k^{\,i} - A_r^{\,i})W(k,i) \end{aligned} \tag{6-81}$$

where $A_k^{\,i}$ and $A_r^{\,i}$ are numbers obtained from the previous stage that are used to compute $A_k^{\,i+1}$ and $A_r^{\,i+1}$, and $W(k,i)$ is $e^{-j2\pi/N}$ raised to some power. After some simple algebra, it can be shown that

$$|A_k^{i+1}|^2 + |A_r^{i+1}|^2 = 2(|A_k^{\,i}|^2 + |A_r^{\,i}|^2) \tag{6-82}$$

It can also be shown straightforwardly that

$$\begin{aligned} \text{Max}\,\{|A_k^{\,i}|, |A_r^{\,i}|\} &\le \text{Max}\,\{|A_k^{\,i+1}|, |A_r^{i+1}|\} \\ &\le 2\,\text{Max}\,\{|A_k^{\,i}|, |A_r^{\,i}|\} \end{aligned} \tag{6-83}$$

Equation 6-82 states that the rms values increase by a factor of 2 with each iteration, and Eq. 6-83 states that the maximum modulus of the numbers in the array are not decreasing with the iteration. Therefore, the numbers in the successive arrays in an FFT would cause overflow of the data registers, unless some scaling is performed. The simplest scaling is a right shift by one bit, this gives a factor of 1/2.

From the above discussion, we may propose the following two scaling policies:

1. *Right shift at each iteration*. The initial data is scaled so that for each n, $|x_n| < 1/2$. The result, after each iteration is right shifted by one bit before performing the Butterfly calculation for the next stage.

2. *Testing for overflow.* The input data is scaled so that $|\mathrm{Re}\, x_n| < 1$ and $|\mathrm{Im}\, x_n| < 1$. When an overflow is detected during the computation, the entire array of that iteration is right shifted by one bit. Note that part of the arrays are the computed result and the rest has yet to be computed.

The first scaling policy obviously does not utilize all the available bits effectively, but is the easy to implement. The second scaling policy gives much better results; its implementation, however, is more involved.

The analysis of the round-off error is complicated by the scaling policy used. While an expression for the first scaling policy can be derived, an accurate analysis of the second scaling policy is not possible because the occurrence of the right shifts are data dependent. We will derive below two expressions for the rms round-off error. One assumes a right shift at each iteration, thus, corresponding to the first scaling policy. The other assuming no shift at all. Thus if the second scaling policy is used, the error must lie between their two values corresponding to a lower bound of the error.

We assume that the data format is B bits plus a sign bit, and that rounding is used after each multiplication. This rounding introduces an error of zero mean and variance $2^{-2B}/12$, as discussed earlier in this chapter. However, this result is obtained for the multiplication of two real numbers. The multiplication of two complex numbers involves four real multiplications, so four such errors are introduced. Assuming the statistical independence of these round-off errors, the variances would add. Thus a complex multiplication will generate a round-off error which has zero mean and variance $4 \times (2^{-2B}/12) = 2^{-2B}/3$.

When a number is right shifted by one bit, the error introduced depends on the lost bit. If it is a zero, no error is introduced. If it is a one, then the error is either 2^{-B} or -2^{-B}, depending on whether the number is positive or negative. We may assume that the last bit is zero or one, each with probability 0.5, and that the number is equally likely to be positive or negative. Under these assumptions, the mean square error due to the one bit right shift can easily be calculated to be $2^{-2B}/2$.

Again, this error characterization is obtained for real numbers. In the FFT computation, the real and imaginary parts of each complex number are stored separately as two real numbers. Assuming the statistical independence of the individual errors introduced due to the right shift, the mean square value of the error in a complex number due to shift is then $2 \times (2^{-2B}/2) = 2^{-2B}$.

We can now start the error analysis for FFT. Reference is made to the flow graph of the 16 point decimation-in-frequency FFT shown in Figure 3.5. Suppose an error of variance Δ^2 is introduced into x_3 of the input array. This error will propagate to all 16 output points. The error at each of the outputs has the same variance, because in the path of propagation, the multipliers all have magnitude 1. So the *total* mean squared error at the output due to this single error is $16\Delta^2$. If the multipliers have magnitude other than 1, modifications are needed. Suppose another error, of the same variance Δ^2 but independent of the first, is introduced to a different input data, say x_{13}. This error will produce the same output errors. Because of the independence of the input errors, the total mean squared error at the output is the sum of these two contributions, namely $32\,\Delta^2$. From the flow graph it is clear that an error of variance Δ^2 introduced to the next array, say $A_6^{\,1}$, will propagate to half of the output points and produce an error of variance Δ^2 in these outputs. In general, for a $N=2^M$ point FFT, an error of variance Δ^2 introduced to a number just *before* the computation of the k-th stage ($1 \leq k \leq M$) will propagate to 2^{M-k+1} output points and produce an error of variance Δ^2 in these outputs.

We now consider separately the error with and without shifting.

(a) *No shift*

Only the errors due to round-off in multiplication need to be considered in this case. Consider the 16 point FFT of Figure 3.5 first. We see from Figure 3.5 and Eq. 3-6 that $(A_0^{\,1},...,A_7^{\,1})$ are computed error free, as there is no multiplication involved. The remaining numbers in the A^1 array $(A_8^{\,1},...,A_{15}^{\,1})$ require multiplication. However, $W_{16}^{\;0}=1$ and $W_{16}^{\;4}=j$, so they do not introduce round-off error. As a result, only 6 of the 16 numbers in the A^1 array are computed with an error. Note that $6=(16/2-2^1)$. The variance of each of these errors is $\Delta^2=2^{-2B}/3$ as we recall from the previous discussion; furthermore, these errors will propagate to the output and produce error at the output in the manner described above. In the computation of numbers in the A^2 array, again, half of them are error free. But here $W_{16}^{\;0}$ and $W_{16}^{\;4}$ each appear twice, so the total numbers in this array computed with error is $(16/2-2^2)=4$.

Consider now the general case of $N=2^M$ point FFT. In the computation of the A^1 array, independent round-off errors, each with variance $\Delta^2=2^{-2B}/3$, are introduced to $(2^M/2-2^1)$ points in the array. Since each of these errors propagates to 2^{M-1} output points, the total mean squared error at the output due to these errors is $(2^M/2-2^1)2^{M-1}\,\Delta^2 = (2^{2M-2}-2^M)\Delta^2$. In the computation of the A^2 array, errors are introduced to $(2^M/2-2^2)$ points and each error is propagated to 2^{M-2} output points. So the total mean squared error due to these is $(2^M/2-2^2)2^{M-2}\,\Delta^2 = (2^{2M-3}-2^M)\Delta^2$. Continuing in this manner, we can determine the output error due to the round-off error at each stage. The grand total mean squared error at the output is the sum of these. It is

$$\Delta^2[(2^{2M-2}-2^M) + (2^{2M-3}-2^M) + \ldots + (2^{2M-M+1}-2^M)] \qquad (6\text{-}84)$$

Notice, only contributions up to the (M−2)-nd stage is included, because at the last two stages, the multiplications are by 1 and j only, hence introduce no error. The summation can be carried out and the result is $\Delta^2[2^{2M-1}-(M+2)2^M]$ or $\Delta^2[N^2/2-(M+2)N]$.

Let S denote the input *signal power*, that is,, $S=x_1^{\,2}+x_2+^2\ldots+x_{N-1}^{\,2}$. Then by the Parseval's relationship, Eq. 6-80, the output signal power is N S. So the total output mean squared error normalized to the mean squared output signal is

$$\frac{\text{Error}}{\text{Signal}} = \frac{[N^2/2-(M+2)N]\Delta^2}{N\,S} = [N/2-(M+2)]\,\frac{\Delta^2}{S} \qquad (6\text{-}85)$$

where $\Delta^2=2^{-2B}/3$ and $N=2^M$.

(b) *Shift at each stage*

In this case, not only do we need to take into account both the error due to multiplication round-off and the error due to right shift at each stage, we also have to consider the effect of the scaling as a result of the right shifts.

Let us take up the multiplication round-off first. Reference is again made to Figure 3.5. As before, an error with variance Δ^2 is introduced into $6=(2^4/2-2^1)$ numbers in the A^1 array. Each of the errors is propagated to $2^4/2$ output points. However, along the path they are scaled by 1/2 at each subsequent stage. So the variance at each output point due to one of the errors is $(1/2^2)^3\Delta^2$ instead of Δ^2 when there is no scaling. Reasoning in this manner, for a $N=2^M$ point FFT, the round-off error in the A^1 array would give rise to an output error variance of $\Delta^2(2^{2M-2}-2^M)\,2^{-2(M-1)}$ or $\Delta^2(2^0-2^{-M+2})$. The round-off error in the A^2 array would give rise to an output error variance of $\Delta^2(2^{2M-3}-2^M)2^{-2(M-2)}$ or $\Delta^2(2^1-2^{M+4})$. Continuing in this manner, the expression Eq. 6-84 is changed to

$$\Delta^2[(2^0-2^{-M+2})+(2^1-2^{-M+4})+\ldots+(2^{M-3}-2^{-M+2(M-2)})] \qquad (6\text{-}86)$$

This summation can be carried out to give the simple result of $\Delta^2(2^M/6-1+2^{-M+2}/3)$ or $\Delta^2(N/6-1+4/3N)$.

Now for the error introduced by the right shifts. The shift before the computation of the k-th stage introduced into each complex number an error of variance $2^{-2B}=3\Delta^2$ as discussed previously. Each of these errors will propagate to 2^{M-k} output points and will result in an error at each of these output points of variance $3\Delta^2 \times 2^{-2(M-k+1)}$ where the factor of $2^{-2(M-k+1)}$ is due to the scaling in the subsequent stages. So the total contribution to the error variance at the output is $2^M \times 3\Delta^2 \times 2^{-2(M-k+1)} \times 2^{M-k}$ or $3 \times 2^{k-2}\Delta^2$. The total output mean squared error due to the right shift at all the stages is simply

$$3\Delta^2 \sum_{k=0}^{M-1} 2^{k-2} = 3\Delta^2(2^M-1)/4 = 3\Delta^2(N-1)/4 \qquad (6\text{-}87)$$

The sum of the contributions due to round-off and due to shifts is

$$\Delta^2(N/6-1+4/3N)+3\Delta^2(N-1)/4 = \Delta^2(11N/12-7/4+4/3N) \qquad (6\text{-}88)$$

Because of the shifts, the output signal power is now $NS/2^{2M} = S/N$, where S again denotes the input signal power. The total output mean squared error normalized to the total output mean squared signal is

$$\frac{\text{Error}}{\text{Signal}} = \left(\frac{11}{12}N^2 - \frac{7}{4}N + \frac{4}{3}\right)\frac{\Delta^2}{S} \qquad (6\text{-}89)$$

where $\Delta^2 = 2^{-2B}/3$. When compared with Eq. 6-85 we see that the shifts for scaling have resulted in an error to signal ratio that grows as N^2 instead of N when there is no shift.

6.6 SUMMARY

The use of finite length words has significant effects on the performance of a digital signal processor. In this chapter, we have treated in detail the more important effects arising in the hardware implementation of these processors. Fixed point arithmetic is usually preferred in the implementation because of its simplicity. The discussion in this chapter is restricted to rounding arithmetic, rather than chopping or truncation, the treatment of which is slightly more involved.

Digital signal processing functions can also be implemented by programming a general purpose computer. In this case it is likely that floating point arithmetic will be used because of better accuracy. The analysis of floating point error is somewhat more complicated than the fixed point counter part. The interested reader should consult the list of references in [6] and [7], especially [18-20].

EXERCISES

6.1 Prove the inequality in Eq. 6-14.

6.2 The transfer characteristic shown in Figure 6.2 is for rounding a number to B bits. It must be modified if chopping is used.

(a) Show that for 2's complement chopping, the transfer characteristic is that shown in Figure 6.2 shifted to the right by $\delta/2$.

(b) Show that for sign-magnitude chopping, the transfer characteristic is symmetric with respect to the origin, and that the positive branch is identical to that in (a).

6.3 A number x with $|x|<1$ is chopped to B bit in 2's complement format. The result is denoted by x^Q.

(a) Show that the error $e=x^Q-x$ satisfies $-2^{-B+1}<e\leq 0$.

(b) Write e' as $e-2^{-B}$. Then $-2^{-B}<e'\leq 2^{-B}$, and e' can be treated as the quantization noise the same as that in rounding.

Suppose the input to a digital filter is quantized using 2's complement chopping. Show that the output error due to input quantization has a D.C. component and a random component. What is the mean squared value of this random component?

6.4 Derive the relationship in Eq. 6-19. Derive a similar relationship for the movement of the zeros due to coefficient quantization.

6.5 Take the fourth order Butterworth lowpass filter designed in Section 2.2 and assume rounding.

(a) Calculate the output error due to input quantization for the direct form.

(b) Calculate the mean squared error at the filter output due to multiplication round-off, using the direct form.

(c) Determine the scaling constant at the input according to the third scaling policy. What effect does this have on the output error?

(d) Repeat (b) and (c) using the parallel form. Draw a complete block diagram, including the scaling multiplier.

(e) Repeat (b) and (c) using cascade form. Take two different orderings of the poles.

6.6 Take the 22-nd order nonrecursive filter designed in Section 2.6 and assume rounding.

(a) Calculate the mean squared error at the output due to both input quantization and multiplication round-off. Ignore the fact that the coefficients are symmetric.

(b) How can the coefficient symmetry be taken into account?

(c) Factor the transfer function (a polynomial). Form two cascade realizations of the filter, one with zeros in successive sections further away from the point $z=1$, and the other getting closer to $z=1$. Calculate the round-off error at the output for these two realizations.

6.7 Consider a second order filter with no zero and with poles at $(1-\delta)e^{\pm jQ}$ $\delta << 1$, and $Q << 1$. Derive an approximate expression of the mean square round-off error at the output. Use the third scaling policy.

6.8 For the filter of exercise 6.7, determine an approximation to the inequalities in Eq. 6-73 and for those in Eq. 6-74.

6.9 Show that, for the filter of Eq. 6-75 to be stable, b_1 and b_2 must lie inside the triangle of Figure 6.12. Determine the region inside the triangle for the filter to have complex roots.

6.10 Analyze the zero input limit cycle of the filter $y_n = x_n + ay_n$ if rounding is used. Repeat the analysis for chopping.

6.11 Determine the error to signal ratio for radix 2 FFT decimation-in-time algorithm. Consider both the no shift case and the case with shift at every stage.

6.12 What is the effect of the input quantization in the FFT calculation?

6.13 Determine the error to signal ratio for radix 4 FFT decimation-in-frequency algorithm, assuming no shift.

6.14 Write a program in Fortran or in PL/1 to simulate the operation of fixed point digital filter, with the number of bits B a varying parameter. Test your program on the fourth order Butterworth filter of exercise 6.4 in direct form. Use a sum of three sinusoids as input. Compare the observed mean squared error with that calculated in exercise 6.4 for several values of B.

REFERENCES

1. J. R. Ragazzini and G. F. Franklin, *Sampled Data Control Systems*, McGraw-Hill, New York, New York, 1958.
2. W. R. Bennett, *Spectra of Quantized Signals*, Bell Sys. Tech. Jr., vol. 27, 1948, pp. 446-472.
3. S. K. Mitra, K. Hirano, and H. Sakaguchi, *A Simple Method of Computing the Input Quantization and Multiplication Round-off Errors in a Digital Filter*, IEEE Trans. Acoust. Speech Sig. Proc., vol. ASSP-22, 1974, pp. 326-329.
4. J. F. Kaiser, *Digital Filters*, in *System Analysis by Digital Computer*; F. F. Kuo and J. F. Kaiser, eds., John Wiley & Sons, New York, 1966.
5. C. S. Weaver *et al.*, *Digital Filtering with Applications to Electrocardiogram Processing*, IEEE Trans. Audio Electroacoust., vol. AU-16, 1968, pp. 350-391.
6. B. Liu, *Effect of Finite Word Length on the Accuracy of Digital Filters—A Review*, IEEE Trans. Circuit Th., vol. CT-18, 1971, pp. 670-677.
7. A. V. Oppenheim and C. J. Weinstein, *Effects of Finite Register Length in Digital Filtering and the Fast Fourier Transform*, Proc. IEEE, vol. 60, 1972, pp. 957-976.
8. J. B. Knowles and R. Edwards, *Effect of Finite-Word-Length Computer in a Sampled-Data Feedback System*, Proc. IEEE (London), vol. 112, 1965, pp. 1197-1207.
9. P. M. Ebert, J. E. Mazo, and M. G. Taylor, *Overflow Oscillations in Digital Filters*, Bell Sys. Tech. Jr., vol. 48, 1969, pp. 2999-3020.
10. L. B. Jackson, *On the Interaction of Round-off Noise and Dynamic Range in Digital Filters*, Bell Sys. Tech. Jr., vol. 49.
11. L. B. Jackson, *Round-off Noise Analysis for Fixed Point Digital Filters Realized in Cascade or Parallel Form*, IEEE Trans. Audio Electroacoust., vol. AU-18, 1970, pp. 107-122.

12. B. Liu and A. Peled, *Heuristic Optimization of the Cascade Realization of Fixed Point Digital Filters*, IEEE Trans. Acoust. Sp. Sig. Proc., vol. ASSP-23, 1975, pp. 464-473.

13. S. Y. Hwang, *On Optimization of Cascade Fixed Point Digital Filters*, IEEE Trans. Cir. & Sys., vol. CAS-21, 1974, pp. 163-166.

14. J. L. Long and T. N. Trick, *An Absolute Bound on Limit Cycles Due to Round-off Errors in Digital Filters*, IEEE Trans. Audio Electroacoust., vol. AU-21, 1973, pp. 27-30.

15. I. W. Sandberg and J. F. Kaiser, *A Bound on Limit Cycles in Fixed-Point Implementations of Digital Filters*, IEEE Trans. Audio Electroacoust., vol. AU-20, 1972, pp. 110-112.

16. P. D. Welch, *A Fixed-Point Fast Fourier Transform Error Analysis*, IEEE Trans. Audio Electroacoust., vol. AU-17, 1969, pp. 151-157.

17. Tran-Thong and B. Liu, *Fixed-Point Fast Fourier Transform Error Analysis*, to appear.

18. I. W. Sandberg, *Floating-Point Round-Off Accumulation in Digital Filter Realization*, Bell Sys. Tech. Jr., vol. 46, 1967, pp. 1775-1791.

19. B. Liu and T. Kaneko, *Error Analysis of Digital Filters Realized with Floating-Point Arithmetic*, Proc. IEEE, vol. 57, 1969, pp. 1735-1747.

20. T. Kaneko and B. Liu, *Accumulation of Round-Off Error in Fast Fourier Transforms*, Jr. Ass. Comp. Mach., vol. 17, 1970, pp. 637-654.

Appendix 6.1

A PL/1 Computer Program for the Heuristic Optimization of the Pole Zero Pairing and Scaling for Minimum Round-off Error

In this appendix we include a PL/1 program for the heuristic optimization of the pole zero pairing and scaling for minimum round-off error. Details on the application of this program, as well as a detailed flowchart describing it logically can be found in [12]. Although the scaling coefficients are not printed out explicitly they are available internally in the procedure ERROR, in array SCALER(I), and can be printed out if so desired.

```
ORDER:PROCEDURE OPTIONS(MAIN) REORDER;
/*THIS IS A PROGRAM FOR HEURISTIC OPTIMIZATION OF CASCADE
 REALIZATIONS OF RECURSIVE DIGITAL FILTERS*/
/*A(I,J) IS ARRAY OF COEFFICIENTS OF ZEROES,I<11,J=3,B(I,J) IS
 ARRAY OF COEFFICIENTS OF POLES,,B(I,1)=1.0,NPER IS NUMBER OF
 SECOND ORDER SECTIONS,A0 IS NORMALIZING CONSTANT. */
DCL(AR(11,3),BO(11,3),S(12)) DECIMAL FLOAT(12);
 DCL(A(11,3),B(11,3),ERR(15)) DECIMAL FLOAT(12);
 DCL(LONG(11),MONG(11),KONG(11),JPERM(15,11),KPERM(15,11))
  BINARY FIXED(15);
 DCL DECI DECIMAL FLOAT(12);
 DCL A0 DECIMAL FLOAT(12);
 DCL ORCOF FILE OUTPUT STREAM PRINT;
 RANPER:PROCEDURE(N,L,M,IKO);
 /*THIS PROCEDURE GENERATES RANDOM PERMUTATIONS OF
  INTEGERS 1 TO N, AND RETURNS THEM IN ARRAY L(*) */
 DCL(L(*),M(*),N) BINARY FIXED(15);
 DCL X DECIMAL FLOAT(12);
 DCL R INITIAL FLOAT DECIMAL (16) STATIC
  INITIAL (1.455191522836685E-11);
 DCL MULT FLOAT DECIMAL (16) STATIC
  INITIAL (1.061774307700000E+10);
 DCL RANDOM FLOAT DECIMAL (16) STATIC;
 IF IKO=1 THEN RANDOM=R INITIAL;
 M=0; KR=N+1;
 DO I=1 TO N;
  KR=KR-1;
  RANDOM=RANDOM*MULT;
  RANDOM=RANDOM-TRUNC(RANDOM);
  K=RANDOM*KR+1; IF K>KR THEN K=KR;
  IO=0;
  DO J=1 TO N;
   IF M(J)¬=0 THEN GOTO LAB52;
   IO=IO+1; IF IO<K THEN GOTO LAB52;
   M(J)=1; L(J)=I; GOTO LAB53;
 LAB52:END;
 LAB53:END;
 RETURN;
```

```
END RANPER;
POLKA:PROCEDURE(A,B,C,KD);
 DCL(A(*),B(*),C(*))DECIMAL FLOAT(12);
 /* THIS PROCEDURE SETS POLYNOMIAL MULTIPLICATION
  C(KD+2) = A(3)*B(KD)          */
  C(1) = A(1)*B(1); C(2) = A(1)*B(2) + A(2)*B(1);
  DO J = 3 TO KD;
   C(J) = A(1)*B(J) + A(2)*B(J-1) + A(3)*B(J-2);
  END;
  C(KD+1) = A(3)*B(KD-1) + A(2)*B(KD);
  C(KD+2) = A(3)*B(KD);
 RETURN;
END POLKA;
MINIK:PROCEDURE(X,K,L);
 /*THIS PROCEDURE FINDS THE MINIMUM VALUE OF ARRAY X(*)
  OF DIMENSION K, AND RETURNS ITS INDEX L        */
 DCL(X(*),XMIN) DECIMAL FLOAT(12);
  XMIN = X(1); L = 1;
  DO ILIA = 1 TO K;
   DO JUL = ILIA+1 TO K;
    IF X(JUL) <= XMIN THEN DO L = JUL; XMIN = X(JUL);
         ILIA = JUL - 1; GOTO CON1;
        END;
   END;
 CON1:END;
 RETURN;
END MINIK;
SALOSS:PROCEDURE(A,B,N,IERR,V);
 /* THIS PROCEDURE EVALUATES COMPLEX INTEGRALS OF TYPE
  1/2*PI*I INT(Q(Z)*Q(1/Z)/P(Z)P(1/Z) DZ/Z)USING THE JURY
 ANGSTROM AGNIEL ALGORITHM*/
  DCL(ALFA,BETA,A0,AS(23),R,V,A(*),B(*))DECIMAL FLOAT(12);
  DCL(SAV(23),SAB(23))DECIMAL FLOAT(12);
  DO I = 1 BY 1 TO (N+1); SAV(I)=A(I); SAB(I)=B(I); END;
  IF A(1) <= 0 THEN GOTO LAB1; R = A(1);
  DO I = 1 BY 1 TO N; R = R + A(I+1);
  END;
  IF R <= 0 THEN GOTO LAB2; R = A(1); N1 = 1 ;
  DO I = 1 BY 1 TO N; N1 = -N1; R=-R+A(I+1); END;
  IF N1 > 0 THEN GOTO LAB3; R = -R ;
LAB3: IF R <= 0 THEN GOTO LAB2 ; A0 = A(1); IERR = 1 ; V = 0.0 ;
 DO K = 1 BY 1 TO N; L = N+1-K; L1=L+1;
 ALFA = A(L1)/A(1);  BETA = B(L1)/A(1);
 V = V + BETA*B(L1);
 DO I = 1 BY 1 TO L; M = L+2-I ;
 AS(I) = A(I)-ALFA*A(M); B(I) = B(I)-BETA*A(M);
 END;
 IF AS(1) <= 0 THEN GOTO LAB2;
 DO I = 1 BY 1 TO L; A(I) = AS(I); END;
 END;
 V = V + B(1)**2/A(1); V = V/A0 ;
 DO I = 1 BY 1 TO (N+1); A(I)=SAV(I); B(I)=SAB(I); END;
 RETURN;
LAB2: IERR = 0 ;
LAB1: RETURN;
END SALOSS;
ERROR:PROCEDURE(N,D,NP,ERB,A0,SCALER);
 /* THIS PROCEDURE COMPUTES THE OUTPUT NOISE GAIN FOR THE D1
 CONFIGURATION OF SECOND ORDER SECTIONS USING L2 SCALING. */
  DCL(N(*,*),D(*,*),ERB,V2(11),V1(11),Q(23),P(23),SCALER(*),
  QN(23),PN(23),DUM(3),SAM(3))DECIMAL FLOAT(12);
  DCL A0 DECIMAL FLOAT(12);
  SAM(*) = N(NP,*) ; DUM(*) = D(NP,*);
  NUM = 2; CALL SALOSS(DUM,SAM,NUM,IERR,V2(NP));
  IF IERR = 0 THEN PUT LIST ('ERROR CONDITION!!!!!!!!!!',NUM) ;
  DO I = 1 BY 1 TO 3; Q(I) = SAM(I) ; P(I) = DUM(I); END;
```

```
DO LIK = (NP-1) BY -1 TO 1;
 KUM = 2*(NP-LIK) + 1; NUM = KUM + 1; SAM(*) = N(LIK,*);
 CALL POLKA(SAM,Q,QN,KUM);
 DUM(*) = D(LIK,*) ; CALL POLKA(DUM,P,PN,KUM);
 CALL SALOSS(PN,QN,NUM,IERR,V2(LIK));
 IF IERR = 0 THEN PUT LIST ('ERROR CONDITION!!!!!!!!!',NUM);
 DO I = 1 BY 1 TO (KUM+2); Q(I)=QN(I); P(I)=PN(I); END;
 END;
 SAM(1)=0; SAM(2)=0; SAM(3)=1.0; DUM(*)=D(1,*);
 NUM=2; CALL SALOSS(DUM,SAM,NUM,IERR,V1(1));
 IF IERR = 0 THEN PUT LIST ('ERROR CONDITION!!!!!!!!!',NUM);
 DO I = 1 BY 1 TO 3; Q(I)=SAM(I); P(I)=DUM(I); END;
 DO LIK = 2 BY 1 TO NP;
  KUM = 2*(LIK-1) + 1; NUM = KUM + 1;
  SAM(*) = N(LIK-1,*); DUM(*) = D(LIK,*);
  CALL POLKA(SAM,Q,QN,KUM); CALL POLKA(DUM,P,PN,KUM);
  CALL SALOSS(PN,QN,NUM,IERR,V1(LIK));
 IF IERR = 0 THEN PUT LIST ('ERROR CONDITION!!!!!!!!!',NUM);
  DO I=1 BY 1 TO (KUM+2); Q(I)=QN(I); P(I) = PN(I); END;
 END;
 ERB = 0;
 DO LIK = 2 BY 1 TO NP; ERB = ERB + V1(LIK)*V2(LIK); END;
 ERB = A0*(5.0*ERB + 3.0*V1(1)*V2(1)) + 2.0;
/* SCALING COEFFICIENTS ARE SCALER(I))  */
 SCALER(1) = 1.0/V1(1);
 DO I = 2 BY 1 TO NP; SCALER(I) = V1(I-1)/V1(I) ; END;
 SCALER(NP+1) = A0*V1(NP);
 RETURN;
END ERROR;
LOCOPT:PROCEDURE(A,B,I1,J1,NP,SER,A0,S);
  DCL(A(*,*),B(*,*),N(11,3),D(11,3),ER(56),SER)DECIMAL FLOAT(12);ORD0135
  DCL(JN(56,11),I1(*),J1(*),NP) BINARY FIXED(15);
  DCL (A0,S(*)) DECIMAL FLOAT(12);
  JN = 0;
  DO J = 1 BY 1 TO NP;
   N(J,*) = A(I1(J),*); JN(1,J) = J1(J);
  END;
  K1 = 1;
  DO NI = 1 BY 1 TO (NP-1);
  DO J = (NI+1) BY 1 TO NP;
   K1 = K1 + 1; JN(K1,*) = JN(1,*);
   JN(K1,J) = JN(1,NI); JN(K1,NI) = JN(1,J);
  END; END;
  KAPA = NP*(NP-1)/2 + 1;
  DO IDA = 1 BY 1 TO KAPA;
  DO J=1 BY 1 TO NP; D(J,*) = B(JN(IDA,J),*) ; END;
  CALL ERROR(N,D,NP,ER(IDA),A0,S);
  END;
  CALL MINIK(ER,KAPA,LIM);
  SER = ER(LIM); J1(*) = JN(LIM,*);
  RETURN;
END LOCOPT;
 PUT PAGE;
 NPER = 4;
 KALKA = 15;
 KUBA = 0;
 /* START RANDOM NUMBER GENERATOR */
 DO IGAF = 1 BY 1 TO 10; CALL RANPER(NPER,LONG,MONG,IGAF); END;
 IGAF = 10;
SALMA: KUBA = KUBA + 1;
 GET EDIT(NPER) (X(1), F(2));
 PUT EDIT
 ('*************************************************************')
 (SKIP,A);
 PUT EDIT(' RECURSIVE FILTER OF ORDER ',2*NPER)( SKIP(3),A,F(4));
 GET SKIP EDIT(A0) (X(1),F(18,12));
```

```
PUTEDIT('AO = ',AO)(SKIP(2),A,F(12,8));
DO I=1 TO NPER;
 GET SKIP EDIT(A(I,1),A(I,2),A(I,3))(X(1),3 F(18,12));
 GET SKIP EDIT(B(I,2),B(I,3))(X(1),2 F(18,12));
 B(I,1)=1.00000000000;
 PUTEDIT('A(',I,',1)=',A(I,1),'A(',I,',2)=',A(I,2),'A(',
 I,',3)=',A(I,3))(SKIP,3(A,F(2),A,F(12,8),X(4)));
 PUTEDIT('B(',I,',1)=',B(I,1),'B(',I,',2)=',B(I,2),'B(',
 I,',3)=',B(I,3))(SKIP,3(A,F(2),A,F(12,8),X(4)));
END;
 KAR = 5;
PUTEDIT('BEST ASSIGNMENT','NOISE GAIN','DB')
 (SKIP(2),X(20),A,X(12),A,X(8),A,X(4),A);
PUTEDIT('FOUND IN 5 RANDOM STARTS')(SKIP,X(20),A);
DO KADI = 1 BY 1 TO KALKA;
LONG = 0; KONG = 0;
DO IKA = 1 BY 1 TO KAR;
CALL RANPER(NPER,LONG,MONG,IGAF);
CALL RANPER(NPER,KONG,MONG,IGAF);
CALL LOCOPT(A,B,LONG,KONG,NPER,ERR(IKA),AO,S);
JPERM(IKA,*)=LONG(*); KPERM(IKA,*)=KONG(*);
END;
CALL MINIK(ERR,KAR,IMAK);
DECI = 10.0*LOG10(ERR(IMAK));
 PUTEDIT(KADI,(KPERM(IMAK,I) DO I=1 TO NPER))
(SKIP,F(4),X(4),11 F(3));
PUTEDIT((JPERM(IMAK,I) DO I=1 TO NPER))(SKIP,X(8),11 F(3));
PUTEDIT(ERR(IMAK),DECI)(COLUMN(56),F(12,1),X(6),F(5,2));
PUTLIST
 ('----------------------------------------------------------------')
SKIP;
END;
END ORDER;
```

Index